Topics in Fluorescence Spectroscopy

Volume 7

DNA Technology

Topics in Fluorescence Spectroscopy

Edited by JOSEPH R. LAKOWICZ

Volume 1: Techniques
Volume 2: Principles
Volume 3: Biochemical Applications
Volume 4: Probe Design and Chemical Sensing
Volume 5: Nonlinear and Two-Photon-Induced Fluorescence
Volume 6: Protein Fluorescence
Volume 7: DNA Technology

Topics in Fluorescence Spectroscopy

Volume 7
DNA Technology

Edited by

JOSEPH R. LAKOWICZ
Center for Fluorescence Spectroscopy and
Department of Biochemistry and Molecular Biology
University of Maryland School of Medicine
Baltimore, Maryland

Springer Science+Business Media, LLC

The Library of Congress cataloged the first volume of this title as follows:

Topics in fluorescence spectroscopy/edited by Joseph R. Lakowicz. p. cm.
Includes bibliographical references and index.
Contents: v. 1. Techniques
1. Fluorescence spectroscopy. I. Lakowicz, Joseph R.

QD96.F56T66 1991 91-32671
543′.0858—dc20 CIP

ISBN 978-1-4757-8210-3 ISBN 978-0-306-47947-2 (eBook)
DOI 10.1007/978-0-306-47947-2

Originally published by Kluwer Academic/Plenum Publishers, New York in 2003
Softcover reprint of the hardcover 1st edition 2003

http://www.wkap.nl

10 9 8 7 6 5 4 3 2 1

A C.I.P. record for this book is available from the Library of Congress

Contributors

W. Patrick Ambrose • Bioscience Division, Los Alamos National Laboratory, Los Alamos, New Mexico 87545 *Current address*: Lawrence Livermore National Laboratory, Livermore, California 94551

Hong Cai • Bioscience Division, Los Alamos National Laboratory, Los Alamos, New Mexico 87545

Sabato D'Auria • Center for Fluorescence Spectroscopy and Department of Biochemistry and Molecular Biology, University of Maryland Medical School, Baltimore, Maryland 21201; and Institute of Protein Biochemistry and Enzymology, National Research Council, Naples, Italy 801215

Michael J. Difilippantonio • Center for Cancer Research, National Cancer Institute, National Institutes of Health, Bethesda, Maryland 20892-8010

Alexander N. Glazer • Department of Molecular and Cell Biology, University of California, Berkeley, California 94720-3200

Peter M. Goodwin • Bioscience Division, Los Alamos National Laboratory, Los Alamos, New Mexico 87545

W. Kevin Grace • Bioscience Division, Los Alamos National Laboratory, Los Alamos, New Mexico 87545

Ignacy Gryczynski • Center for Fluorescence Spectroscopy and Department of Biochemistry and Molecular Biology, University of Maryland Medical School, Baltimore, Maryland 21201

Zygmunt Gryczynski • Center for Fluorescence Spectroscopy and Department of Biochemistry and Molecular Biology, University of Maryland Medical School, Baltimore, Maryland 21201

Robert C. Habbersett • Bioscience Division, Los Alamos National Laboratory, Los Alamos, New Mexico 87545

Mary E. Hawkins • Pharmacology and Experimental Therapeutics Section, National Cancer Institute, Bethesda, Maryland 20892

Su-Chun Hung • Department of Molecular and Cell Biology, University of California, Berkeley, California 94720-3200

James H. Jett • Bioscience Division, Los Alamos National Laboratory, Los Alamos, New Mexico 87545

Richard A. Keller • Bioscience Division, Los Alamos National Laboratory, Los Alamos, New Mexico 87545

Ulrich J. Krull • Chemical Sensors Group, Department of Chemistry, University of Toronto at Mississauga, Mississauga, Ontario, Canada L5L 1C6

Erica J. Larson • Bioscience Division, Los Alamos National Laboratory, Los Alamos, New Mexico 87545 *Current address*: Engineering Science and Applications Division, Los Alamos National Laboratory, Los Alamos, New Mexico 87545

Suzanne Lassiter • Department of Chemistry, Louisiana State University, Baton Rouge, Louisiana 70803

Joanna Malicka • Center for Fluorescence Spectroscopy and Department of Biochemistry and Molecular Biology, University of Maryland Medical School, Baltimore, Maryland 21201

Richard A. Mathies • Department of Chemistry, University of California, Berkeley, California 94720-1460

Linda B. McGown • Department of Chemistry, Duke University, Durham, North Carolina 27708

Larry E. Morrison • Vysis Inc., Downers Grove, Illinois 60515-5400

Clyde Owens • Department of Chemistry, Louisiana State University, Baton Rouge, Louisiana 70803

Paul A. E. Piunno • FONA Technologies, Inc., Waterloo, Ontario, Canada, N2V 2K1

Thomas Ried • Center for Cancer Research, National Cancer Institute, National Institutes of Health, Bethesda, Maryland 20892-8010

Mosè Rossi • Institute of Protein Biochemistry and Enzymology, National Research Council, 810215 Naples, Italy

Paul R. Selvin • Physics Department and Biophysics Center, University of Illinois, Urbana, Illinois 61801

Steven A. Soper • Department of Chemistry, Louisiana State University, Baton Rouge, Louisiana 70803

Emanuel Waddell • Department of Chemistry, Louisiana State University, Baton Rouge, Louisiana 70803

James H. Werner • Bioscience Division, Los Alamos National Laboratory, Los Alamos, New Mexico 87545

Jin Xie • Department of Molecular and Cell Biology, University of California, Berkeley, California 94720-3200

Yichuan Xu • Department of Chemistry, Louisiana State University, Baton Rouge, Louisiana 70803

Preface

During the past fifteen years, DNA technology has advanced at an astounding pace. The pioneering papers on the use of fluorescence in sequencing appeared in 1986–1987, resulting in publication of the human genome sequence in February of 2001. Fluorescence detection has played a dominant role not only in sequencing, but in genetics research and diagnostics. In addition to sequencing, fluorescence is used in essentially all measurements of DNA amplification by PCR and other methods. Molecular beacons are used to light up specific sequences in mixtures of DNA, even at the intracellular level. In the past several years, DNA arrays or gene chips have become an important research tool in developmental biology, and drug discovery, and targeted treatments. It is now possible to place an excess of 30,000 DNA sequences on a single microscope slide, and to identify the amounts of complimentary DNA by fluorescence ratiometric measurements. In this volume, we attempted to provide an overview of the fundamental principles associated with these modern applications of fluorescence to nucleic acids. Because of the rapid rate of development in this field, it is not possible for any volume to be completely up to date on the latest details. Hopefully the fundamental principles will remain useful to scientists using modern DNA methodology.

Joseph R. Lakowicz
Center for Fluorescence Spectroscopy
March 2002

Preface

Contents

Topics in Fluorescence Spectroscopy

Volume 7

DNA Technology

1

DNA Sequencing Using Fluorescence Detection

Steven A. Soper, Clyde Owens, Suzanne Lassiter, Yichuan Xu, and Emanuel Waddell

1.1. General Considerations

1.1.1. What Is DNA?

1.1.1.1. Organization of Genome

The blueprint for all cellular structures and functions is encoded in the genome of any organism. The genome consists of deoxyribonucleic acid (DNA) which is tightly coiled into narrow threads that, when completely stretched, reach a length of approximately 1.5 m yet possesses a width of only ~2.0 nm (2×10^{-9} m). The threads of DNA are typically associated with many different types of proteins and are organized into structures called chromosomes that are housed within the nucleus of cells in most eukaryotic organisms. Within the human genome, there are 23 pairs of chromosomes.

DNA is comprised of several different chemical units, which are a deoxyribose sugar unit, a phosphate group and one of four different nucleotide bases (adenine (A), guanine (G), cytosine (C), or thymine (T)). At a molecular level, it is the order of these bases which carries the code to build proteins within the cell that inevitably control the function of various cells and also determine the organism's physical characteristics. In the human genome, 3 billion bases are contained within the 23 pairs of chromosomes. The length of the chromosomes (in base pairs) varies greatly, with the smallest chromosome containing 50 million bases (chromosome Y) and the longest containing 250 million bases (chromosome 1). It is the primary function of DNA sequencing to determine the order of these nucleotide bases.

Steven A. Soper, Clyde Owens, Suzanne Lassiter, Yichuan Xu, and Emanuel Waddell • Department of Chemistry, Louisiana State University, Baton Rouge, Louisiana, 70803.

Topics in Fluorescence Spectroscopy, Volume 7: DNA Technology, edited by Joseph R. Lakowicz, Kluwer Academic / Plenum Publishers, New York, 2003

1.1.1.2. Functions of Genes

It should be noted that the coding regions (regions which carry the information for the construction of proteins) of the genome are contained within genes. Each gene consists of a specific sequence of nucleotide bases, with three nucleotide bases (codon) directing the cells' protein-synthesizing machinery to add a specific amino acid to the target protein. It is estimated that within the human genome, there are approximately 30,000 genes, with the length of the genes highly variable. However, only approximately 10% of the human genome is thought to contain protein-coding sequences.

1.1.2. What Is DNA Sequencing?

A flow chart depicting the important steps involved in the process of DNA sequencing is shown in Figure 1.1. As can be seen, there are three primary steps: mapping, sequencing, and finally, assembly. Within each of these three general steps are a number of sub-steps, which include such processes as cloning and subcloning (mapping), template preparation and gel electrophoresis (sequencing), and the computer algorithms required to assemble the small bits of sequencing data into contiguous strings which comprise the intact chromosome. Each chromosome may consist of hundreds of millions of nucleotide bases and, unfortunately, the sequencing phase of the intricate process can only handle pieces of DNA that vary from 1000 base pairs (bp) to 2000 bp in length. Therefore, the chromosome must typically be broken down into manageable pieces using either restriction enzymes or mechanical shearing, and, then cloned into bacterium in order to increase the copy number of the individual pieces of DNA. Following sequencing of each cloned fragment, the individual pieces must be reassembled into a contiguous strand representing the entire chromosome. This process typically involves sophisticated computer algorithms to look for commonalties in the small fragments and overlap them to build the sequence of the entire chromosome.

1.1.2.1. DNA Sequencing Factories

In order to accomplish the lofty goal of sequencing the entire human genome, sequencing factories have been assembled to produce large amounts of data and deposit this data into public databases for easy accessibility by the general scientific and medical communities. One such production-scale sequencing center is located at the Baylor College of Medicine (www.hgsc.bcm.tmc.edu). To put the sequencing demands into perspective, it is informative to give some statistics on this particular sequencing center. As of May, 1999, the Baylor sequencing center

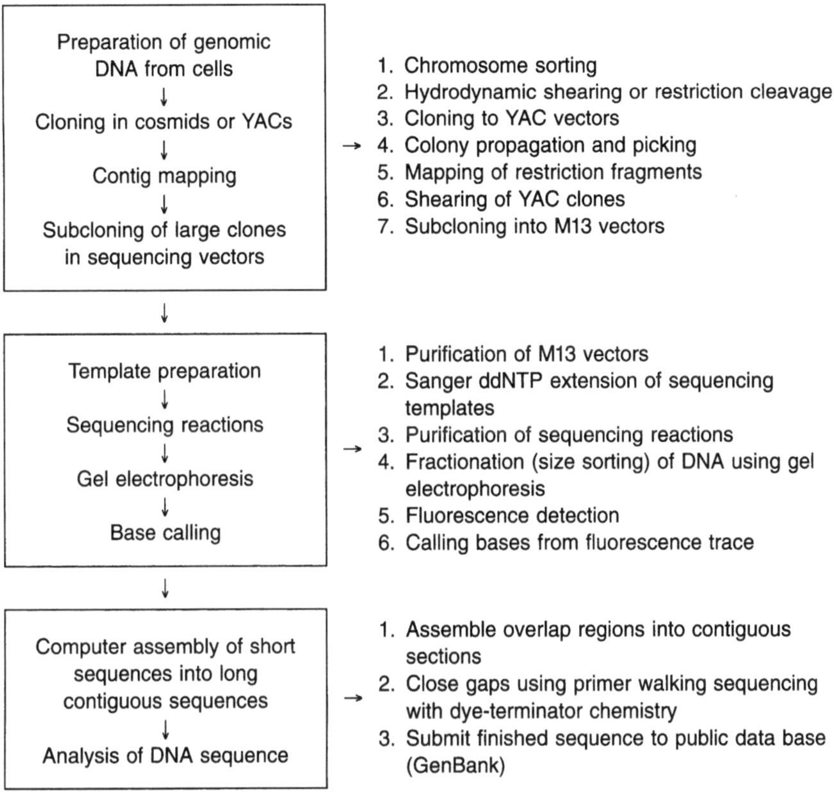

Figure 1.1. Flow chart showing the processing steps typically involved in sequencing DNA starting with isolating genomic DNA from cells.

had deposited over 26 Mbp (26 × 10^6 base pairs, 0.7% of the human genome) of sequence data into the public data bases and typically runs approximately 14 automated fluorescence-based DNA sequencing machines 12 hours a day. This amounts to performing 50,000 sequencing reactions a month.

In this chapter, we will focus on the sequencing phase of the genome processing steps. It is in this particular step that fluorescence—both hardware and probe development—has had a profound impact on augmenting the throughput of acquiring sequencing data. We will begin by briefly introducing (or refreshing) the reader on the molecular and geometrical structure of DNA and review the common schemes used to sequence DNA. In addition, we will discuss the various electrophoretic modes for fractionating DNA based on size, an integral component in high throughput sequencing. We will then discuss common strategies of

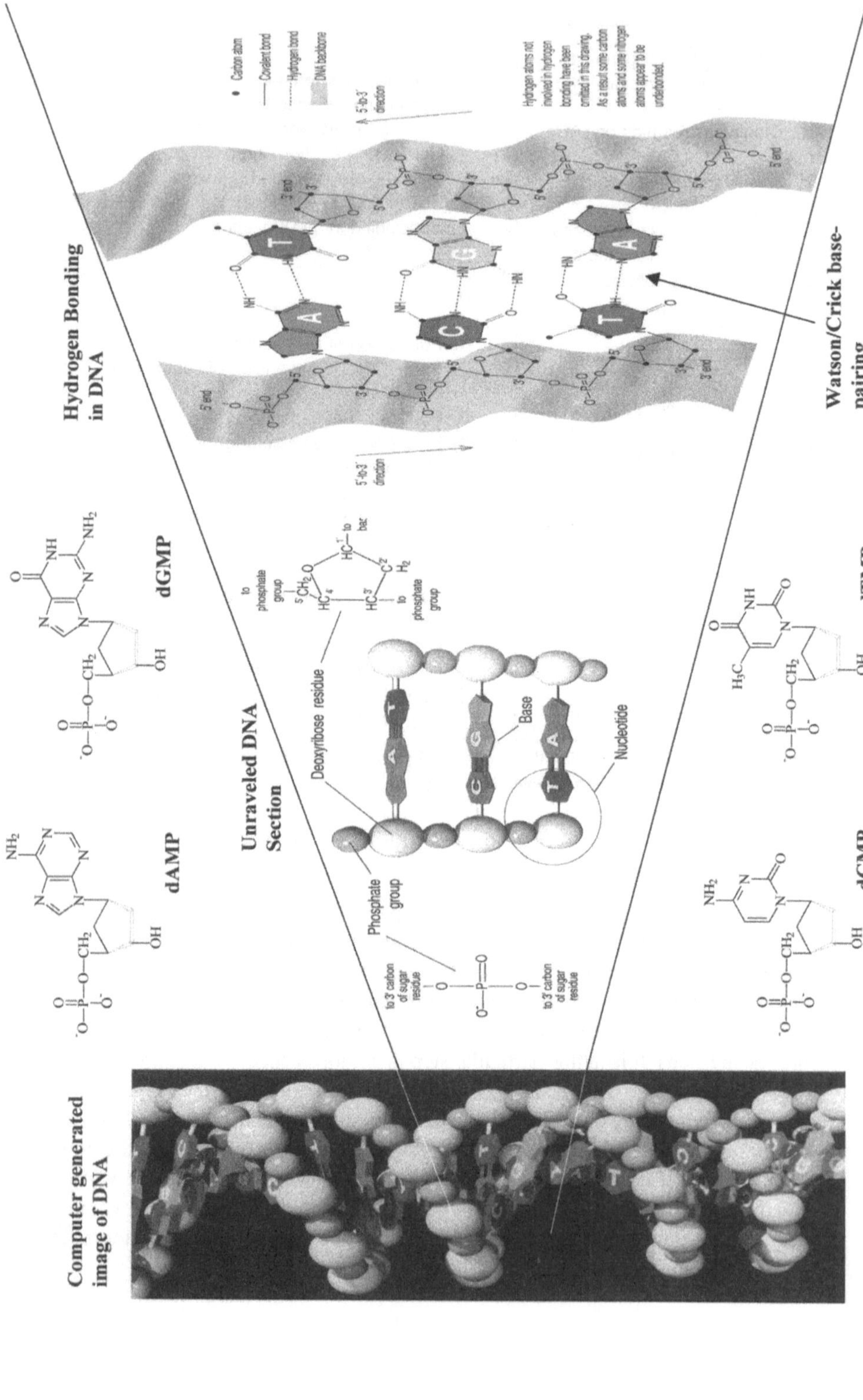
Computer generated image of DNA
Unraveled DNA Section
Hydrogen Bonding in DNA
Watson/Crick base-pairing
dAMP
dGMP
dCMP
dTMP
Phosphate group
Deoxyribose residue
Base
Nucleotide
to 3' carbon of sugar residue
to phosphate group
Carbon atom
Covalent bond
Hydrogen bond
DNA backbone
5'-to-3' direction
5' end
3' end
Hydrogen atoms not involved in hydrogen bonding have been omitted in this drawing. As a result some carbon atoms and some nitrogen atoms appear to be unbonded.

fluorescence detection used in many sequencing machines and include in this discussion, hardware developments as well as probe (labeling dye) developments.

1.1.3. Structure of DNA

1.1.3.1. Nucleotide Bases

As stated in the previous section, the basic building blocks of DNA are a deoxyribose sugar unit, a phosphate group, and the nucleotide base, of which there are four, and when combined chemically they form an individual nucleotide. In Figure 1.2 is shown the chemical structure of the four nucleotide units which comprise DNA. The bases are grouped into two different classes: the pyrimidines (T, C) and the purines (A, G). In the case of RNA (ribonucleic acid) the structural differences include the incorporation of a ribose sugar (inclusion of a hydroxyl group at the 2′ position) and also the substitution of uracil for thymine.

Since DNA, more specifically chromosomes, are composed of many of these individual nucleotide units strung together, the nucleotides are covalently attached via phosphodiester linkages, which occur between the phosphate group on the 5′ site of one sugar unit and the 3′ hydroxyl group on another nucleotide (see Figure 1.2). Therefore, DNA exists as a biopolymer, with the repeating units being deoxyribose and phosphate residues that are always linked together by the same type of linkage and form the backbone of the DNA molecule. However, the order of bases along the biopolymer backbone can vary greatly and impart a high degree of individuality to any particular DNA molecule.

1.1.3.2. Watson–Crick Base Pairing

The three-dimensional structure of DNA is known to differ greatly from that of proteins, which are also biopolymers composed of different amino acid residues. It was discovered by James Watson and Francis Crick in 1953 that DNA exists in a double helical structure (see Figure 1.2) with the sugar–phosphate backbone oriented on the outside of the molecule and the bases positioned on the inside of the double helix.[1] They also surmised that the two strands of the double-stranded molecule were held together via hydrogen bonds between a pair of bases on opposing strands. From modeling, they found that A could pair only with T and C with G (see Figure 1.2). Each of these base pairs possesses a symmetry that permits it to be placed into the double helix in two different ways (A to T and T

Figure 1.2. Chemical structures of deoxyribose nucleic acid (DNA). The computer image shows the double helical nature of DNA. Also shown are the hydrogen bonding between nucleotide bases (Watson–Crick base pairing) and the structures of the nucleotide building blocks.

to A; C to G and G to C). Thus, all possible permutations (4) of sequence can exist for all four bases. In spite of the irregular sequence of bases along each strand, the sugar–phosphate backbone assumes a very regular helical structure, with each turn of the double helix comprised of 10 nucleotide units.

1.1.4. Methods for Determining the Primary Structure of DNA

The important factor to consider when developing a sequencing strategy for whole genomes is to remember that one can sequence only small sections of DNA (1000–2000 bp) and that entire chromosomes are comprised of well over 1×10^6 bp. The actual process of generating DNAs that can be handled by sequencing machines is typically involved and requires a number of cloning and purification steps followed by actual sequencing and then assembly of the pieces into contiguous regions of the target chromosome. While these specific processes will not be covered in this chapter, Figure 1.3 gives a schematic diagram of a typical strategy that is used in many sequencing laboratories. This strategy is termed an ordered shotgun approach and starts with the mechanical shearing (breaking apart) of

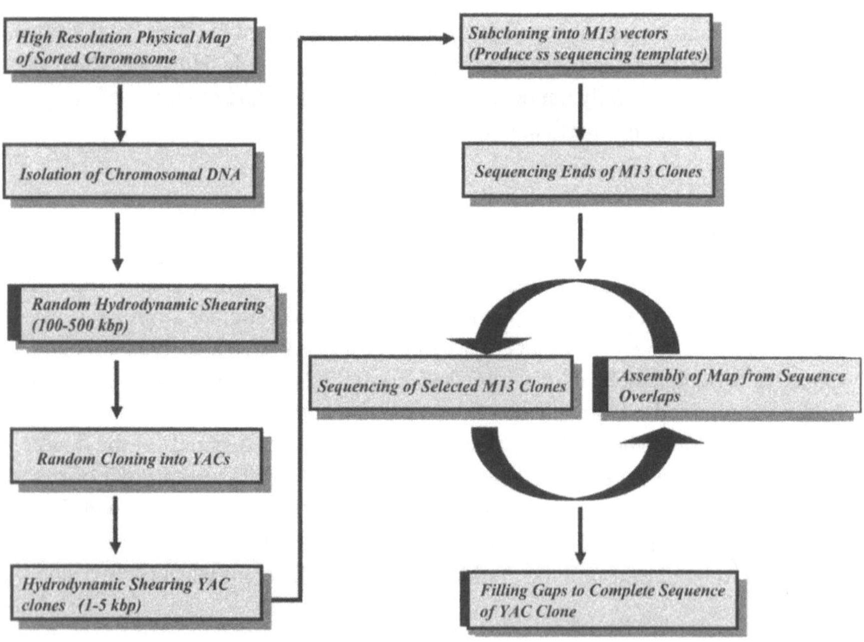

Figure 1.3. Processing flow chart of ordered shotgun sequencing of DNA.

intact chromosomes into pieces composed of 100,000 to 200,000 bp.[2] These sheared DNAs are then cloned into yeast artificial chromosomes (YACs), one sheared section per YAC that allows amplification of the number of copies of the insert. Cloning provides an unlimited amount of material for sequencing and serves as the basis for construction of libraries, which are random sets of cloned DNA fragments. Genomic libraries are sets of overlapping fragments encompassing an entire genome. Once these libraries have been constructed, single inserts are extracted from the library and further sheared into fragments ranging from 1000–2000 bp. These fragments are then sub-cloned into M13 vectors to produce high quality single-stranded DNAs appropriate for actual sequencing. Typically, M13 sub-clones are sequenced from both ends to allow construction of maps of the single YAC inserts, which are then used to provide a scaffold for the complete sequence analysis of the YAC insert. The important procedures that will be the focus of this chapter will be the procedures that are actually used to produce the sequence data of the M13 sub-clones. The two most common procedures used are the Maxam–Gilbert chemical degradation method and the Sanger dideoxy-chain termination method.[3,4] The commonality in both of these methods is the production of a nested set of fragments that are all terminated (cleaved) at a common base(s).

1.1.4.1. Maxam–Gilbert Sequencing

The Maxam–Gilbert sequencing method uses chemical cleavage methods to break single-stranded DNA molecules at either one or two bases, which is followed by a size fractionation step to sort the cleaved products. The cleavage reaction involves two different reactions; one that cleaves at G and A (purine) residues and the other that cleaves at C and T (pyrimidine) residues. The first reaction can be slightly modified to cleave at G only and the second at T only. Therefore, one can run four separate cleavage reactions, G only, A + G, T only, T + C, from which the sequence can be deduced. In each cleavage reaction, the general process involves chemically modifying a single base, removing the modified base from its sugar, and finally breaking the bond of the exposed sugar in the DNA backbone.

The chemical steps involved in G cleavage are shown in Figure 1.4a. In this step, dimethylsulfate is used to methylate G. After eviction of the modified base via heating, the strand is broken at the exposed sugar by subjecting the DNA to alkaline conditions. To cleave at both A and G residues, the procedure is identical to the G cleavage reaction except that a dilute acid is added after the methylation step (see Figure 1.4b). The reaction that cleaves at either a C or C and T residue is carried out by subjecting the DNA to hydrazine to remove the base and piperidine to cleave the sugar–phosphate backbone. The extent of each reaction can be carefully limited so that each strand is cleaved at only one site.

The entire Maxam–Gilbert process is depicted in Figure 1.4c. As can be seen,

(a) Cleavage Reaction for Guanine

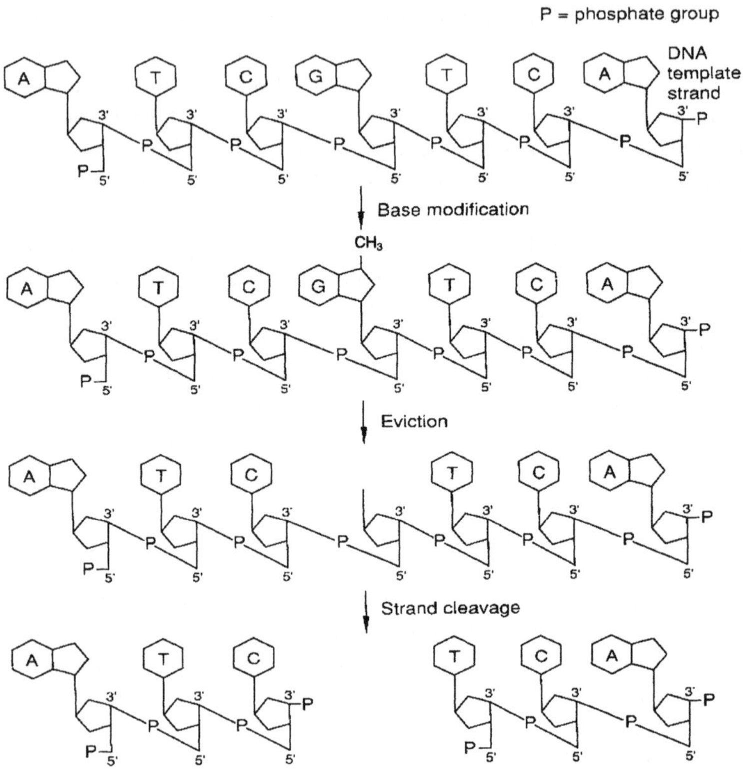

Dimethylsulfate is used to methylate guanine. After eviction of the modified base, the exposed sugar, deoxyribose, is then removed from the backbone. Thus the strand is cleaved in two.

(b) Fragments from Single Cleavage at G

5'-^{32}P-ATGACCGATTTGC-3' — Labeled template strand

5'-^{32}P-AT-3' 5'-ACCGATTTGC-3'

5'-^{32}P-ATGACC-3' 5'-ATTTGC-3'

5'-^{32}P-ATGACCGATT-3' 5'-C-3'

Six different types of fragments are produced. Only three of those include the labeled 5' end of the original strand.

Figure 1.4. Maxam–Gilbert chemical degradation method for DNA sequencing. In (a) is shown the chemical cleavage method for a guanine residue using dimethylsulfate to methylate the G residue. In (b), the fragments generated from chemical cleavage at only G are shown, while in (c) the entire Maxam–Gilbert process is displayed. (*continued on next page.*)

(c) Steps in Maxam-Gilbert Sequencing

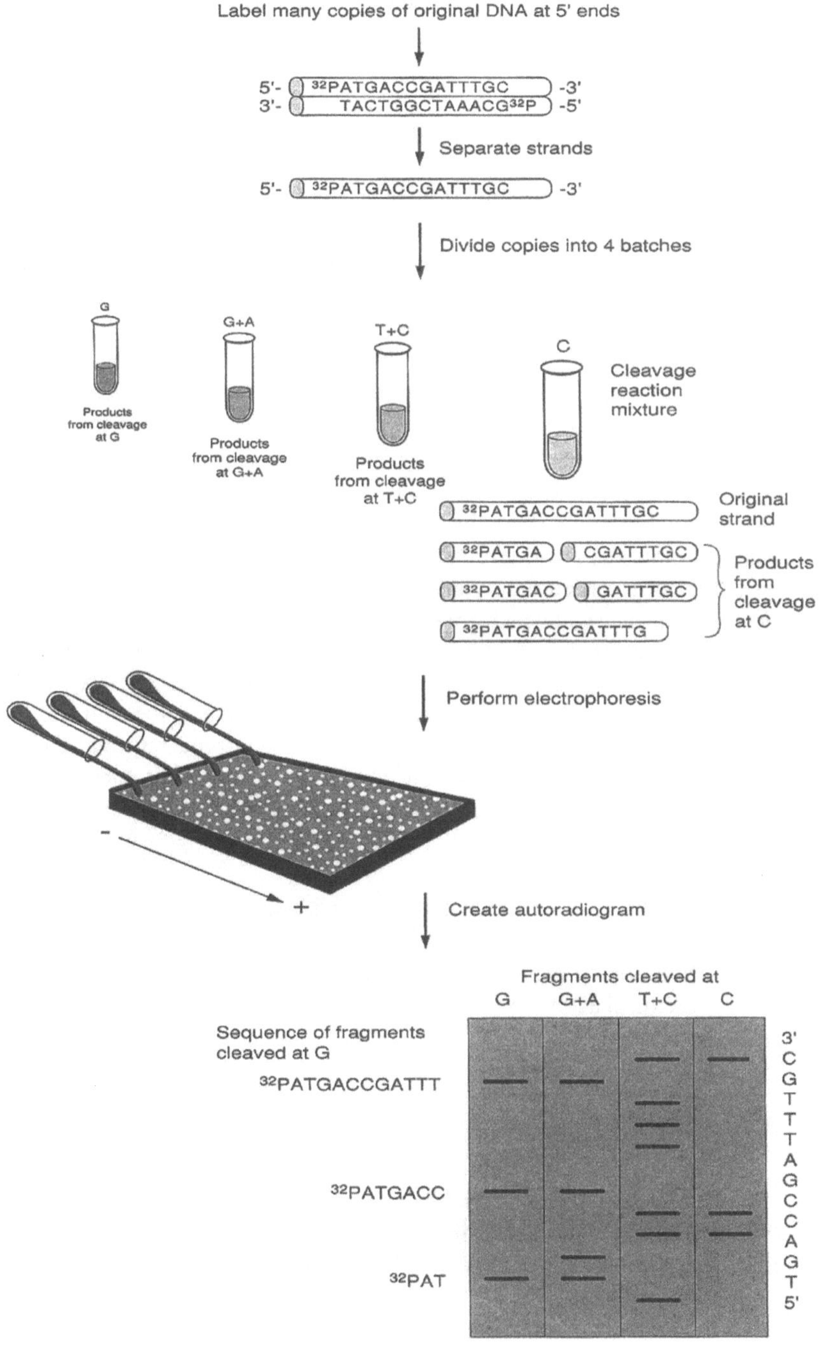

four cleavage reactions are run: G, G + A, T, T + C. Prior to chemical cleavage, the intact DNA strand is labeled (typically with a ^{32}P radiolabel for detection at the 5′ end). Following chemical cleavage, the reactions are run in an electrophoresis gel (polyacrylamide) and separated based on size. The actual sequence of the strand is then deduced from the generated gel pattern.

1.1.4.2. Sanger Chain Termination Method

Contrary to the Maxam–Gilbert method, the Sanger procedure is an enzymatic method and involves construction of a DNA complement to the template whose sequence is to be determined. The complement is a strand of DNA that is constructed with a polymerase enzyme, which incorporates single nucleotide bases according to Watson–Crick base pairing rules (A–T; G–C) (see Figure 1.5). The nested set of fragments is produced by interrupting the polymerization by inserting into the reaction cocktail a base that has a structural modification, that modification being the lack of a hydroxyl group at the 3′ position of the deoxyribose sugar (dideoxy nucleotide, ddNTP). Figure 1.5A shows the chemical structure of a ddNTP. When mixed with the deoxynucleotides (dNTP), the polymerization proceeds until the ddNTP is incorporated.

The entire Sanger, chain-termination protocol is shown in Figure 1.5C. The process is typically carried out by adding to the template DNA a primer which has a known sequence and which anneals (binds) to a complementary site on the unknown template. This primer can carry some type of label, for example a covalently attached fluorochrome. However, situating the fluorescent label on the ddNTP can be done as well. Following annealing of the primer to the template, the reaction is carried out by the addition of a polymerase enzyme, all four dNTPs, and one particular ddNTP. Therefore, four reactions are carried out, each containing a particular ddNTP. Following polymerization, the reactions are loaded onto an electrophoresis gel and size fractionated. The final step involves reading the sequence from the gel.

The Sanger method has been the preferred sequencing method for most large-scale sequencing projects due to its ease in preparing the nested set of fragments. Many times reactions can be run under standard conditions, without the need for chemical additions at timed intervals. In addition, the process is very conducive to automation.

1.1.5. Modes of Electrophoresis

Whatever fluorescence detection protocol is used for calling bases, the important analytical technique that is required is a fractionation step, in which the

DNA molecules are sorted by size. While the focus of this chapter is on fluorescence-based detection in sequencing applications, it is informative to briefly discuss some of the common gel electrophoresis platforms that are used, since the fluorescence detector is integrated with the sizing step to provide on-line detection.

The commonality in all electrophoresis formats is the use of an electric field to shuttle the DNAs through a maze which consists of a polymer, either static or dynamic, that possesses pores of various sizes. In all electrophoresis experiments, the mobility of the molecule (μ, cm^2/Vs, defined as the steady state velocity per unit electric field strength) in an electric field is determined by

$$\mu = \frac{q}{f} \quad (1.1)$$

where q is the net charge on the molecule and f is the frictional property of the molecule and is related to its conformational state as well as the molecular weight (MW) of the molecule. For example, proteins can be considered solid spheres and thus, $f \sim (MW)^{1/3}$. In the case of DNAs, either single stranded or double stranded, $f \sim (MW)^1$ because the DNA molecule acts as a free-draining coil. Since q is also related to the length of the DNA molecule, one finds that $\mu \sim N_b^o$, where N_b^o indicates that the mobility (in free solution) is independent of the number of bases comprising the DNA molecule. Because of this property of DNAs, the electrophoresis step must include some type of sieving medium, which can be a polymer consisting of pores with a definitive size.

For any type of analytical separation, resolution is a key parameter which is optimized to improve the performance of the separation. The resolution (R) for electrophoretic separations can be calculated from the simple relationship

$$R = \frac{1}{4}\frac{\Delta\mu_{app}}{\mu_{app,avg}}\ {}^{1/2} \quad (1.2)$$

where μ_{app} is the difference in mobility between two neighboring bands, $\mu_{app,avg}$ is the average electrophoretic mobility for the same neighboring bands, and N is the plate number which represents the efficiency (bandwidth) for the electrophoresis. As can be seen from this equation, the resolution can be improved by increasing the difference in the mobility of the two bands (increase selectivity, gel property) or increasing the plate numbers (narrower bands). As a matter of reference, when $R = 0.75$, two bands are baseline resolved. For DNA sequencing, the accuracy in the base call depends intimately on the resolution obtained during gel fractionation.

Common polymers used for DNA sequencing are linear or crosslinked polyacrylamides or polyethylene oxides, both of which possess the appropriate pore size for sorting single-stranded DNAs. Polyacrylamides are prepared from the acrylamide monomer ($-CH_2{=}CHCONH_2$), which is copolymerized with a certain per-

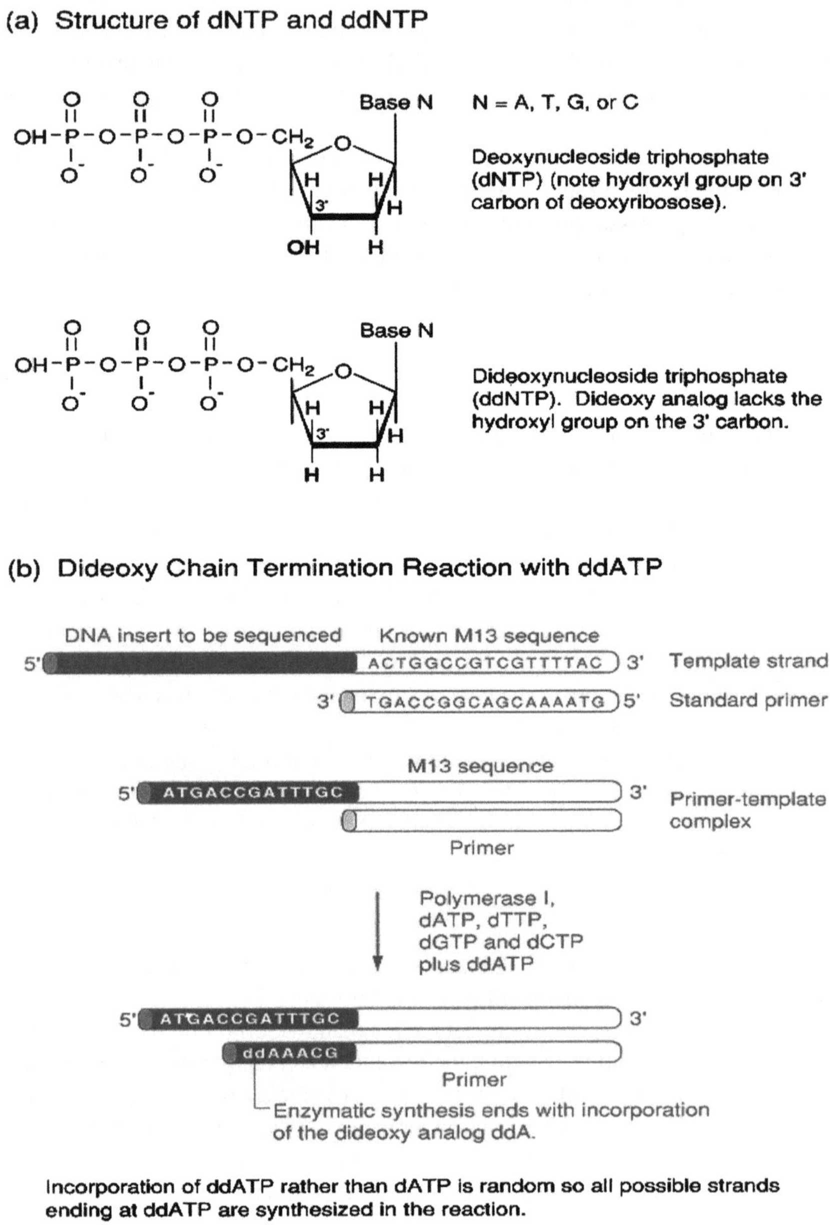

Figure 1.5. Sanger chain termination DNA sequencing method. (a) Chemical structures of a dNTP and also a ddNTP, which lacks a hydroxyl group on the 3′ site of the deoxyribose sugar. (b) Chain termination reaction with ddATP. Also shown is primer annealing to the template, which is directed by Watson–Crick base paring rules. (c) Entire Sanger chain termination process showing primer annealing, chain extension and termination and finally, gel fractionation.

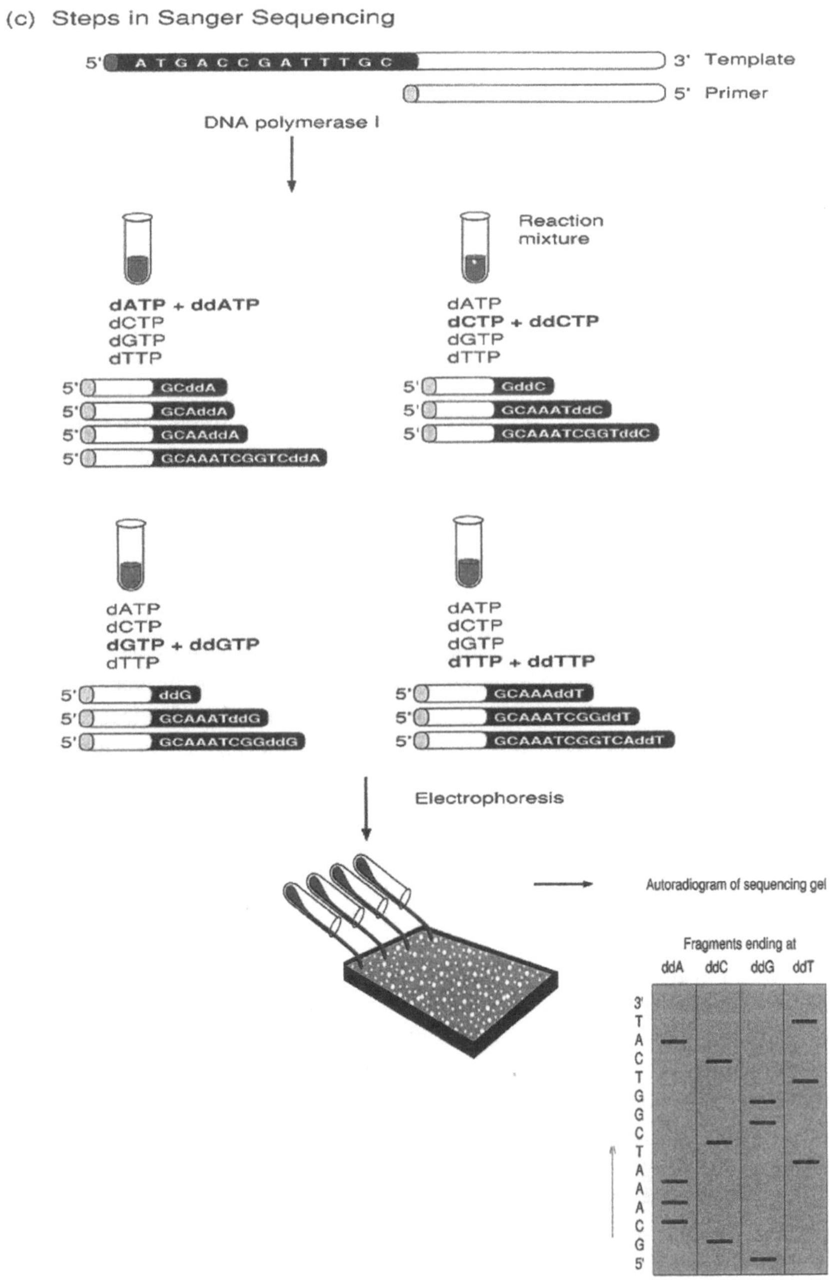

Figure 1.5. *(Continued)*

centage of crosslinker, N,N′-methylenebisacrylamide ($-CH_2(CHCOCH{=}CH_2)_2-$), in the presence of a catalyst accelerator-chain initiator mixture. The porosity of the gel is determined by the relative proportion of acrylamide monomer to crosslinking agent. For reproducible fractionation of DNAs, they must be maintained in a single-strand conformation, which is accomplished by adding denaturants, such as urea or formamide, to the sieving gel.

The existence of a gel network contributes significantly to the electrophoretic migration pattern (i.e., electrophoretic mobility) observed in the sequencing process. Any DNA fragments to be fractionated will inevitably encounter the gel network of polymer threads. This encounter increases the effective friction and consequently lowers the velocity of movement of the molecules. It is obvious that the retardation will be most pronounced when the mean diameter of gel pores is comparable to the size of the DNA fragments. Size therefore plays a critical role in determining the relative electrophoresis mobility and the degree of separation of different DNA fragments. It is this sieving effect that partly determines the resolution obtained with polyacrylamide gels. The gel network also minimizes convection currents caused by small temperature gradients.

The main characteristic property of DNA is that the value of the negative charge on the molecule is almost independent of the pH of the medium. Therefore, their electrophoretic mobility is due mainly to differences in molecular size, not in charge. For a particular gel, the electrophoretic mobility of a DNA fragment is inversely proportional to the logarithm of the number of bases up to a certain limit. Since each DNA fragment is expected to possess a unique size, its electrophoretic mobility is unique and it migrates to a unique position within the electric field in a given length of time. Therefore, if a mixture of DNA fragments is subjected to electrophoresis, each of the fragments would be expected to concentrate into a tight migrating band at unique positions in the electric field.

1.1.5.1. Slab Gel Electrophoresis

DNA sequencing is usually performed with slab gels. In slab gel electrophoresis, polymerization of acrylamide as well as the electrophoresis are conducted in a mold formed by two glass plates with a thin spacer. When analytical electrophoresis is performed, several samples are usually run simultaneously in the same slab. Since all samples are present in the same gel, the conditions of electrophoresis are quite constant from sample to sample. In slab gel electrophoresis, the sample is loaded using a pipette into wells formed into the gel during polymerization. A loading volume of 1–10 μL is typical for slab gel electrophoresis. The field strength that can be applied to these types of gels range from 50–80 V/cm, with the upper limit determined by heating (Joule) caused by current flow through the gel. Due to the thick nature of the gel, this heat is not efficiently dissipated, causing convective mixing and thus, zone broadening, which limits the upper level on the electric field strength that can be effectively used.

1.1.5.2. Capillary Gel Electrophoresis

In order to reduce the development time in the electrophoresis, many sequencing applications now use capillary gel electrophoresis, in which a glass tube (i.d. = 50–100 μm; length = 30–50 cm) contains the gel sieving matrix and an electric field is applied to this capillary column. Due to the higher surface-to-volume ratio afforded by the thin glass capillary, large electric fields can be applied (~300 V/cm), which results in shorter electrophoresis development times (2–3 hrs) and also enhanced plate numbers compared to slab gel electrophoresis. The glass capillary contains silanol groups on its surface, which at high pH are deprotonated. As such, the glass capillary is negatively charged above pH = 7.0. Cations in the electrolyte solution build up at the wall of the glass tube, producing an electrical double layer. When a voltage is applied across the capillary, the cations migrate to the wall and exert a force on the surrounding fluid causing a bulk flow of solution toward the cathode (negative terminal). This electrically induced flow is called an electroosmotic flow and, unfortunately, it can interfere with the electrophoresis of the DNAs. Therefore, in DNA separations using glass capillaries, the wall is coated with some type of polymer (for example a linear polyacrylamide) to suppress this electroosmostic flow. After coating the wall, the capillary can then be filled with the sieving gel, which can be a linear polyacrylamide (no crosslinking) or some other type of gel, such as a hydroxyl cellulose of poly (ethylene oxide).[5] In addition, crosslinked gels may also be used in these small capillaries, but since crosslinked gels are not free flowing like their non-crosslinked counterparts, they must be polymerized directly within the capillary tube. Most capillary-based DNA sequencers use linear gels, since they can be easily replaced using high pressure pumping allowing the capillary to be used for multiple sequencing runs. The crosslinked gels cannot be removed from the capillary once polymerized.

The electrophoresis is performed following filling the column with the gel by inserting one end of the capillary into a sample containing the sequencing mix and applying an electric field to the capillary tube. The injection end is typically cathodic (negative), with the opposite end being anodic (positive). By applying a fixed voltage for a certain time period, a controlled amount of sample can be inserted into the column with the injection volume ranging from 1 to 10 nL. Since the capillary is made of fused silica, fluorescence detection can be performed directly from within the capillary tube. In most cases, the capillary column is coated with a polyimide coating (non-transparent) to give it strength and the optical detection window can be produced by simply burning off a section of the polymer using a low temperature flame.

The higher plate numbers (i.e., better resolution) that is obtained in capillary gel electrophoresis compared to slab gel is a direct consequence of the ability to use higher electric fields. Since the development time is significantly shorter in capillary gel electrophoresis due to the ability to apply higher electric fields, band spreading due to longitudinal diffusion is reduced resulting in higher plate numbers. The ability to use higher electric fields in capillary gel electrophoresis results from the fact that Joule heating is suppressed since the heat can be effectively dissipated by the high surface-to-volume ratio capillary.

1.1.6. Detection Methods for DNA Sequencing

1.1.6.1. Autoradiographic Detection

Following electrophoresis, the individual DNA bands separated on the gel must be detected and subsequently analyzed. One of the earlier methods implemented to detect DNA bands in gels was autoradiography. In this mode, one of the phosphates of an individual nucleotide is replaced with a radioisotope, typically ^{32}P ($\tau_{1/2}$ = 14 days) or ^{35}S ($\tau_{1/2}$ = 87 days), both or which are radioprobes that emit β-particles. When Sanger methods are used to prepare the sequencing reactions, either the primer or the dideoxynucleotide can contain the radiolabel. The labeling is done using an enzyme (T4 polynucleotide kinase), which catalyzes the transfer of a γ-phosphate group from ATP to the 5′-hydroxy terminus of a sequencing primer. After the electrophoresis has been run, the gel is dried and then situated on an X-ray film. The film is developed (exposure to radiation from radioprobes) and dark bands are produced on the film where the DNA was resident. This is then followed by reading the sequence from the gel manually.

The primary advantage of this approach is the inexpensive nature of the required equipment to perform the measurement. It basically requires only a gel dryer, film holder, and film. The difficulties associated with this approach are numerous. One important issue is the fact that radioisotopes are used, and therefore waste disposal becomes a difficult problem to contend with. Throughput issues (data production rates) are also a primary concern when using autoradiographic detection. For example, radiography can sometimes require several days to expose the film in order to get strong signals to read the bases from the gel. In addition, the detection is done after the electrophoresis and not during the electrophoretic run. Also, since there is no means to identify the individual bases using radioprobes, each base must be analyzed in a different lane of the gel. And finally, the bases must be called manually, which many times leads to frequent errors in the sequence reconstruction. Therefore, the inability to obtain data production rates sufficient to accommodate large sequencing projects has made radiographic detection obsolete for high throughput applications.

1.1.6.2. Fluorescence Detection

For most DNA sequencing applications, irrespective of the separation platform used, fluorescence is the accepted detection protocol for several important reasons. Fluorescence allows one to perform the base calling and detection in an automated fashion and alleviates the need for manual base calling. In addition, fluorescence can be carried out during the separation, eliminating long film development times. More importantly, due to the ability to implement multiple probes possessing unique spectral properties, the four bases comprising the DNA molecule can be identified in a single gel lane, potentially increasing throughput by a factor of four compared to radiographic detection. All of these important advantages associated with fluorescence allow for higher throughputs in DNA sequencing applications. As such, fluorescence can be considered as one of the most important recent technical innovations in DNA sequencing and has made it feasible to consider tackling large genome sequencing projects, such as the human genome.

The first demonstrations on the use of fluorescence in DNA sequencing came with the work of Smith *et al.*,[6] Prober *et al.*,[7] and Ansorge *et al.*[8] In these works, slab gels or large gel tubes were used to fractionate the DNA ladders produced during enzymatic polymerization using Sanger sequencing strategies. The fluorescence detection was accomplished using four spectroscopically unique probes, which allowed the DNA sequence reconstruction to be done in a single electrophoresis lane of the gel. The chemical structures of the dye labels used in the Smith *et al.*[6] and Prober *et al.*[7] experiments are shown in Figures 1.6 and 1.7,

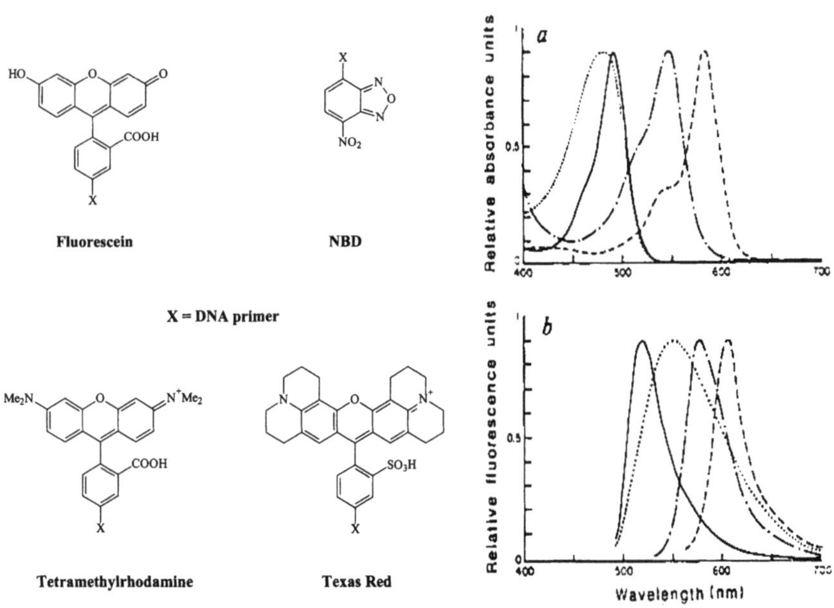

Figure 1.6. Chemical structures and absorbance and emission spectra of the dyes used for labeling primers. (Adapted with permission from reference 6.)

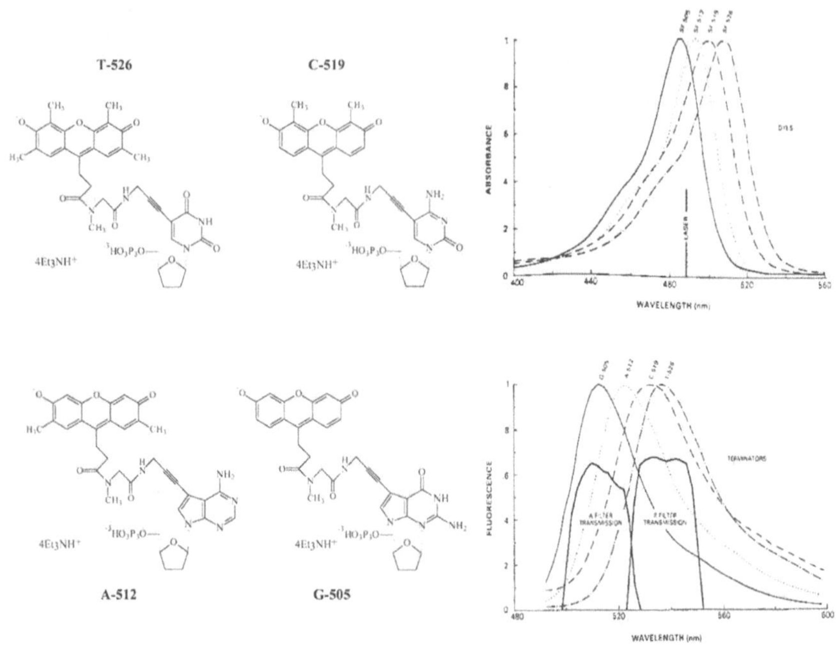

Figure 1.7. Chemical structures and the absorption and emission spectra of dyes conjugated to ddNTPs. Also shown in the figure are the excitation laser wavelength (488 nm) and the filter set used to isolate the fluorescence onto the two detection channels. (Adapted with permission from reference 7.)

respectively. As can be seen, the dyes were either attached (covalently) to the sequencing primer or to the dideoxynucleotides. The advantage of using dye-labeled dideoxynucleotides is that the sequencing reactions can be performed in a single reaction tube, whereas the dye-labeled primer reactions must be performed in four separate tubes and pooled prior to electrophoresis. In the case of the dye-labeled terminators, succinylfluorescein analogs were used with slight structural modifications to alter the absorption/emission maxima. The dyes were attached either to the 5 position of the pyrimidine bases or the 7 position of the 7-deaza-purines, both of which are non-hydrogen bonding sites on the nucleotide base. The linker structure is also important, which in this case was a propargylamine, since the presence of the dye onto the terminator radically affects its ability to be incorporated by the polymerase enzyme. For dye-labeled primers, the oligonucleotides possessing the appropriate sequence were prepared on a standard DNA synthesizer. For Smith *et al.*,[6] a thymidine derivative was prepared, which contained a phosphoramidite at the 3′ carbon and a protected alkyl amino group at the 5′ carbon (typically a 6-carbon linker structure). During the final addition cycle of the oligonucleotide prepared via solid-phase synthesis using phosphoramidite chemistry the thymidine residue is added and, following deprotection of the alkyl amino group and cleavage from the support, a free primary amine group results which can be reacted with any amino-reactive fluorescent dye to produce the oligonucleotide derivative.

Also shown in Figures 1.6 and 1.7 are the absorption and emission profiles for the dye sets used in these experiments. The major attributes of the dye sets are that they can be efficiently excited with either 488 and/or 514.5 nm lines of the Ar ion laser. In addition, there is minimal separation between the emission maxima of the dyes, which allows processing of the fluorescence on as few detection channels as possible. However, there is significant overlap in the emission spectra of the four dyes, producing severe spectral leakage into other detection channels, which must be corrected by software.

The optical hardware used to process the four-color fluorescence for both systems are depicted in Figure 1.8. For the Smith *et al.* experiment,[6] a single laser was used as well as a single detection channel, which in this case consisted of a multiline Ar ion laser and a conventional photomultiplier tube (PMT). Placed in front of the laser and PMT were filter wheels to select the appropriate excitation wavelength (488 or 514.5 nm) and emission color. The filter pairs used during fluorescence readout were 488/520 nm; 488/550 nm; 514/580 nm; 514/610 nm. In the case of the Prober *et al.* experiment,[7] due to the narrow distribution between the excitation maxima of the dyes, only excitation using the 488 nm line from the Ar ion laser was required as well as two PMT tubes to process the emission from the four colors. Discrimination of the four colors was accomplished by monitoring the intensity of each dye on both detectors simultaneously. By histograming the ratio of the fluorescence intensity of each dye (produced from an electrophoresis band) on the two detection channels, a discrete value was obtained that allowed facile discrimination of the four different fluorescent dyes (i.e., terminal base). In order to determine the limit of detection of these fluorescence systems, injections of known concentrations of dye labeled sequencing primers were electrophoresed. In both cases, the mass detection limit was estimated to be 10^{-17}–10^{-18} moles.

In Figure 1.9 is shown the data output from these systems. Unfortunately, reading the sequence directly from the raw gel data becomes problematic due to several non-idealities, including signal from a single dye appearing on multiple detection channels due to the broad and closely spaced emission bands, dye-dependent electrophoretic mobility shifts, and non-uniformity in the intensity of the electrophoresis bands due to the enzymatic reaction used to construct the individual DNA size-ladders. As such, several post-electrophoresis processing steps were required to augment sequence reconstruction of the test template. In the case of the Smith *et al.* example,[6] these steps involved:

(A)

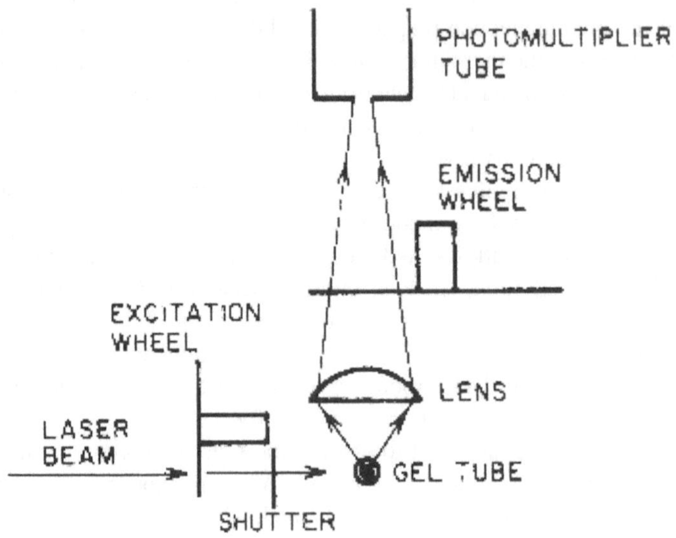
PHOTOMULTIPLIER TUBE
EMISSION WHEEL
EXCITATION WHEEL
LENS
LASER BEAM
GEL TUBE
SHUTTER

(B)

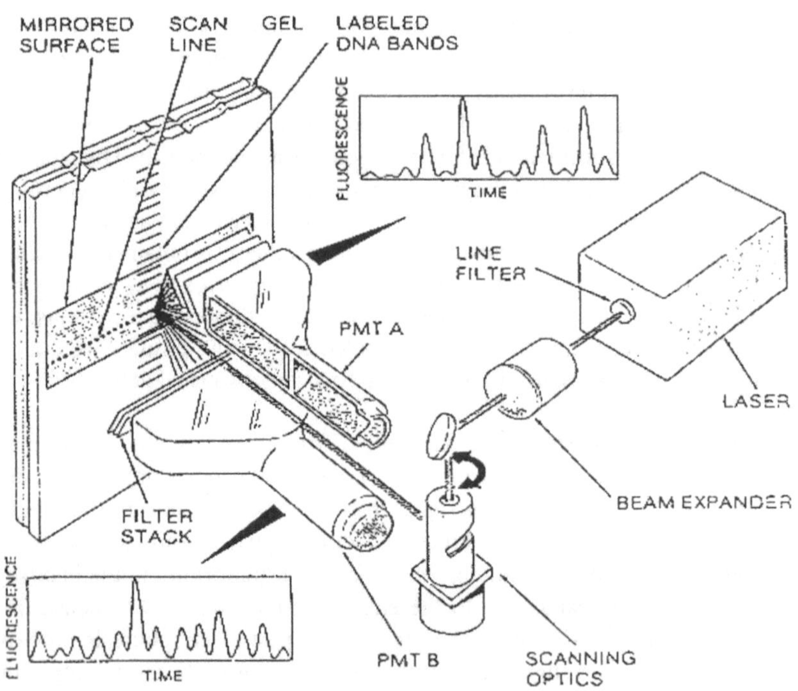
MIRRORED SURFACE
SCAN LINE
GEL
LABELED DNA BANDS
FLUORESCENCE
TIME
LINE FILTER
PMT A
LASER
BEAM EXPANDER
FILTER STACK
FLUORESCENCE
TIME
PMT B
SCANNING OPTICS

1. High frequency noise removal using a low-pass Fourier filter.
2. A time-delay between measurements at different wavelengths corrected by linear interpolation between successive measurements.
3. A multicomponent analysis performed on each set of four data points, which produced the amount of the four dyes present in the detector as a function of time.
4. The peaks present in the data stream located.
5. The mobility corrected for the dye attached to each DNA fragment. In this case, it was empirically determined that fluorescein and rhodamine-labeled DNA fragments moved as if they were 1 base longer than the NBD-labeled fragments and the Texas Red fragments moved as if they were 1¼ bases longer.

The important performance criteria in any type of automated DNA sequencer is its throughput, the number of bases it can process in a single gel read and its accuracy in calling bases. In terms of base calling accuracy, these early instruments demonstrated an error rate of approximately 1% with a read length approaching 500 bases. The throughput of the instrument described by Prober *et al.*[7] was estimated to give a raw throughput of 600 bases per hour (12 electrophoresis lanes). Interestingly, many present-day commercial automated sequencers still use similar technology in their machines and the throughput can be as high as 16,000 bases per hour (96 electrophoresis lanes).

1.2. Fluorescent Dyes for DNA Labeling and Sequencing

As stated above, fluorescence detection has had a tremendous impact in the area of DNA sequencing because of the speed of the readout process as well as the ability to discriminate amongst the four nucleotide bases in a single gel lane due to the unique spectral properties of the target dye molecules. There are a variety of dye sets (typically four dyes per set, one for each nucleotide base) that have been developed for DNA sequencing applications and whichever dye set one wishes to

Figure 1.8. Fluorescence detector systems for DNA sequencing. (A) The laser consisted of an Ar ion laser operating at 488 and 514 nm. The wavelength was selected using a rotating filter wheel. The emission was collected by a lens and sent through a filter wheel that contained four different filters, one for each of the multi-color dyes used to label the sequencing primers (see Figure 1.6 for their structures and fluorescence properties). (Adapted with permission from reference 6.) (B) The laser used for excitation is an Ar ion laser operated at 488 nm only. The beam was rastered over the slab gel plate using scanning optics. The emission was then processed on one of two elongated, stationary photomultiplier tubes (PMTs) (see Figure 1.7 for dye set used with this fluorescence detector). (Adapted with permission from reference 7.)

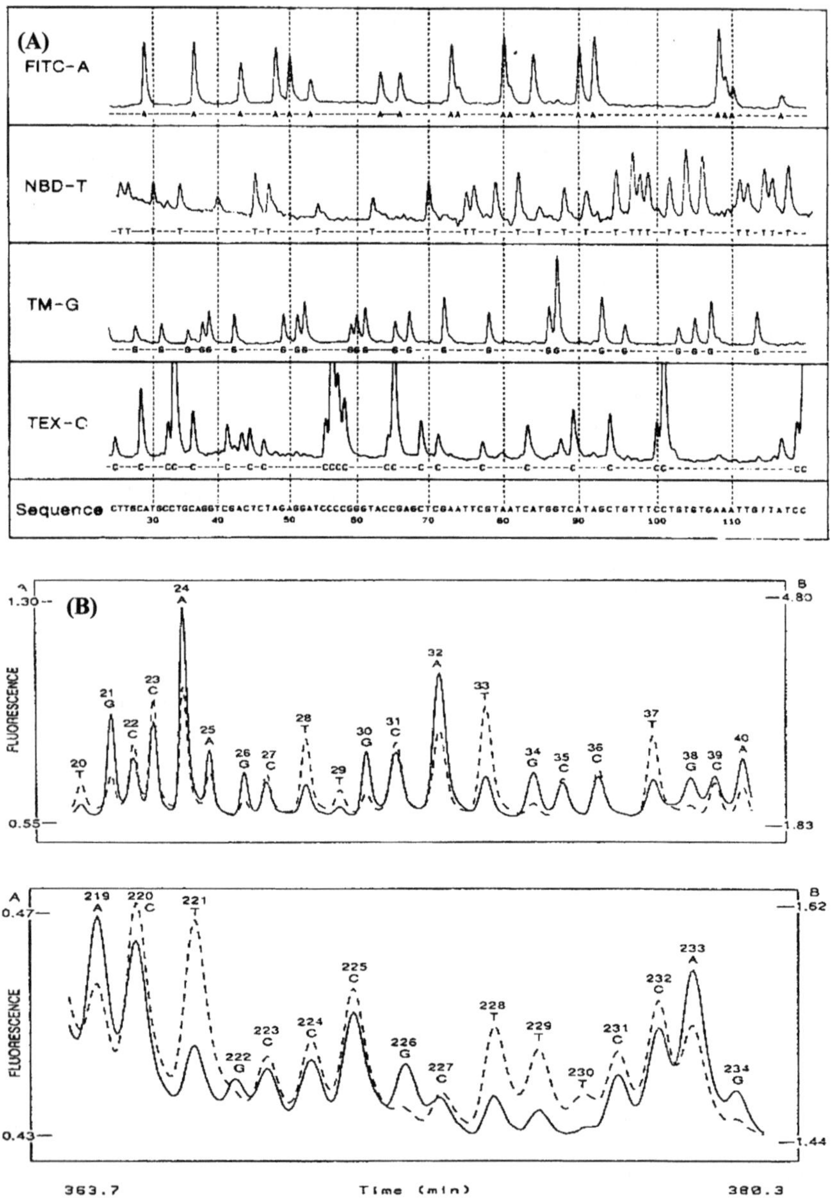
(A)
FITC-A
NBD-T
TM-G
TEX-C
Sequence
(B)
FLUORESCENCE
Time (min)
363.7
380.3

consider, certain properties associated with the dyes for DNA sequencing applications are important. These properties are listed below:

1. *Each dye in the set must be spectroscopically distinct.* The ideal situation from a throughput point of view, is to identify each base in a single gel lane instead of four gel lanes. In most cases, the discrimination is based on differences in the emission properties of each dye (distinct emission maxima), however, other fluorescence properties can be used as well, such as fluorescence lifetimes.
2. *The dye set should preferably be excited by a single laser source.* It becomes instrumentally difficult to implement multiple excitation sources since lasers are typically used as the source of excitation to improve the limit of detection in the measurement and the upkeep on multiple lasers becomes problematic.
3. *The dye set should possess high extinction at the excitation frequency and also large quantum yields in the gel matrix used to fractionate the DNAs.* Good photophysical properties are necessary in order to improve detectability. In addition, the dye set should show reasonable quantum yields in denaturing gels, which consist of high concentrations of urea and/or formamide.
4. *The dye set should show favorable chemical stability at high temperatures.* A common procedure in most Sanger sequencing strategies is to implement a thermostable polymerase and then subject the sequencing reactions to multiple temperature cycles (55°C–95°C) in order to amplify the amount of product generated (cycle sequencing). Therefore, the dyes must be able to withstand high temperature conditions for extended periods of time.

Figure 1.9. Fluorescence sequencing data obtained from the two fluorescence systems described in Figure 1.8. (A) The dyes (see Figure 1.6) were conjugated to a primer with a sequence, 5′-CCCAGTCACGACGTT-3′, complementary to a site on the M13 phage vector. The reactions were run using Sanger chain termination methods and run in four different vessels, one for each terminator. The polymerase enzyme was a T7 polymerase and the reactions were pooled prior to electrophoresis. (Adapted with permission from reference 6.) (B) Sequence of the M13mp8 template using the dye-labeled terminator set shown in Figure 1.7. The base assignment is given as a letter above each electrophoretic peak. The reactions were prepared using Sanger chain termination methods and prepared in a two-stage fashion. In the first stage, the template (M13mp18) was added to a reaction tube along with the primer, heated to 95°C for 2 mins and then cooled in ice water for 5 mins. In the second stage, dNTPs and the dye-labeled ddNTPs were added along with an AMV reverse transcriptase enzyme. Following chain extension, the reactions were purified (removal of excess terminators) using gel filtration (Sephadex spin column) and loaded onto the gel. The gel was 8% polyacrylamide containing 7 M urea as the denaturant. In these traces, the solid line represents the fluorescence signal from PMT A and the dashed line is from PMT B. (Adapted with permission from reference 7.)

5. *The dyes must induce minimal mobility shifts during the electrophoresis analysis of the sequencing ladders.* Many times, the mobility shifts induced by individual dyes can cause misordering of the individual bases during sequence reconstruction. As such, post-electrophoresis corrections are many times employed to rectify this perturbation in the DNA's mobility.
6. *The dyes must not significantly perturb the activity of the polymerase enzyme.* This is especially true in dye-terminator chemistry, since the proximity of the dye to the polymerase enzyme can dramatically influence the ability of the enzyme to incorporate that dye-ddNTP conjugate into the polymerized DNA molecule.

1.2.1. Visible Fluorescence Dyes for DNA Labeling

The common dye set that is frequently used in many automated, fluorescence-based DNA sequencers is the FAM, JOE, TAMRA, and ROX series (see Figure 1.10), which consist of fluorescein and rhodamine analogs containing a succinimidyl ester for facile conjugation to amine terminated sequencing primers or terminators. These dyes can be efficiently excited by the 488 and 514.5 nm lines from an Ar ion laser and also possess emission profiles that are fairly well resolved. Many of the Applied Biosystems automated DNA sequencers use this particular dye set (for example, ABI 373 and 377 series, see www2.perkin-elmer.com). Shown in Figure 1.10 are emission spectra for this dye set as well as the filter set that is used to isolate the emission from the dyes onto the appropriate detection channel. While this dye set is fairly robust and works well with typical DNA cycle sequencing conditions, there are some difficulties in using this series, namely the broad emission profiles, dye dependent mobility shifts, and inefficient excitation of TAMRA and ROX with the 514.5 nm line from the Ar ion laser.

In order to eliminate dye-dependent mobility shifts and to minimize cross-talk between detection channels using four-color processing, a bodipy dye set (4,4-difluoro-4-bora-3,4α-diaza-s-indacene-3-proprionic acid) have been used for DNA sequencing applications.[9] The structures of the dyes are presented in Figure 1.11. Inspection of the fluorescence emission profiles for this particular dye set indicated that the bandwidths were less than those associated with the FAM/JOE/TAMRA/ROX dye set, which resulted in less spectral leakage between detection channels in the fluorescence readout hardware. Inspection of Figure 1.11 reveals that all of these dyes are neutral and have very similar molecular structures, thereby minimizing dye-dependent mobility shifts. In order to fully compensate for any mobility shift differences within the dye set, a modified linker structure was synthesized to account for this difference (see Figure 1.11).

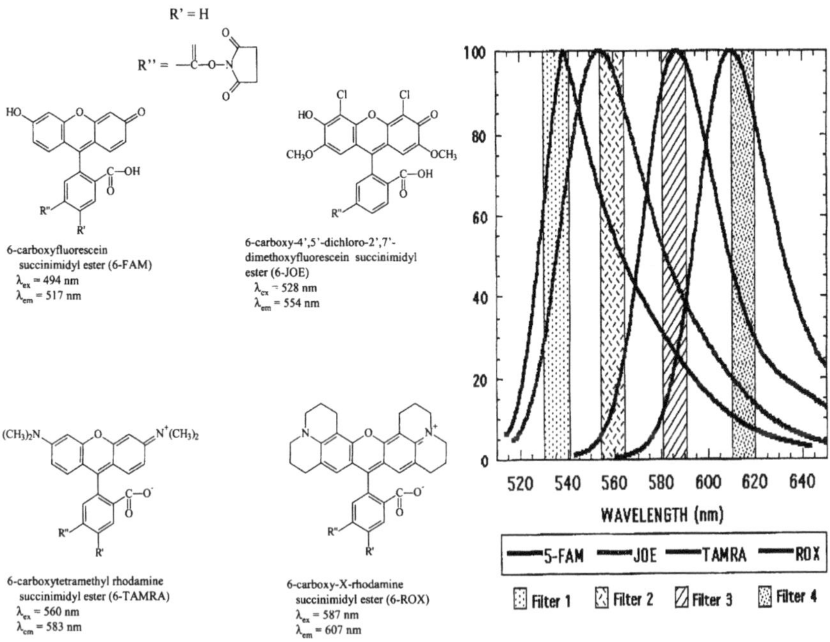

Figure 1.10. Common fluorescence dye set used for labeling primers for DNA sequencing applications. The functional group on each dye is a succinimidyl ester, which readily conjugates to primary amine groups. Also shown in this figure are the emission spectra of this dye set as well as the filters used to select the appropriate dye color processed by each detection channel.

A particular shortcoming associated with this dye set is the chemical instabilities they display when subjected to extended heating at high temperatures. An example of this is shown in Figure 1.12, in which Bodipy-labeled primers were used with cycle sequencing conditions. In cycle sequencing, a linear amplification of the ddNTP terminated fragments can be generated by subjecting the reaction cocktail to multiple temperature cycles consisting of 95°C (thermal denaturation of the double-stranded DNA molecule); 55°C (annealing the sequencing primer to the DNA template); and finally 72°C (chain extension using a thermostable polymerase enzyme, such as *Taq* polymerase). The difficulty arises during the 95°C temperature step, where significant dye decomposition can result. In Figure 1.12 is shown fluorescence traces of sequencing ladders prepared using either a 95°C step time of 30 s or 5 s. As can be seen, there is a significant loss in signal when the cycling time (at 95°C) is 30 s, but the signal is partially restored when this cycling time is reduced to 5 s.

Many of the dyes discussed above have been employed in dye-primer sequencing applications, in which the primer used for selecting the DNA polymer-

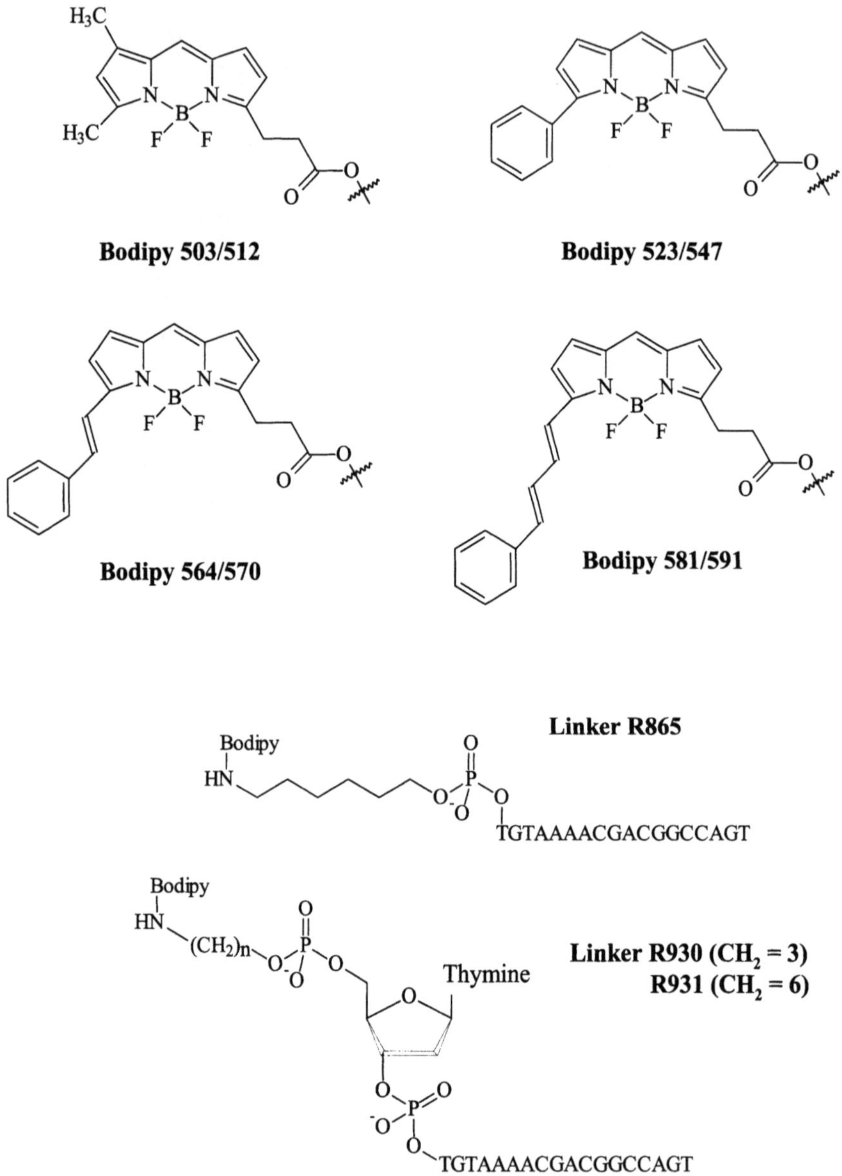

Figure 1.11. Bodipy dye set used for DNA sequencing applications. Below each structure is shown the absorption/emission maxima (nm) of the particular dye. Also shown is the universal sequencing primer and the modification required to attach the dye to the primer. The linkers were either 3 or 6 carbon linkers with primary amine groups. The linkers are designated as R865 (C-6 linker, U-C_6), R930 (C-3 linker, U^+-C_3) or R931 (C-6 linker, U^+-C_6). (Adapted with permission from reference 9.)

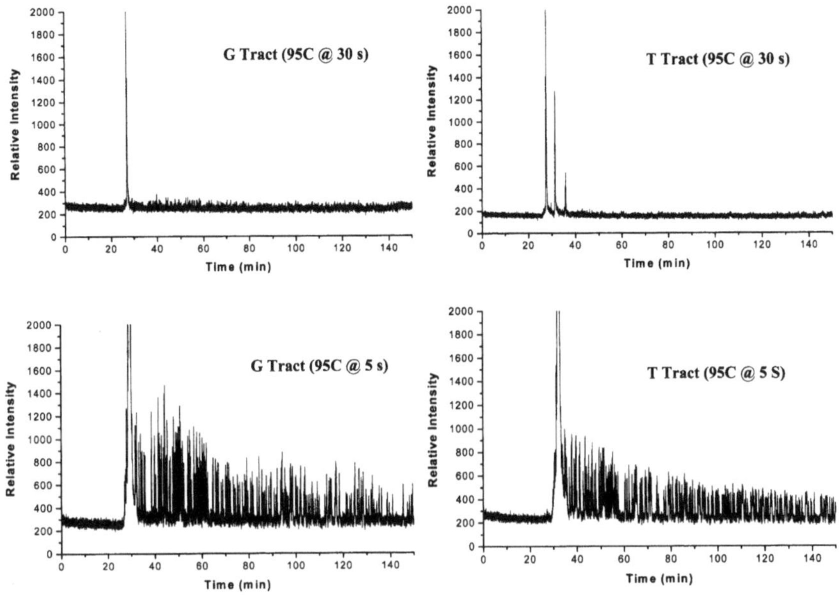

Figure 1.12. Cycle sequencing effects on bodipy-labeled DNA sequencing primers. The sequencing reactions were prepared using Sanger chain termination methods and a *Taq* DNA polymerase. The cycling conditions consisted of 25 cycles with 95°C for 30 or 5 s; 55°C for 10 s; 72°C for 60 s. The reactions were run with a single terminator (ddGTP, ddTTP) and an M13mp18 template. The sequencing reactions were analyzed in a capillary gel electrophoresis DNA sequencer using laser excitation at 532 nm. The dyes were Bodipy 564/570 (ddGTP) and Bodipy 581/591 (ddTTP) (see Figure 1.11 for structures of dyes). The large peak at ~28 mins is unextended dye-labeled primer.

ization site on the unknown template is determined by Watson–Crick base pairing. An alternative is dye-terminator chemistry, where the fluorescent dyes are covalently attached to the ddNTP used in Sanger sequencing strategies. The advantages of this are twofold: (1) The sequencing reactions can be carried out in a single tube. In dye-primer chemistry, the sequencing reactions are carried out in four separate tubes and then pooled prior to the electrophoresis step. If one implements four spectroscopically unique fluorescent probes attached to the ddNTP, then the reactions can be carried out in a single tube, which reduces reagent consumption and also minimizes sample transfer steps. (2) Primers of known sequence do not need to be synthesized. In primer sequencing, making oligonucleotides of 17–23 bases in length can be costly and time consuming, especially if the dye must be chemically tethered to the primer. The use of dye-terminator chemistry eliminates the need for synthesizing dye-labeled primers. Unfortunately, dye-terminators themselves can be quite expensive, and many polymerase enzymes are very sensitive to the type of dye attached to the ddNTP.

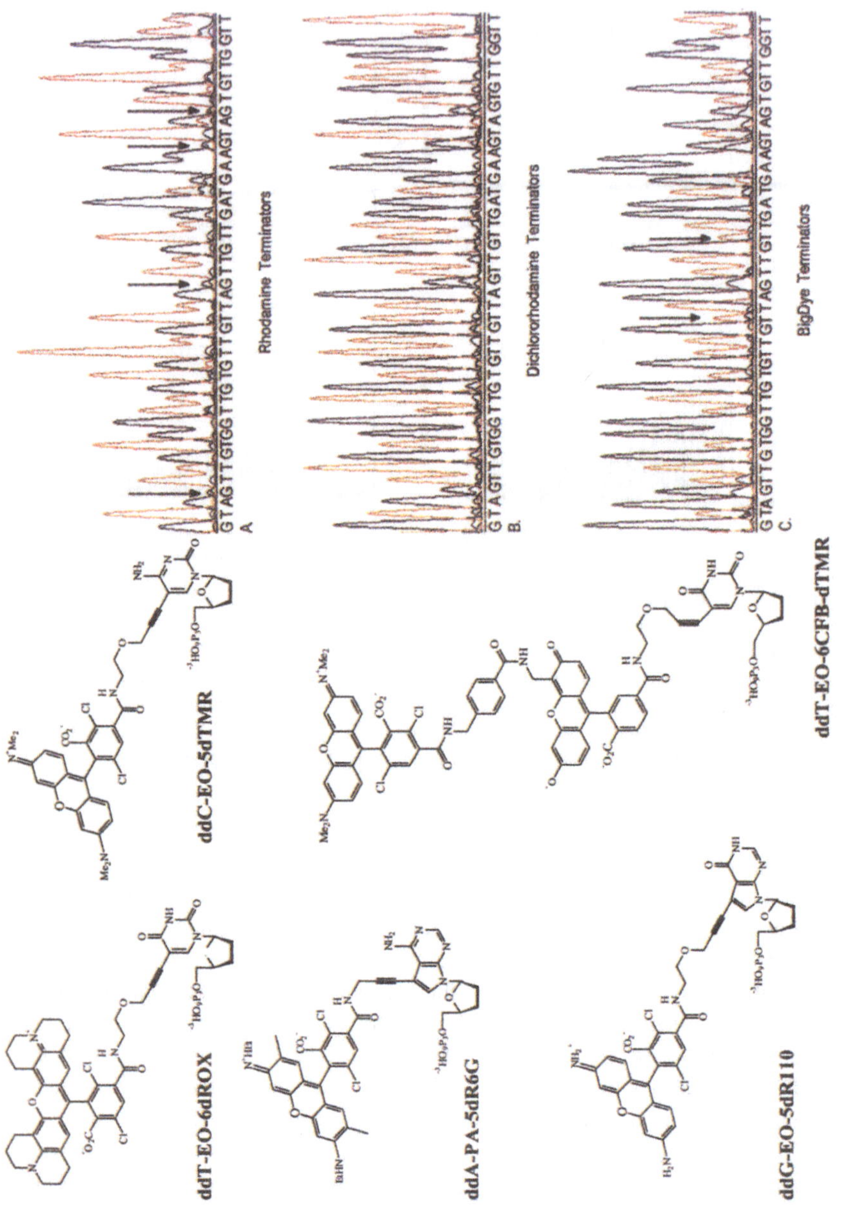
ddT-EO-6dROX
ddC-EO-5dTMR
ddA-PA-5dR6G
ddG-EO-5dR110
ddT-EO-6CFB-dTMR
A.
Rhodamine Terminators
B.
Dichlororhodamine Terminators
C.
BigDye Terminators

For example, fluorescein dyes are poor labels for terminators when using the *Taq* polymerase due to its incompatibility with the binding pocket of *Taq*, while the rhodamine dyes are more hydrophobic and as such, are more suitable for use with *Taq* polymerase. The result is that the peak heights for the electrophoresis bands can vary tremendously due to differences in incorporation of the dye-modified ddNTPs by the particular polymerase enzyme. In dye-primer chemistry, this disparity is absent due to the large displacement of the dye from the polymerization site.[10] It is interesting to note that several mutant forms of *Taq* polymerase have been prepared to allow more facile incorporation of dye-labeled terminators.[11] For example *Taq* Pol I (AmpliTaq FS) has two modifications in it, a substitution which eliminates the 3′→5′ nuclease activity and also a substitution that improves 2′,3′-ddNTP incorporation.

The nature of the dye on the terminator can also influence the mobility of the polymerized DNA fragment as well. For example, rhodamine dyes are typically zwitterionic and, as such, appear to stabilize hairpin (secondary) structures in the DNA fragment causing compressions in the electrophoresis data, especially in GC-rich regions of the template. On the other hand, the fluorescein dyes, which are negatively charged, do not suffer from such anomalies.[10]

Slight structural modifications on the base chromophore can influence its incorporation during DNA polymerization. Also, the linker structure can influence the incorporation efficiency as well. A set of dye-labeled terminators that have been found to give fairly even peak heights in sequencing patterns and produce minimal mobility shifts are shown in Figure 1.13 along with the sequencing patterns that were generated with both the rhodamine dyes and the dichlororhodamine (d-rhodamine) analogs.[12] The linker structures chosen for this set were either the propargylamine linker developed by Prober *et al.*[7] or a propargyl ethoxyamino linker. The choice of linker was selected to accommodate the polymerase enzyme and to minimize the mobility differences within the dye set. For the rhodamine terminators, very weak G-peaks which appeared after A-peaks were observed. However, for the d-rhodamine terminators, this disparity in peak

Figure 1.13. Chemical structures of the d-rhodamine dyes used for labeling terminators. Also shown is the structure of an energy transfer terminator (BigDye terminator). The accompanying panels show a comparison of sequencing data obtained using rhodamine (A), d-rhodamine (B), and BigDye (C) labeled terminators. The arrows in (A) are G-peaks with weak signal. In (C), the arrows indicate weaker T-peaks following G-peaks. The sequencing reactions were run with an AmpliTaq DNA polymerase and cycle sequencing conditions (30 cycles with a temperature program of 95°C for 30 s; 55°C for 20 s; 60°C for 4 mins). The DNA template was isolated from a bacterial artificial chromosome (BAC). Following chain extension, the reactions were purified using a Centri-Sep spin column to remove excess terminators and then dried in a vacuum and resuspended in 2–4 μL of formamide. The reactions were electrophoresed in a slab gel with laser-induced fluorescence detection. (Adapted with permission from reference 12.)

intensity was alleviated, with the peak heights in the pattern being much more uniform.

1.2.2 ET Dyes for DNA Sequencing

One of the major problems associated with many of the single dye sets mentioned previously is that their absorption spectra are dispersed over a relatively large spectral range, which provides poor excitation efficiency even for dual laser (488 and 514 or 543) systems. As such, the red-dyes are used at higher concentrations during DNA polymerization to circumvent poor excitation. To overcome this problem without sacrificing spectral dispersion in the emission profiles, the phenomena of Förster energy transfer has been used to design sequencing primers, which can be efficiently excited with a single laser line.[13–18] While the focus of a subsequent chapter in this book will deal with this topic, it is informative to briefly introduce these energy transfer (ET) primers at this time in the context of DNA sequencing.

The chemical structures of the ET primers developed by Richard Mathies and his group[13] are shown in Figure 1.14. The donor dye in this case was FAM, which could be excited with the 488 nm line of an Ar ion laser. The acceptor dyes were either FAM, JOE, TAMRA, or ROX. The donor (FAM) was attached to the sequencing primer on the 5′ end during the solid-phase DNA synthetic preparation of the M13 (-40) sequencing primer using phosphoramidite chemistry. The sequencing primer also contained a modified base (T*) that possessed a linker structure with a primary amine. The appropriate acceptor was conjugated to the primary amine group off the modified base following cleavage from the solid support via a succinimidyl ester functional group. The spacer distance between the donor and acceptor was selected by positioning T* within the M13 (-40) primer sequence during solid phase synthesis. The naming of these ET primers followed the convention, donor-spacer (bp)-acceptor. In the case of the FAM and JOE ET primers, a 10 base spacer was used and for TAMRA and ROX, a 3 base spacer was selected. The choice of spacer size was primarily determined by producing ET primers, which showed uniform electrophoretic mobilities.

The absorption and emission profiles of the ET primer series are showin in Figure 1.14. While the absorption spectra show bands from both the acceptor and donor dyes, the emission profiles are dominated by fluorescence from the acceptor dyes. In fact, the energy transfer efficiency has been determined to be 65% for F10J, 96% for F3R, and 97% for F3T.[13] As can be seen from the emission profiles displayed in Figure 1.14, the emission intensity was found to be significantly higher for the ET primers compared to their single dye primer partners when excited with 488 nm laser light from the Ar ion laser due to higher efficiency in

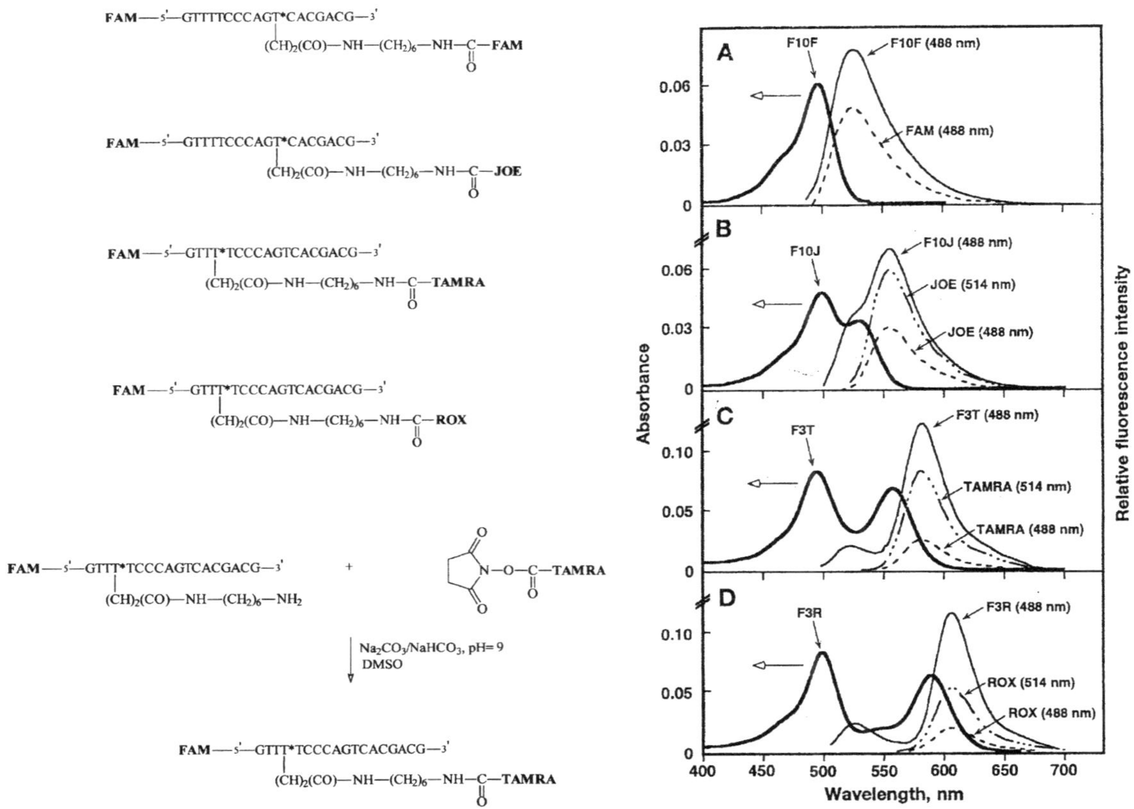

Figure 1.14. Chemical structures and absorption and fluorescence emission spectra of energy transfer (ET) labeled DNA sequencing primers. The bottom panel shows the reaction of a dye-labeled FAM primer (5′-end with $-NH_2$ T modified residue) and a succinimidyl TAMRA dye (F3T). The absorption (dark line) and emission spectra for both the ET primers and the single dye-labeled primers are also shown for comparison purposes. The number in parenthesis is the excitation wavelength used for collecting the emission profile. (Adapted with permission from reference 13.)

excitation. This translates into improved fluorescence sensitivity of these ET primers during electrophoresis.

Using these ET primers does offer some advantages due to their improved detection sensitivities, namely eliminating the need for adjusting the concentrations of the dye primers used during polymerization and also the need for smaller amounts of template in the sequence analysis. In fact, the use of ET primers required about one quarter the amount of template compared to the single dye primers.

Energy transfer dye pairs can also be situated on terminators and an example is shown in Figure 1.13.[12] In this example, d-rhodamine and rhodamine dyes are used with a propargyl ethoxyamino linker. This dye set has been called Big Dye terminators. Again, the dyes were selected so as to provide fairly uniform peak heights in the electrophoresis, and also uniform mobility shifts within the series. A sample of a sequence run using these Big Dye terminators is also shown in Figure 1.13.

1.2.3. Near-Infrared Dyes for DNA Sequencing

The attractive feature associated with fluorescence in the near-IR ($\lambda_{ex} > 700$ nm) includes smaller backgrounds observed during signal collection and the rather simple instrumentation required for carrying out ultrasensitive detection. In most cases, the limit of detection for fluorescence measurements is determined primarily by the magnitude of the background produced from scattering and/or impurity fluorescence. This is particularly true in DNA sequencing since detection occurs within the gel matrix, which can be a significant contributor of scattering photons. In addition, the use of denaturants in the gel matrix, such as urea (7 M) or formamide, can produce large amounts of background fluorescence. The lower background that is typically observed in the near-IR can be attributed to the fact that few species fluoresce in the near-IR. In addition, the $1/\lambda^4$ dependence of the Raman cross section also provides a lower scattering contribution at these longer excitation wavelengths.

An added advantage of near-IR fluorescence is the fact that the instrumentation required for detection can be rather simple and easy to use. A typical near-IR fluorescence detection apparatus can consist of an inexpensive diode laser and single photon avalanche diode (SPAD). These components are solid-state allowing the detector to be run for extended periods of time requiring little maintenance or operator expertise.

Near-IR fluorescence can be a very attractive detection strategy in gel sequencing because of the highly scattering medium that the separation must be performed in. Due to the intrinsically lower backgrounds that are expected in the near-IR compared to the visible, on-column detection can be performed without sacrificing detection sensitivity. To highlight the intrinsic advantages associated with the use of near-IR fluorescence detection in capillary gel DNA sequencing applications, a direct comparison between laser-induced fluorescence detection at 488 nm excitation and 780 nm excitation has been reported.[19] In this study, a sequencing primer labeled with FAM or a near-IR dye were electrophoresed in a capillary gel column and the detection limits calculated for both systems. The results indicated that the limits of detection for the near-IR case were found to be 3.4×10^{-20} moles, while for 488 nm excitation the limit of detection was 1.5×10^{-18} moles. The improvement in the limit of detection for the near-IR case was observed in spite of the fact that the fluorescence quantum yield associated with the near-IR dye was only 0.07, while the quantum yield for the FAM dye was ~0.9. The improved detection limit resulted primarily from the significantly lower background observed in the near-IR. Near-IR has also been demonstrated in sequencing applications using slab gel electrophoresis where the detection sensitivity has been reported to be ~2000 molecules.[20,21]

The fluorophores that are typically used in the near-IR include the tricarbocyanine (heptamethine) dyes, which consist of heteroaromatic fragments linked by a polymethine chain (see Table 1.1). The absorption maxima can be altered by changing the length of this polymethine chain or changing the heteroatom within the heteroaromatic fragments. These near-IR dyes typically possess large extinction coefficients and relatively low quantum yields in aqueous solvents.[22] The low fluorescence quantum yields result primarily from high rates of internal conversion. An additional disadvantage associated with these dyes are their poor water solubility. These dyes show a high propensity toward aggregation forming aggregates with poor fluorescence properties. This aggregation can be alleviated to a certain degree by inserting charged groups within the molecular framework of the fluorophore, for example alkyl-sulfonate groups. In addition, these dyes have short fluorescence lifetimes and are very susceptible to photobleaching compared to the visible dyes typically used in DNA sequencing applications. In most cases, the near-IR chromophore is covalently attached to either a sequencing primer or ddNTP via an isothiocyanate functional group. As can be seen in Table 1.1, this functional group can be placed on the heteroatom (N in this case) or in the para-position of the bridging phenyl ring. When situating the isothiocyanate group on the heteroatom, the net charge on the dye becomes neutral and as such, has limited water solubility. In the case of placing the isothiocyanate group on the bridging phenyl ring, this gives a net -1 charge to the dye and improves its water solubility. Unfortunately, the ether linkage is susceptible to nucleophilic attack, especially by dithiothriotol (DTT), which can release the dye from the moiety (sequencing primer or ddNTP) to which it is attached. This can produce a large peak in the electropherogram which can mask some of the sequencing fragments that comigrate with the free dye. This problem can be alleviate to a certain extent using an ethanol precipitation step following DNA polymerization.

Table 1.1. Chemical Structures of Some Typical Near-IR Fluorescent Dyes and Their Photophyiscal Properties[a]

Sulfonated, heavy-atom modified near-IR dyes (X = I, Br, Cl, F)	Dye substitution	Absorbance (nm)	Emission (nm)	ε (M^{-1} cm^{-1})	ϕ	τ_f (ps)
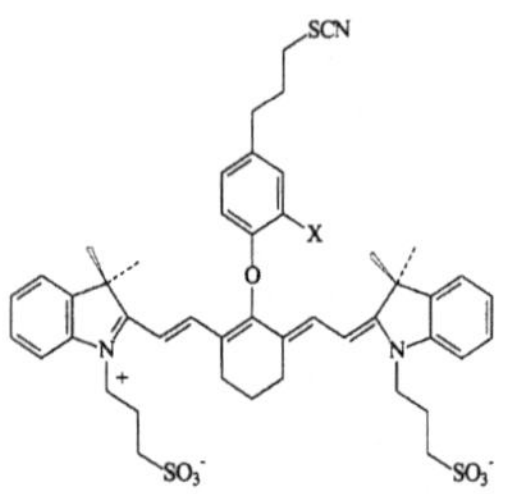	I	766	796	216,000	0.15	947
	Br	768	798	254,000	0.14	912
	Cl	768	797	239,000	0.14	880
	F	768	796	221,000	0.14	843

IRD 41	Absorbance (nm)	Emission (nm)	ϕ	ε (M^{-1} cm^{-1})
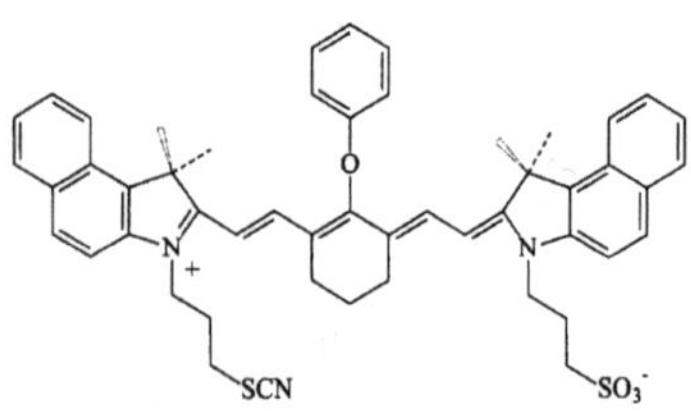	787	807	0.16	200,000

[a]In the first panel of this table, the dyes were modified with an intramolecular heavy atom so as to produce a set of dyes with four distinct lifetime values. In both cases, the dyes contained an isothiocyanate to allow conjugation to amine-containing molecules.

1.3. Instrumental Formats for Fluorescence Detection in DNA Sequencing

The ability to read the fluorescence during the electrophoresis fractionation of the DNA ladders and also, accurately identify the terminal base (Sanger sequencing) is a challenging task due to a number of technical issues. As such, a number of fluorescence readout devices have been developed for reading such data. When considering the design of a fluorescence detector for sequencing applications, several instrumental constraints must be enveloped into the design, including:

1. *High sensitivity.* As pointed out previously, the amount of material (fluorescently-labeled DNA) loaded onto the gel can be in the low atto-mole range (10^{-18} moles) and the detector must be able to read this

fluorescence signature with reasonably high signal-to-noise in order to accurately call the base. Therefore, in most cases the source of excitation, irrespective of the separation platform, is a laser that is well matched to the excitation maxima of the dye set used in the sequencing experiment.

2. *Identify the bases using a spectral marker.* The instrument must be able to identify one of the four bases terminating the sequencing fragments by accurately processing the fluorescence via spectral discrimination (wavelength) or some other fluorescence property, such as the lifetime. For spectral discrimination, this would require sorting the fluorescence by wavelength using either filters or gratings. In addition, multiple detection channels would be required.
3. *Process the fluorescence from many electrophoresis lanes.* In most sequencing instruments the fluorescence must be read from multiple gel lanes (slab gel electrophoresis) or multiple capillaries (capillary gel electrophoresis). This can be done by either using a scanning system, in which the relay optic (collection optic) is rastered over the gel lanes or capillaries or an imaging system, in which the fluorescence from the multiple gel lanes or capillaries are imaged onto some type of multichannel detector, such as a charge coupled device (CCD) or image intensified photodiode array.
4. *Robust instrumentation.* Since many sequencing devices are run by novice operators and are run for extended periods of time, the detector format must be able to operate on a dependable basis and be turn-key in operation.

This short list of requirements for any type of fluorescence readout device appropriate for sequencing presents itself with many challenges that often times are non-complementary with the sequencing requirements. For example, high sensitivity is particularly demanding since the separation platforms used to fractionate the DNA are becoming smaller and therefore, smaller amounts of material must be inserted into the device. In addition, detecting material directly within the gel matrix (typically a polyacrylamide gel) can be problematic due to the intense scattering photons that it produces. Also, the signal is transitory in that the DNA fragment resides within the probing volume (defined by the laser beam size) for a few seconds. Another issue is that many separation channels or lanes must be interrogated for high throughput applications. On top of these considerations, high signal-to-noise is required to obtain high accuracy in the base-calling phases of the readout process. As such, significant design considerations go into fabricating a fluorescence detector system for DNA sequencing applications.

There are two general types of fluorescence detector formats: scanning and imaging type devices. In most high throughput sequencing devices, multiple lanes of the slab gel or multiple capillaries are run in parallel to increase system throughput. For example, many machines run in a 96-lane format since micro-titer plates, standard plates used to prepare sequencing reactions, come in a 96-well format. In the scanning systems, the excitation beam is tightly focused and irradiates only a single lane, while the relay optic is rastered over the lanes of the gel or the capillary array and the fluorescence from each lane processed sequentially on one set of detection channels. For the imaging systems, all of the electrophoresis lanes are irradiated by a laser(s) simultaneously, with the fluorescence readout accomplished using a multichannel detector, such as a CCD.

1.3.1. Fluorescence Scanning Instruments

A typical example of a scanning system is depicted in Figure 1.15.[23,24] In this system, a confocal geometry with epi-illumination is used, in which the objective used to collect the emitted fluorescence also serves to focus the laser beam into individual capillaries (or lanes of the slab gel) used for the electrophoresis. Following collection, the emission is focused onto a spatial filter at the secondary image plan of the collection objective. The laser light (488 nm, Ar ion laser, 1 mW) is directed into the objective using a dichroic filter. The capillaries are held into a linear array and, in this particular example, the capillary array is translated beneath the microscope objective. As can be seen, the laser only irradiates one capillary at a time, but since the beam is tightly focused (diameter = 10 μ) the electronic transition can be saturated at relatively low laser powers, improving signal-to-noise in the fluorescence measurement.[25] In addition, the noise can be significantly suppressed in this system, since a pinhole is used in the secondary image plane of the collection microscope objective preventing scattered out-of-focus light generated at the walls of the capillary from passing through the optical system. The capillary array is scanned at a rate of 20 mm/s with the fluorescence sampled at 1500 Hz/channel (color channel) resulting in a pixel image size of 13.3 μ. The fluorescence is collected by a 32× microscope objective (numerical aperture = 0.4) resulting in a geometrical collection efficiency of approximately 12%. Once the fluorescence has been collected by the objective, it is passed through a series of dichroics to sort the color and then processed on one of four different PMTs, with each PMT sampling a different color (spectral discrimination). While the present system is configured with four color channels, the system could easily be configured to do two-color processing as well by removing the last dichroic filter in the optical train and two of the PMTs. The concentration limit of detection of this system has been estimated to be 2×10^{-12} M (SNR = 3), which was determined by flowing a solution of fluorescein through an open capillary.[23] The detection limit would be expected to degrade in a gel filled capillary due to the higher background that would be generated by the gel matrix.

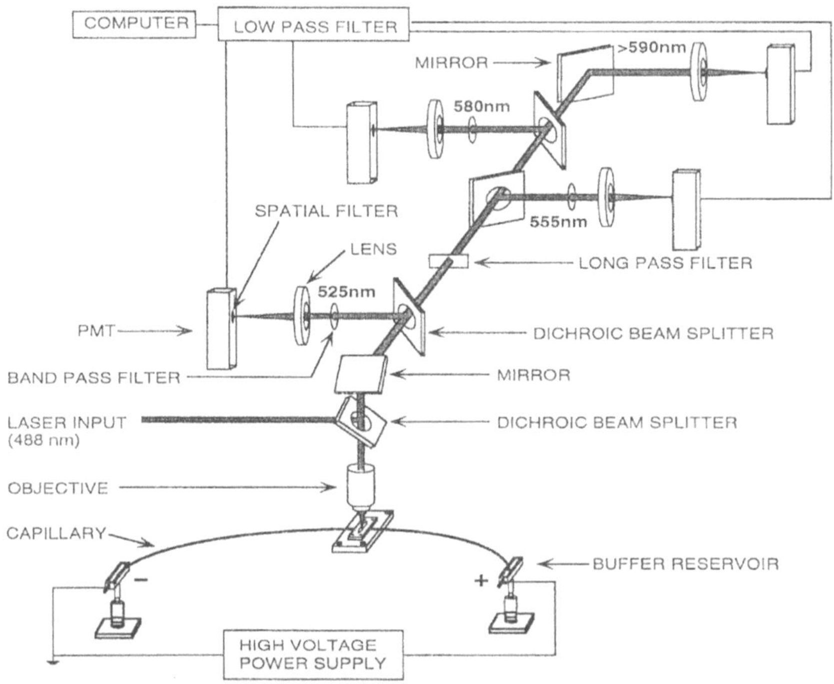

Figure 1.15. Four-color, laser scanning fluorescence system for DNA sequencing applications using capillary gel electrophoresis. The excitation source was an air-cooled Ar ion laser operating at 488 nm (1 mW average power). The collection and focusing optic consisted of a microscope objective (20×, NA = 0.5). The emission was directed onto one of four different PMTs using dichroic filters and further isolated from background photons using a bandpass filter. The fluorescence was sampled at 2 Hz with output filtered using a low-pass filter with a time-constant of 1 s. (Adapted with permission from reference 14.)

1.3.2. Fluorescence Imaging Systems for DNA Sequencing

An example of an imaging system for reading fluorescence from multichannel capillary systems for DNA sequencing is shown in Figure 1.16A.[26,27] In this example, a sheath flow cell is used with the laser beam(s) traversing the sheath flow and the capillary output dumping into the sheath stream (see Figure 1.16C). This geometry allows simultaneous irradiation of all the material migrating from the capillaries without requiring the laser beam to travel through each capillary, which would cause significant scattering and reduce the intensity of the beam as it travels through the array. The sheath flow cell also causes contraction of the fluid

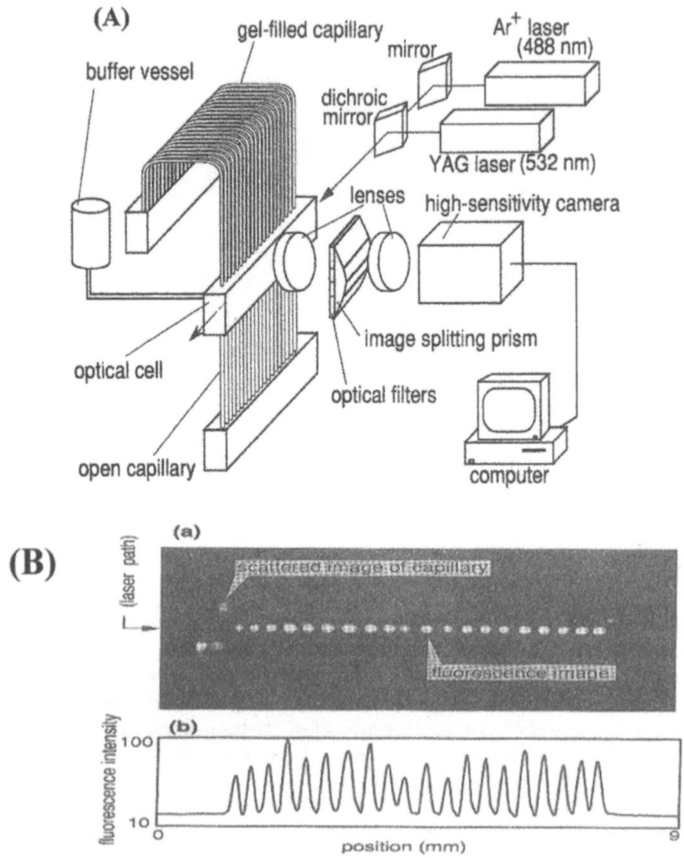

Figure 1.16. (A) Schematic of an imaging laser-induced fluorescence detector for reading four-color fluorescence from capillary gel arrays. The lasers used for excitation were an Ar ion (488 nm) and a YAG laser (532 nm). The lasers were allowed to traverse below the output of the gel columns and the fluorescence collected with a lens system. The fluorescence line image was split into four different color images using a polyhedral image-splitting prism coupled to optical filters. The filtered fluorescence was detected using a 2-dimensional CCD. (B) Fluorescence image in the one-color mode across an array of 20 capillaries. On the bottom panel, the integrated fluorescence intensity is shown. (C) Schematic view of the multiple sheath flow cell using gravity feed for the sheath flow. Twenty-gel filled capillaries were aligned at a 0.35 mm pitch in an optical cell (26 mm × 26 mm × 4 mm). (D) Photograph of the capillary array aligned in the sheath flow cell. (Adapted with permission from reference 26.) *(continued)*

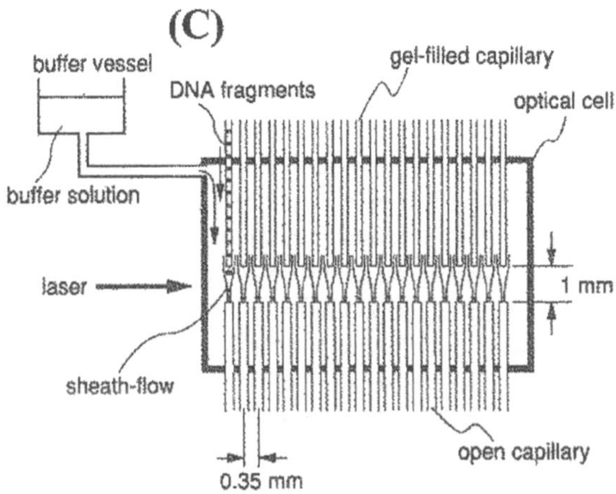

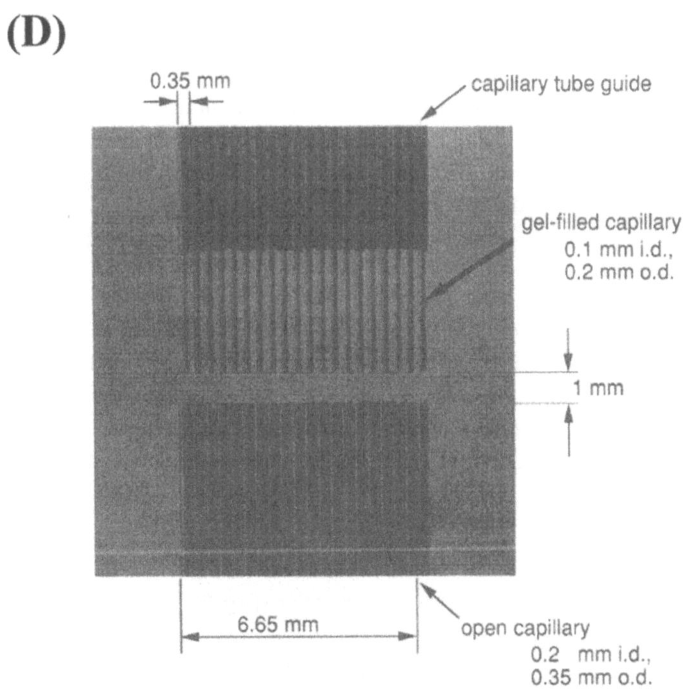

Figure 1.16. (*Continued*)

output of the capillary since the sheath flow runs at a greater linear velocity compared to the sample (capillary) stream. A fluorescence image of the output from the capillary array is shown in Figure 1.16B, indicating that the sample stream diameter at the probing point was ~0.18 mm and also demonstrating minimal cross-talk between individual capillaries in the array. The laser beams (488 nm from Ar ion, 6 mW; 532 nm from frequency doubled YAG, 6 mW) were positioned slightly below the exit end of the capillaries (see Figure 1.16A) with the beams brought colinear using a dichroic filter. The collection optic and focusing optic produced a total magnification of 1 and resulted in a geometrical collection efficiency of 1%. In order to achieve multicolor processing capabilities, the collected radiation was sent through an image-splitting prism to produce four separated (spectrally) line images on the array detector (one for each dye used to label the terminal bases). In addition, a series of narrow bandpass filters were placed in front of the image-splitting prism to assist in isolating the appropriate colors for data processing. The detector that was used for this system was a 2-dimensional CCD camera with a cooled image intensifier. Interestingly, the detection limit reported for this system was found to be 2×10^{-12} M when operated in the four-color mode, comparable to that seen for the scanning system discussed above. However, it should be pointed out that the presence or absence of the gel matrix will not affect the sensitivity of the fluorescence measurement in this case, since the fluorescence interrogation is done off column in the sheath flow. In fact, researchers have reported that the implementation of the sheath flow geometry in gel electrophoresis can offer a significant improvement in limits of detection by minimizing scattering contributions to the background.[28]

When comparing these two fluorescence readout systems, several issues should be highlighted. One is the duty cycle, which takes into account the loss in signal due to multiple lane sampling. For any type of scanning system, the sampling of the electrophoresis lanes is done in a sequential fashion. For a scanning system sampling 96 capillaries, the duty cycle is approximately 1%. However, in the imaging system, all capillaries are sampled continuously and the duty cycle is nearly 100%. Therefore, comparisons of detection limits for any system must include a term for the duty cycle, since lower duty cycles will degrade the limit of detection.

1.4. Dye Primer/Terminator Chemistry and Fluorescence Detection Formats

In most sequencing applications, dye-labeled primers are used for accumulating sequencing data using automated instruments. This stems from the fact that dye-labeled primers are typically less expensive to use compared to their dye-labeled terminator counterparts. Also, in most applications, small pieces of DNA (1–2 kbp in length) are cloned into bacterial vectors for propagation (to increase copy number), such as M13s, which have a known sequence and serve as ideal priming sites. However, dye-labeled primers do present one with problems, for example, the sequencing reactions must be run in four separate tubes during polymerization and then pooled prior to the gel electrophoresis. In addition, unextended primer can result in a large electrophoretic peak (i.e., high intensity) which can often times mask the ability to call bases close to the primer annealing site.

Dye-labeled terminators can be appealing to use in certain applications, for example, cases where high quality sequencing data is required and in primer walking strategies. In primer walking, the sequence of the DNA template is initiated at a common priming site using a primer that is complementary to that site. After reading the sequence at that site, the template is subjected to another round of sequencing, with the priming site occurring at the end of the first read. In this way, a long DNA can be sequenced by walking in a systematic fashion down the template. Dye terminators are particularly attractive since primers need to be synthesized frequently in primer-walking strategies and the need for non-labeled primers simplifies the synthetic preparation of these primers. Dye terminators improve the quality of sequencing data in many cases since the excess terminators are removed prior to electrophoresis (using size exclusion chromatography) and as such, give clean gel reads free from intense primer peaks. However, it should be noted that in most cases terminators can produce uneven peak heights (broad distribution of fluorescence intensities) due to the poor incorporation efficiency of dye-terminators by polymerase enzymes.

When dye-labeled primers are used, several different formats can be implemented to reconstruct the sequence of the template when using fluorescence detection. In the case of spectral discrimination, these formats may vary in terms of the number of dyes used, the number of detection channels required, or the need for running 1–4 parallel electrophoresis lanes. For example, if the sequencing instrument possesses no spectral discrimination capabilities, the electrophoresis must be run in four different lanes, one for each base comprising the DNA molecule. However, if four different dyes are used the electrophoresis can be reduced to one lane and as a consequence, the production rate of the instrument goes up by a factor of 4. The fluorescence-based formats that will be discussed here includes (# dyes/# electrophoresis lanes) single dye/four lane; single dye/single lane; two color/single lane, and four color/single lane strategies.

The most pressing issue in any type of DNA sequencing format is the accuracy associated with the base call, which is intimately related to a number of experimental details, for example the number of spectral channels used in the instrument as well as the signal-to-noise ratio in the measurement. The information content of a signal, I, can be determined from the simple relation[29]

$$I = n \log_2 (\text{SNR}) \tag{1.3}$$

where n is the number of spectral channels and SNR is the signal-to-noise ratio associated with the measurement. The term I is expressed in bits and typically 2 bits are necessary to distinguish between four different signals, but only if there is no spectral overlap between the dyes used for identifying the bases. Unfortunately, in most multicolor systems, the spectra of the dyes used in the sequencing device show significant overlap and as such, many more bits will be required to call bases during the sequencing run.

While the above equation can provide information on how to improve the accuracy of the base call, it does not provide the sequencer with information on the identity of the individual electrophoretic peaks (base call) nor the quality of a base call within a single gel read. For example, if four color sequencing is used with dye-primer chemistry, how should one process the data and what is the confidence to which an electrophoretic peak is called an A, T, G or C? In order to provide such information, an algorithm has been developed to not only correct for anomalies associated with fluorescence-based sequencing but also to assign a quality score to each called base. The typical algorithm that is used is called the *Phred* scale and it uses several steps to process the sequencing data obtained from fluorescence-based, automated DNA sequencers.[30]

The data input into *Phred* consists of a trace, which is electrophoretic data processed into four spectral channels, one for each base. The algorithm consists of four basic steps:

1. *Idealized electrophoretic peak locations (predicted peaks) are determined.* This is based on the premise that most peaks are evenly spaced throughout the gel. In regions where this is not the case, typically during the early and late phases of the electrophoresis, predictions are made as to the number of correct bases and their idealized locations. This step is carried out using Fourier methods as well as the peak spacing criterion and helps to discriminate noise peaks from true peaks.
2. *Observed peaks are identified in the trace.* Peaks are identified by summing trace values to estimate the area in regions which satisfy the criterion, $2 \times v(i) \geqslant v(i+1) + v(i-1)$, where $v(i)$ is the intensity value at point i. If the peak area exceeds 10% of the average area of the preceding 10 accepted peaks, and 5% of the area of the immediate preceding peak, it is accepted as a true peak.
3. *Observed peaks are matched to the predicted peak locations, omitting some peaks and splitting others.* In this phase of the algorithm, the observed peak arises from one of four spectral channels, and thus, can be associated with one of the four bases. It is this ordered list of matched observed peaks which determines a base sequence for the DNA template in question.
4. *Observed peaks that are uncalled (unmatched to predicted peaks) are*

processed. In this step, an observed peak that did not have a complement in the predicted trace is called and assigned a base and finally inserted into the read sequence.

As can be seen, this algorithm mainly deals with sorting out difficulties associated with the electrophoresis by identifying peaks in the gel traces, especially in areas where the peaks are compressed (poor resolving power) or where multiple peaks are convolved due to significant band broadening produced by diffusional artifacts.

Many times, pre-processing of the traces is carried out prior to *Phred* analysis to correct for dye-dependent mobility shifts. In most cases, these mobility shifts are empirically determined by running an electropherogram of a single dye-labeled DNA ladder (for example, T-terminated ladder), and comparing the mobilities to the same ladder, but labeled with another dye of the set. This type of analysis can be very complex and involved, since the mobility shift is not only dependent upon the dye and linker structure, but also upon the separation platform used. For example, dyes which show uniform mobility shifts in slab gel electrophoresis may not show the same effect in capillary gel electrophoresis. In addition, these mobility shifts can be dependent upon the length of the DNA to which the dye is attached.[31] An example of this phenomenon is shown in Figure 1.17. In this particular example, cyanine dyes were covalently anchored to an M13 (-40) sequencing primer and annealed to an M13 template followed by extension with a single terminator (ddT). The tracts were electrophoresed using capillary gel electrophoresis. In Figure 1.17A, comparison between a Cy5T7 (-2 charge) and Cy5.5T (-1 charge) tracts indicate that the Cy5T7-labeled fragments migrate faster in the beginning of the run (smaller DNA fragments), but after 300 bp the two dye-labeled fragments co-migrate. In Figure 1.17B, one notices that for the dyes Sq5T4 (neutral charge) and Cy5.5T12 (-2 charge), a mobility crossover occurs at ~125 bp, with the shorter fragments migrating faster with the Cy5.5T12 label and after this, the Sq5T4 labeled fragments migrate faster. These types of mobility shifts have been ascribed to not only differences in the net charge of the dye-label but also potential dye–DNA base interactions. These interactions, predominately driven by hydrophobic interactions, may cause loops or hairpin structures on the 5′ end of the dye–DNA complex. These structures would cause a faster migration rate compared to fully extended structures produced by most dye–DNA complexes.

1.4.1. Single Color/Four Lane

In this processing format, only a single fluorescence detection channel is required to analyze the signal from the labeling dye since only a single dye is used

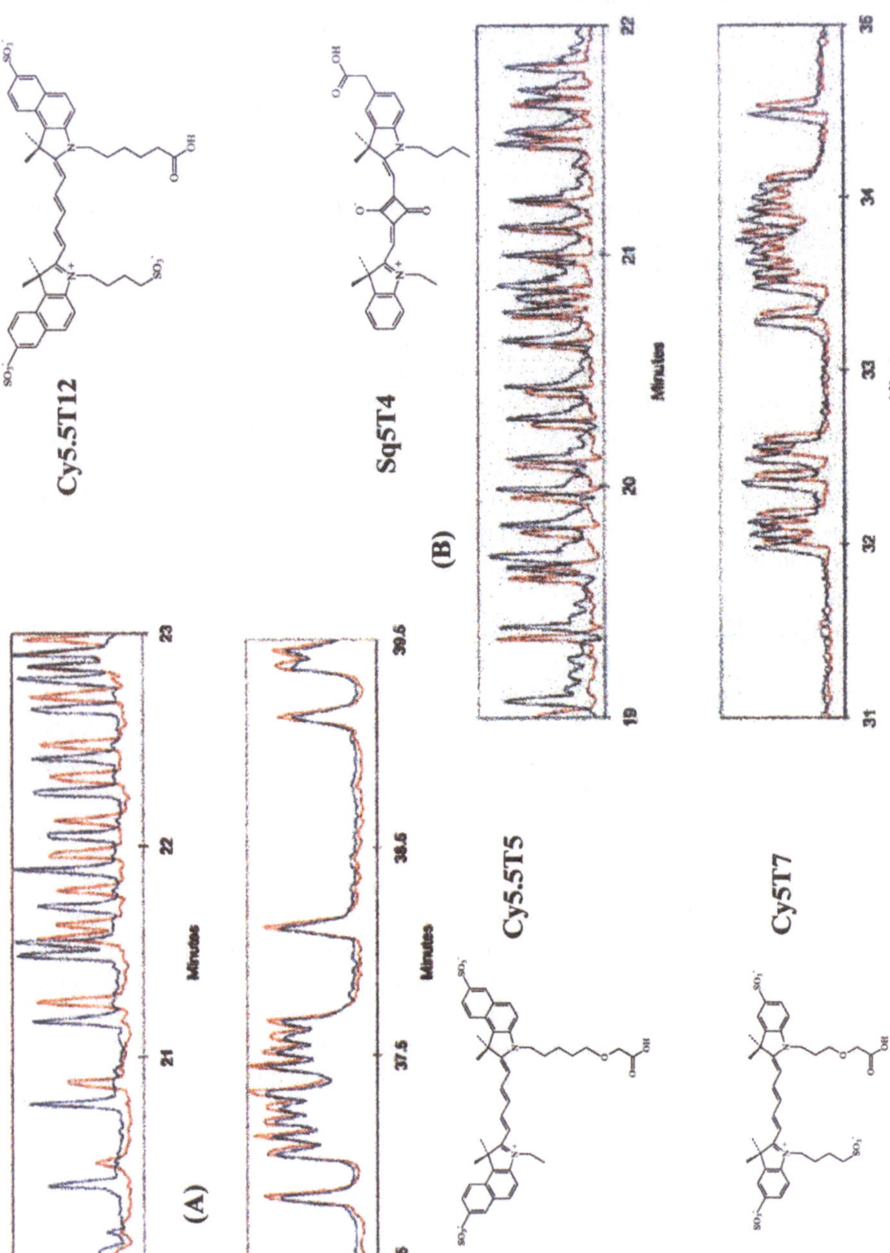

Cy5.5T12
Sq5T4
Cy5.5T5
Cy5T7
(A)
(B)
Minutes

to detect the sequencing fragments produced following chain extension. However, since no color discrimination is implemented, the electrophoresis must be run in four lanes, one for each base, similar to the format used in traditional autoradiographic detection. While this is a reasonable approach for slab gel separations, it is not a viable strategy in capillary gel applications due to the poor run-to-run reproducibility in the migration rates of the fragments traveling through the different capillaries. This is due to differences in the gel from capillary to capillary as well as differences in the integrity of the wall coatings used to suppress the electroosmotic flow. In the slab gel format, reproducibility in the migration times becomes less of a problem since all of the lanes are run in the same gel matrix. The output of a typical, single color/four lane sequencing device is shown in Figure 1.18 along with the called bases. In this example, each different color trace represents an electropherogram from an individual lane of the slab gel, which are overlaid to allow reconstruction of the sequence of the template. In this case, the device used a single microscope head containing the collection optics, a diode laser, filters and avalanche photodiode to read the fluorescence from the gel.[32] The microscope scanner is rastered over the gel at a rate of ~0.15 cm/s and monitors fluorescence along a single axis of the gel. The gel is approximately 20 cm in width and 42 cm in length and can accommodate 48 separate lanes. The time required to secure this data was 6 hours, with the extended time due primarily to the limited electric field that can be applied to the thick slab gel.

1.4.2. Single Color/Single Lane

In this sequencing approach, only a single fluor is used and as such, only a single laser is required to excite the fluorescence and only a single detection channel is needed to process the fluorescence. The advantage of this approach is that instrumentally it is very simple since the hardware required for detection is simple. In addition, since the sequence is reconstructed from a single electrophoresis lane and not four, the throughput can be substantially higher compared to a single fluor/four lane method.

The bases are identified by adjusting the concentration ratio of the terminators used during DNA enzymatic polymerization to alter the intensity of the

Figure 1.17. Electrophoresis traces and chemical structures of four different cyanine dyes used to end-label DNA sequencing primers. (A) Gel trace of Cy5.5T5 (red trace) and Cy5T7 (blue trace). (B) Gel trace of Cy5.5T12 (red trace) and Sq5T4 (blue trace). In all cases, the ladders were prepared using a single terminator (ddT) and an M13mp18 template. The electrophoresis was carried out in a capillary gel column (field strength = 185 V/cm) using a hydroxyethyl cellulose sieving buffer. (Adapted with permission from reference 31.)

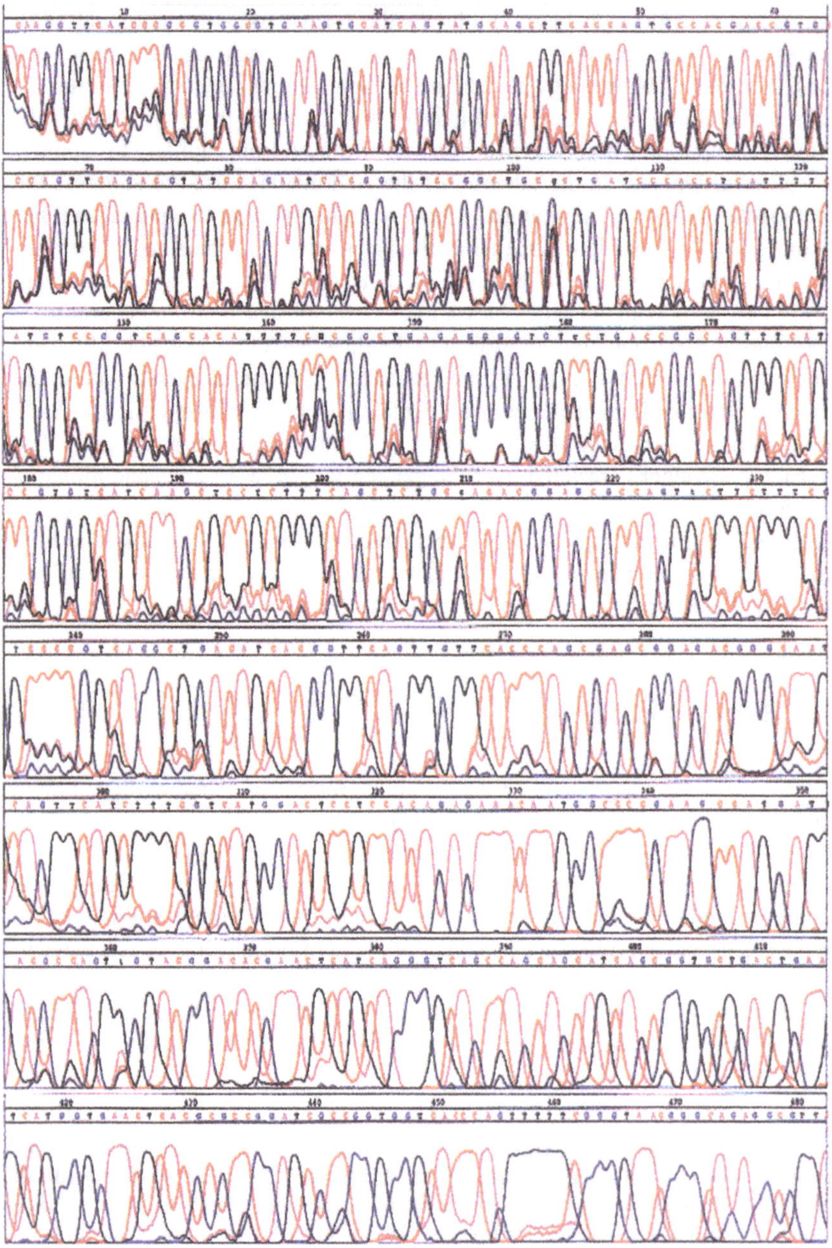

resulting electrophoretic bands.[19,29,33,34] Therefore, if the concentration of the terminators used during DNA polymerization was 4:2:1:0 (A:C:G:T), a series of fluorescence peaks would be generated following electrophoretic sizing with an intensity ratio of 4:2:1:0, and the identification of the terminal bases would be carried out by categorizing the peaks according to their heights. In order to accomplish this with some degree of accuracy in the base calling, the ability of the DNA polymerase enzyme to incorporate the terminators must be nearly uniform. This can be achieved using a special DNA polymerase, which in this case is a modified T7 DNA polymerase.[35,36] This enzyme has been modified so as to remove its proof reading capabilities by eliminating its 3′→5′ exonuclease activity. Since this method requires uniform incorporation of the terminators, it is restricted to the use of dye-primer chemistry. In addition, since the T7 enzyme is not a thermostable enzyme, cycle sequencing cannot be used.

An example of sequencing data accumulated using this base-calling strategy is presented in Figure 1.19, in which the terminator concentration was adjusted at a concentration ratio of 4:2:1:0 (A:C:G:T).[19] The accuracy in calling bases was estimated to be 84% up to 250 bases from the primer annealing site, with readable bases up to 400 but the accuracy deteriorating to 60%. This is a common artifact in this approach; that being, poor base calling accuracy. The poor base calling results from variations in the activity of the T7 DNA polymerase and in addition, the null signal used to identify Ts. Ambiguities are present when one must identify multiple null signals, which can lead to insertions or deletions.

1.4.3. Two Color/Single Lane

In order to improve on the base calling accuracy associated with the single color/single lane strategy without having to increase the instrumental complexity of the fluorescence readout device associated with sequencing instruments, one may employ a two color format to identify the four terminal bases in sequencing applications. In this approach, one or two lasers are used to excite one of two spectrally distinct dyes used for labeling the sequencing primers, and the fluorescence is processed on one of two detection channels consisting of bandpass filters and photon transducers.

A schematic of a two-color scanning instrument is shown in Figure 1.15 (last

Figure 1.18. Single-color/four-lane sequencing trace (called bases 1–480) of a PCR amplified λ-bacteriophage template. The sequencing was performed using an IRD-800 labeled primer (21mer) and slab gel electrophoresis instrument (Li-COR 4000). The gel consisted of 8% polyacrylamide with 7 M urea. The dimensions of the gel were 25 cm (width) by 41 cm (length). The traces from the four lanes were overlaid to reconstruct the sequence of the template.

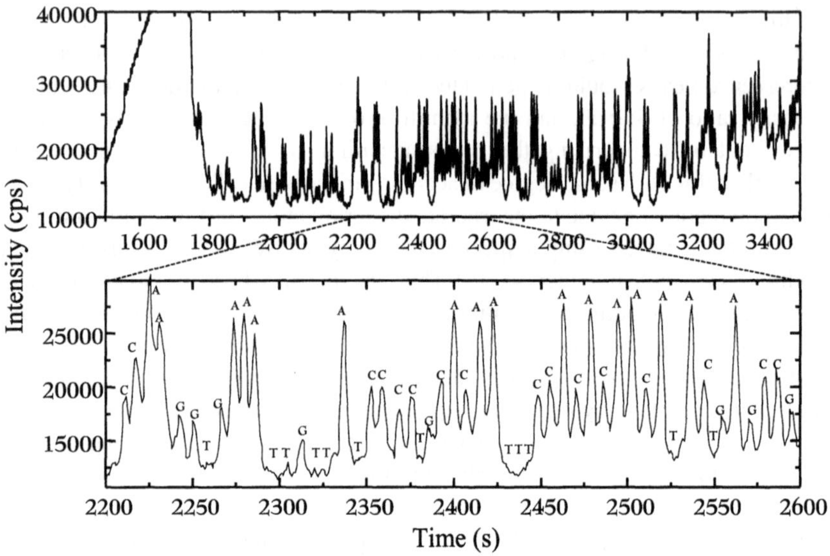

Figure 1.19. Single-color/single-lane capillary gel sequencing data. The template was an M13mp18 phage and the primer was labeled with a near-IR fluorescent dye and detected using near-IR, laser-induced fluorescence. The terminators were adjusted to a concentration ratio of 4:2:1:0 during extension to allow identification based on fluorescence intensities. The separation was performed at a field strength of 250 V/cm and the gel column contained a 3% T/3% C crosslinked polyarcylamide gel. (Adapted with permission from reference 19.)

dichroic was not used and two PMTs were removed). It was used to excite the fluorescence of the labeling dyes, FAM and JOE. Due to their similar absorption maxima, a single laser (488 nm) could efficiently excite the fluorophores. In addition, this dye-pair was selected since they produced sequencing fragments which co-migrated and thus, no mobility correction was required. The bases were identified using a binary coding scheme, which is shown in Table 1.2.[24] During DNA polymerization, four separate reactions were run with the A-reaction containing an equimolar mixture of the FAM and JOE labeled sequencing primers. In the case of G, only the JOE labeled primer was present, while for T only the FAM labeled primer was present, and for C no dye labeled primer was used. By ratioing the signal in the red (JOE) to green (FAM) channel, a value was obtained which could be used to identify the terminal base. The attractive feature associated with this protocol is that while the absolute intensity of the bands present in the electropherogram may vary by a factor of 20 due to sequence dependent termination, the ratio only varies by a factor of 1.7. The read length using this binary coding system was up to approximately 350 bases, with the number of errors ~15 (accuracy = 95.7%). The majority of the errors were attributed to C determinations, since a null signal was used to indicate the presence of this base.

Table 1.2. Binary Coding Scheme for Two-Color DNA Sequencing[a]

	FAM	JOE
A	1	1
C	0	1
G	1	0
T	0	0

[a]1 indicates the presence of the dye-labeled primer during DNA polymerization and 0 indicates the absence of the dye label.

In order to alleviate the errors in the base calling associated with identifying bases using a null signal, a two-dye, two-level approach can be implemented.[37,38] In this method, the bases using a common dye-label have the concentration of the ddNTPs adjusted during chain extension to alter the intensity of the fluorescence peaks developed during the electrophoresis. Also required in this approach is uniform peak heights, requiring the use of the modified T7 DNA polymerase in the presence of manganese ions. For example, Chen et al.[37] used a FAM-labeled primer for marking Ts and Gs, with the concentration ratio of the ddNTPs adjusted to 2:1 (T:G). Likewise, the As and Cs were identified using a 2:1 concentration ratio of the terminators with the labeling dye in this case being TAMRA. Sequencing data produced an effective read length of 350 bases, with an accuracy of 97.5%. When the concentration ratios of the terminators sharing a common labeling dye was increased to 3:1, the read length was extended to 400 bases with a base-calling accuracy 97%. As can be seen, the elimination of null signals to identify bases can improve the base calling accuracy in these type of sequence determinations.

While most fluorescence labeling strategies for DNA sequencing which depend on differences in intensities of the electrophoretic peaks to identify bases use dye-labeled primers, internal labeling, where the fluorescent dye is situated on the dNTP, can also be used.[39] The advantages associated with using dye-labeled dNTPs are: (1) ability to use a wide range of primers since no dye-labeled primer is required, (2) incorporation of the dye-labeled dNTP can be much more uniform than dye-labeled ddNTPs, and (3) dye-labeled dNTPs are much less expensive compared to dye-labeled primers and terminators. Using a tetramethylrhodamine-labeled dATP and a fluorescein labeled dATP, a two-color/single lane sequencing assay has been reported.[39] For internal labeling, a two step polymerization reaction was used, in which the template was annealed to the sequencing primer (unlabeled) along with the dye-dATP and the four unlabeled dNTPs, as well as the polymerase enzyme. The extension reaction was incubated at 37°C for 10 mins, after which the appropriate terminator was added and the reaction allowed to proceed for an additional 10 mins. The initial extension reaction extended 6–8 nucleotides to a quartet of As, with 80–90% of the fragments containing a single dye-labeled dATP. Since only two dyes were used in this particular example, the concentration ratio for a pair of terminators sharing a common dye was adjusted (3:1) so as to allow discrimination based upon the intensity of the resulting electrophoretic peaks. Analysis of the sequencing data indicated that the read length was found to be 500 bases with an accuracy of 97%.

1.4.4. Four Color/Single Lane

The commonly used approach in most commercial DNA sequencing instruments using fluorescence detection is the use of a four color/single lane strategy for identifying the terminal bases in sequencing applications. The primary reasons for using a four color/single lane approach is that it provides high accuracy in the base calling, especially for long reads, and the throughput can be high due to the fact that all bases comprising the template DNA can be called in a single gel tract. Unfortunately, a four color detector requires extensive optical components to sort the fluorescence and also, in some cases, multiple excitation sources are needed in order to efficiently excite the fluorophores used to label the individual sequencing ladders. In addition, post-electrophoresis software corrections may be required to account for spectral leakage into detection channels.

Most dye-terminator reads are used with this four color strategy, since the data analysis (base calling) does not depend on uniform incorporation efficiencies, which are hard to achieve using dye-labeled teminators. The same type of instrumentation that is used for four color/dye-primer reads can also be used for four color/dye-terminator reads as well. The only difference one finds is in terms of the sample preparation protocols and the software corrections in the sequencing data, such as different mobility correction factors. In most cases, a size exclusion step is used following DNA polymerization to remove excess dye-labeled terminators since they are negatively charged and can mask the sequencing data due to the presence of a large dye-terminator band in the gel.

An example of a four-color detector for capillary gel electrophoresis is shown in Figure 1.20A, in which two laser sources (Ar ion laser, 488 nm, and green helium-neon laser, 543 nm) were used to excite the dye set: FAM, JOE, TAMRA and ROX. In order to process the emission on a single detection channel, a four stage filter wheel was used, which was synchronized to a sector wheel situated in front of the two lasers. The synchronization was set so as to pass 488 nm excitation for FAM and JOE and simultaneously place the band pass filters for

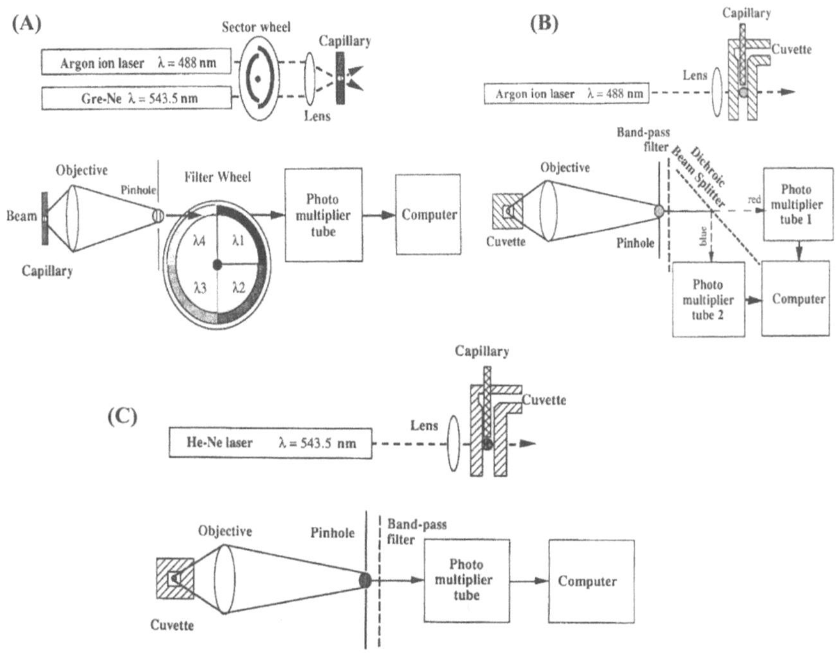

Figure 1.20. Laser-induced fluorescence detection apparati for processing DNA sequencing data obtained using capillary gel electrophoresis. The formats are four-color/single lane (A), two-color/single-lane (B) and one-color/single lane (C).

FAM and JOE in the optical path. Following this, the 543 nm laser light was passed and the filters for TAMRA and ROX were situated within the optical path. Typical traces produced from this system are shown in Figure 1.21, in which the sequence of an M13mp18 phage test template was analyzed. A read length exceeding 550 bases was obtained at an accuracy of 97%.

1.4.5. So Which Sequencing Format Is Better?

With a variety of different fluorescence detection formats available, the question becomes, which configuration is better in terms of base calling, both from a read length and an accuracy point of view. In addition, what detection format produces the best signal-to-noise ratio in the measurement. Other issues require attention as well, namely the complexity of the instrumentation required for

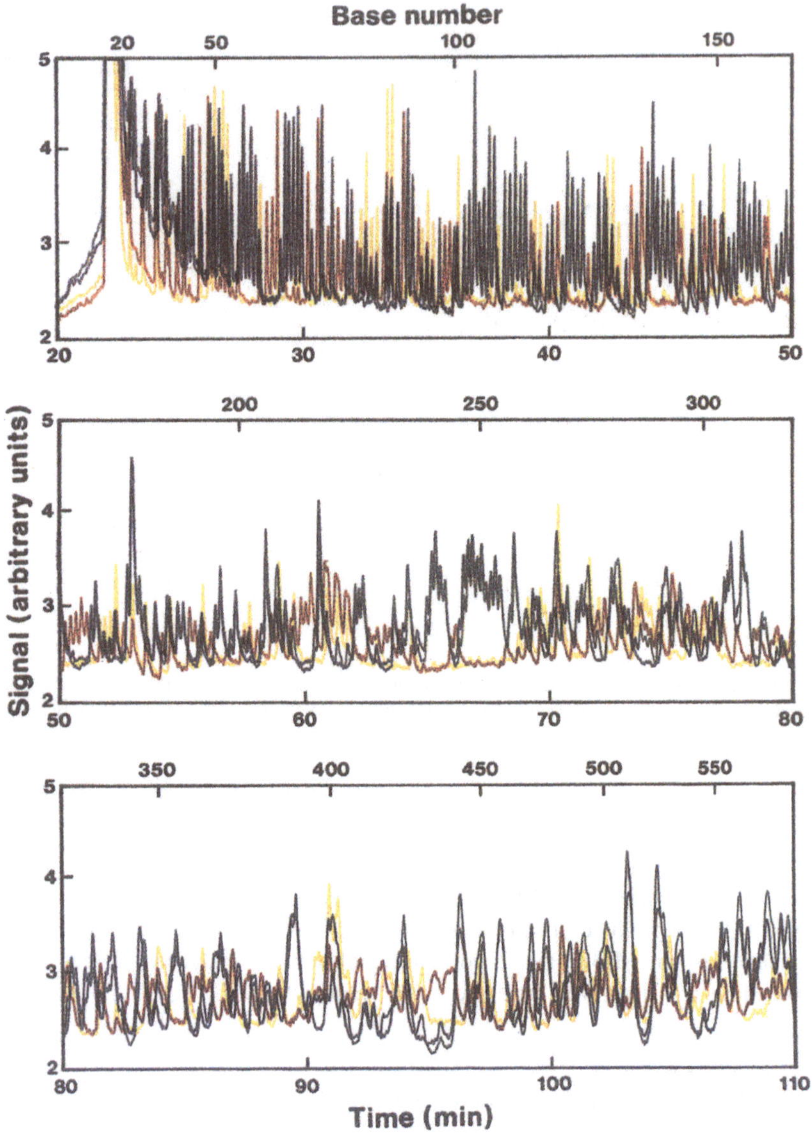

Figure 1.21. Four-color/single-lane sequencing of an M13mp18 template with a histidine tRNA insert. The numbers along the top of the electropherogram represent cumulative bases from the primer annealing site. The electrophoresis was performed in a capillary column with a length of 41 cm and an i.d. of 50 μ. The sieving gel consisted of a crosslinked polyarcylamide (6% T/5% C) and was run at a field strength of 150 V/cm. The dyes used for the labeling of the sequencing primers were FAM, JOE, TAMRA, and ROX. Each color represents fluorescence from a different wavelength region (blue—540 nm; green—560 nm; yellow—580 nm; red—610 nm). (Adapted with permission from reference 29.)

Table 1.3. Comparison of Figures of Merit for Various Detector Formats for Fluorescence-Based DNA Sequencing[a]

Detection mode	Noise (mass)[c]	Detection limit (moles)[c]	Read length[d]	Base calling accuracy	Information content[b]	
					10 amol	100 zmol
One color	700 ymol	2 zmol	400	85%	14	7
Two color	7 zmol	20 zmol	500	97%	21	8
Four color	70 zmol	200 zmol	550	97%	29	2

[a]The sequencing assays used capillary gel electrophoresis for fractionating the DNA ladders. The gel consisted of a crosslinked polyacrylamide gel with with formamide or urea as the denaturant. In all cases, the template was an M13mp18 phage with dye-primer chemistry used for fluorescence labeling. In the four-color experiment, FAM, JOE, TAMRA, and ROX were the labeling dyes, while for the two-color experiment, FAM and TAMRA were used, and for the one-color example, only TAMRA was used. The sequencing primer was an M13 (−40) primer and the polymerase was a Sequenase enzyme.

[b]Calculated from equation (1.3).

[c]ymol = 10^{-24} moles; zmol = 10^{-21} moles.

[d]The read lengths were determined from sequencing data with high signal-to-noise ratios that were comparable across the series.

detection. A rigorous comparison between the various detector configurations for sequencing applications has been carried out.[28] In this study, three different detector formats were used and they are shown in Figure 1.20. These consisted of a four color/single lane format, a two color/single lane format, and finally a single color/single lane format.

A summary of the data collected from this work is shown in Table 1.3. As can be seen, the single color experiment, in which there is a single detection channel, provides the lowest limits of detection, followed by the two color system, and finally, the four color system. The significant improvement in the detection limit for the two color and single color systems was partly due to the use of the sheath flow detector that was employed in these formats. However, in spite of the use of the sheath flow cell, the general trend is that the lower number of spectral channels typically results in better signal-to-noise in the fluorescence measurement due to the fact that spectral sorting is not required. Spectral sorting causes emission losses due to reflection or inefficient filtering by the bandpass filters used in the optical train. If a filter wheel is used, as in this particular example, a reduced duty cycle will degrade the signal-to-noise ratio in the measurement. This does not necessarily mean that the lower number of spectral channels will give better sequencing data, however. As can be seen from the results of Table 1.3, the four color format produced better read lengths and favorable base calling accuracies compared to the other formats. This is a consequence of the fact that since one is using four spectral channels, the information content in the signal goes up, but only at reasonable loading levels of sample into the sequencing instrument. It is clear that at low loading levels of DNA sequencing ladders, the one color or two color approach may be better due to improved limits of detection.

1.4.6. Single Color/Four Lifetime Sequencing

While most sequencing applications using fluorescence require spectral discrimination to identify the terminal base during electrophoretic sizing, an alternative approach is to use the fluorescence lifetime of the labeling dye to identify the terminal base. In this method, either time-resolved or phase-resolved techniques can be used to measure the fluorescence lifetime of the labeling dye during the gel electrophoresis separation.

The monitoring and identification of multiple dyes by lifetime discrimination during a gel separation can allow for improved identification efficiency when compared to that of spectral wavelength discrimination. When the identity of the terminal nucleotide base is accomplished through differences in spectral emission wavelengths, errors in the base call can arise from broad, overlapping emission profiles, which results in cross talk between detection channels. Lifetime discrimination eliminates the problem of cross talk between detection channels and also, can potentially allow processing of the data on a single readout channel. Several other advantages are associated with fluorescence lifetime identification protocols, including:[40]

- The calculated lifetime is immune to concentration differences.
- The fluorescence lifetime can be determined with higher precision than fluorescence intensities.
- Only one excitation source is required to efficiently excite the fluorescent probes and only one detection channel is needed to process the fluorescence for appropriately selected dye sets.

One potential difficulty associated with this approach is the poor photon statistics (limited number of photocounts) that can result when making such a measurement. This results from the need to make a dynamic measurement (the chromophore is resident in the excitation beam for only 1–5 s) and the low mass loading levels associated with many DNA electrophoresis formats. Basically, the low number of photocounts acquired to construct the decay profile from which the lifetime is extracted can produce low precision in the measurement, which would affect the accuracy in the base call. In addition, the high scattering medium in which the fluorescence is measured (polyacrylamide gel) can produce large backgrounds, again lowering the precision in the measurement. An additional concern with lifetime measurements for calling bases in DNA sequencing applications is the heavy demand on the instrumentation required for such a measurement. However, the increased availability of pulsed diode lasers and simple avalanche photodiode detectors have had a tremendous impact on the ability to assemble a time-resolved instrument appropriate for sequencing applications.

There are two different formats for measuring fluorescence lifetimes: time-resolved[40–47] and frequency-resolved.[48–52] Since the time-resolved mode is a digital (photon counting) method, it typically shows better signal-to-noise than a frequency-resolved measurement, making it more attractive for separation platforms that deal with minute amounts of sample. In addition, the use of time-resolved methods allow for the use of time-gated detection in which background photons, which are coincident with the laser pulse (scattered photons) can be gated out electronically, improving the signal-to-noise ratio in the measurement.

A typical time-correlated single photon counting (TCSPC) device consists of a pulsed excitation source, a fast detector, and timing electronics. A device that has been used for making time-resolved measurements during capillary gel electrophoresis is shown in Figure 1.22.[43] The light source consisted of an actively pulsed solid-state GaAlAs diode laser with a repetition rate of 80 MHz and an average power of 5.0 mW at a lasing wavelength of 780 nm. The pulse width of the laser was determined to be ~50 ps (FWHM). The detector selected for this instrument was a single photon avalanche diode (SPAD), which has an active area of 150 m

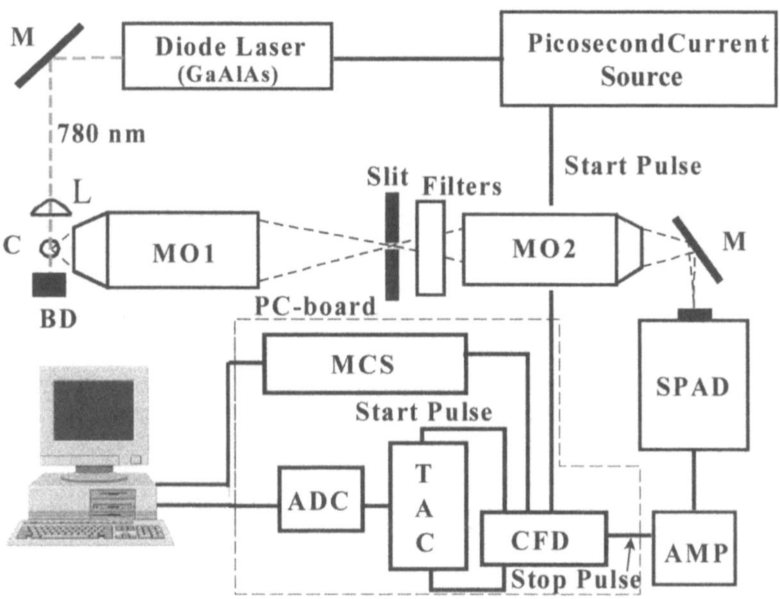

Figure 1.22. Time-correlated single photon counting detector for capillary gel electrophoresis. The laser source was a pulsed diode laser that operated at repetition rate of 80 MHz and lased at 780 nm (average power = 5 mW). The laser was focused onto a capillary gel column with the emission collected using a 40× microscope objective (NA = 0.85). The fluorescence was imaged onto a slit and then spectrally filtered and finally focused onto the photoactive area of a single photon avalanche diode. The electronics for processing the time-resolved data was situated on a single PC-board, which was resident on the PC-bus. (Adapted with permission from reference 43.)

and is actively cooled. In addition, the SPAD has a high single photon detection efficiency (>60% above 700 nm). The counting electronics (constant fraction discriminator (CFD), analog-to-digital converter (ADC), time-to-amplitude converter (TAC), and pulse-height analyzer) are situated on a single TCSPC board. The board plugs directly into a PC-bus and exhibits a dead time of <260 ns, allowing efficient processing of single photon events at counting rates exceeding 2×10^6 counts/s. This set of electronics allows for the collection of 128 sequential decay profiles with a timing resolution of 9.77 ps per channel. The instrument possesses a response function of approximately 275 ps (FWHM), adequate for measuring fluorescence lifetimes in the sub-nanosecond regime.

Probably the most important aspect associated with lifetime determinations in sequencing applications is considerations of the processing or calculation algorithms used to extract the lifetime value from the resulting decay. Since the photon statistics are poor and the accuracy in the base call depends directly on the lifetime differences between fluors in the dye set and the relative precision in the measurement, algorithms which deal with this situation are required, as well as those that can be performed on-line during the electrophoresis. Two algorithms for on-the-fly fluorescence lifetime determinations have been used, namely the maximum likelihood estimator (MLE) and the rapid lifetime determination method (RLD). MLE calculates the lifetime via the following relation[53]

$$1 + (e^{T/\tau_f} - 1)^{-1} - m(e^{mT/\tau_f} - 1)^{-1} = N_t^{-1} \sum_{i-1}^{m} iN_i \tag{1.4}$$

where m is the number of time bins within the decay profile, N_t is the number of photocounts in the decay spectrum, N_i is the number of photocounts in time bin i, and T is the width of each time bin. A table of values using the left-hand side (LHS) of the equation is calculated by setting m and T to experimental values and using lifetime values (τ_f) ranging over the anticipated values. The right-hand side (RHS) of the equation is constructed from CE decay data over the appropriate time range. The fluorescence lifetime may then be determined by matching the value of the RHS obtained from the data with the table entry. The relative standard deviation in the MLE may be determined from $N^{-1/2}$.

Fluorescence lifetimes calculated using the RLD method is performed by integrating the number of counts within the decay profile over a specified time interval and using the following relationship[54]

$$\tau_f = -\Delta t / \ln (D_1/D_0) \tag{1.5}$$

where Δt is the time range over which the counts are integrated and D_0 is the integrated counts in the early time interval of the decay spectrum, while D_1 represents the integrated number of counts in the later time interval. Both the MLE and RLD methods can extract only a single lifetime value from the decay, which in the case of multiexponential profiles would represent a weighted average of the various components comprising the decay.

Wolfrum and co-workers have also implemented a special pattern recognition technique.[42] Basically, the method involves comparing a pattern to the measured decay, and searches for a pattern that best fits the measurement. This algorithm is equivalent to the minimization of a log-likelihood ratio where fluorescent decay profiles serve as the pattern. Since the pattern recognition algorithm uses the full amount of information present in the data, it potentially has the lowest error or misclassification probability.

To demonstrate the feasibility of acquiring lifetimes on-the-fly during the capillary gel electrophoresis (CGE) separation of sequencing ladders, C-terminated fragments produced from Sanger chain-terminating protocols and labeled with a near-IR fluorophore on the 5′ end of a sequencing primer were electrophoresed and the lifetimes of various components within the electropherogram determined.[40] An example of the data produced from this detection format is shown in Figure 1.23. The average lifetimes determined using the MLE method was found to be 843 ps, with a standard deviation of ±9 ps (RSD = 1.9%). The lifetime values calculated here compared favorably to a static measurement performed on the same dye.

Since the base calling is done with lifetime discrimination as opposed to wavelength discrimination, new types of dye sets can and need to be used that suit the identification method. For example, it is not necessary to use dyes with discrete emission maxima and as such, structural variations in the dye set can be relaxed. A dye set developed for lifetime discrimination has been prepared and consists of a near-IR chromophore, which have unique fluorescence lifetimes, with the lifetime altered via the addition of an intramolecular heavy atom.[55] Each of these dyes possesses the same absorbance maximum and fluorescence emission maximum, but different fluorescence lifetimes (see Table 1.1). Since these were tricarbocyanine dyes, the lifetimes were found to be <1.0 ns, with the lifetimes for the dye set ranging from 947 ps to 843 ps when measured in a polyacrylamide gel containing urea, with the observed lifetimes less than what was observed in methanol but still exhibiting single exponential behavior.

A dye set appropriate for lifetime identification purposes has been prepared and used in four lifetime DNA sequencing applications.[46] The dyes absorb radiation from 624 to 669 nm and possess lifetimes which range from 1.6 to 3.7 ns (see Figure 1.24). Unfortunately, the dye set shows multiexponential behavior in sequencing gels containing denaturants (urea, 7 M). In addition, to correct for dye-dependent mobility shifts, unique linker structures were used. Using this dye set (dye-primer chemistry) and a pulsed diode laser operating at 630 nm, the sequence of a M13mp18 phage was evaluated. The lifetimes were extracted from the decays using a pattern recognition algorithm. The read length was found to be 660 bases with a calling accuracy of 90%.

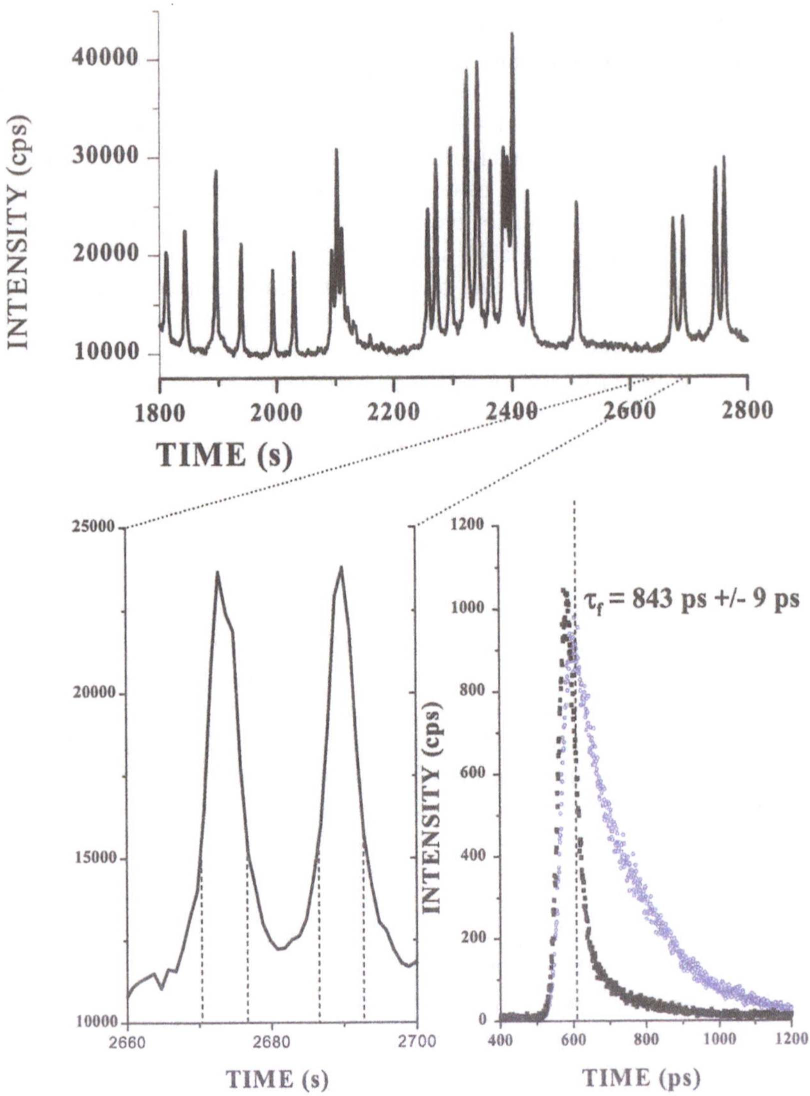

Figure 1.23. Capillary gel electropherogram of C-terminated fragments produced from an M13mp18 template with time-resolved fluorescence detection (A). The labeling dye (F-substituted, heavy atom modified near-IR fluorescent label, see Table 1.1) was attached to the 5′ end of a sequencing primer. In (B) is shown an expanded view of two peaks selected from the electropherogram showing the time area (dashed lines) over which photoelectrons were used to construct the decay profiles. The decay profiles shown in (C) consist of the gel blank (black) and the dye-labeled DNA fragment (blue). The dashed line shows the start channel in which the calculation was initiated. The lifetime was calculated using equation 3. The electrophoresis was carried at a field strength of 200 V/cm using a crosslinked polyacrylamide gel. (Adapted with permission from reference 40.)

1.5. Single Molecule DNA Sequencing Using Fluorescence Detection

The sequencing strategies discussed to this point depend on a fractionation step (electrophoresis) to sort the DNAs by size. While much progress has been made in reducing the time required to develop the electropherogram, resulting in increases in the throughput of acquiring sequencing data, many problems arise when using gel electrophoresis. These problems include: the ability to sequence DNA pieces that are only 1000–2000 bps in length, the requirement of gels to fractionate the DNA, the relatively slow speed associated with the process, and the data processing requirements. The most pervasive problem is the ability to work with only DNA pieces that are 1000–2000 bases in length. Since the chromosomes exceed 1×10^6 bases, the assembly of the sequence of the entire chromosome with extremely short pieces makes the task daunting. What would relieve some the technical challenges associated with assembly is to directly sequence larger strands of DNA. In addition to this, since YAC clones are ~100,000 bps, this requires the shearing and then sub-cloning of these DNAs into M13s to produce templates for sequencing. The ability to work with longer DNAs would eliminate the need for this secondary cloning step.

A very attractive approach has been suggested to rapidly sequence DNA strands that are >40,000 bp in length. The process is based upon the principle of single molecule detection using fluorescence photon burst analysis.[56] The process is depicted in Figure 1.25. Basically, the process involves immobilizing a long strand of DNA in a flow stream and then clipping the terminal base using an exonuclease enzyme, releasing the nucleotide from the original strand. The single nucleotide (either fluorescently labeled or non-labeled) is then carried to a laser beam, which excites the fluorescence with the color used to identify the base clipped from the DNA strand. As can be seen, the single molecule sequencing approach does not involve a gel fractionation step, potentially significantly speeding up the sequencing rate. In essence, the sequencing rate is determined by the rate at which the enzyme clips nucleotide bases from the strand, which for many exonuclease enzymes is on the order of 1000 bases per second. There are three key technical challenges associated with utilizing this technique: (1) tethering the DNA strand to a bead for holding it stationary within a flow stream, (2) creating a complement of the original DNA strand using dye-labeled dNTPs, and (3) detecting single dye-labeled nucleotides using laser-induced fluorescence detection.

Since the premise of this sequencing protocol is to analyze single DNA molecules, it is necessary to select a single DNA molecule and then subsequently trap it within a flow stream for processing. In order to accomplish this, a DNA molecule can be prepared which contains a biotin molecule on its 5′ end. The biotin-modified DNA molecule can then be attached to a microbead (i.d. = 1–10 m), which is coated with streptavidin. Streptavidin is a protein that contains binding sites for biotin. The association of biotin to streptavidin is very strong

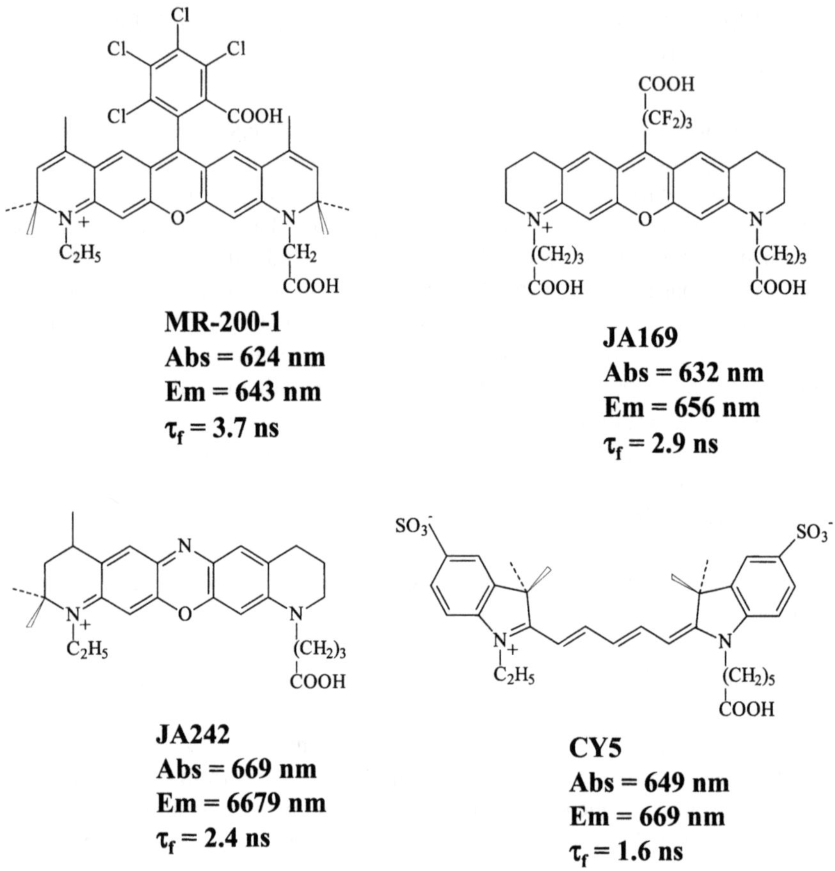

Figure 1.24. Chemical structures of dyes used for multiplex, time-resolved DNA sequencing as well as the fluorescence properties of this dye set. Also shown is the sequencing data obtained when using the dye set shown above conjugated to sequencing primers. The sequencing was performed in a 5% linear polyacrylamide gel containing 7 M urea at a field strength of 160 V/cm. The fluorescence detector consisted of a pulsed diode laser (630 nm) with the optics configured in a confocal geometry. (Adapted with permission from reference 46.) (*continued*)

($K_{assoc.} = 10^{15}\ M^{-1}$), and in addition, is stable to heat and the reagents used in most enzymatic reactions. In this case, it is necessary to select a bead that has one DNA molecule attached to it, since a bead with multiple DNA strands would create a registry problem because the exonucleases work at different rates. Therefore, it is necessary to select conditions to produce a sufficient population of beads containing a single DNA molecule. This can be done statistically by incubating the biotinylated DNA with a large excess of microbeads. This produces a large

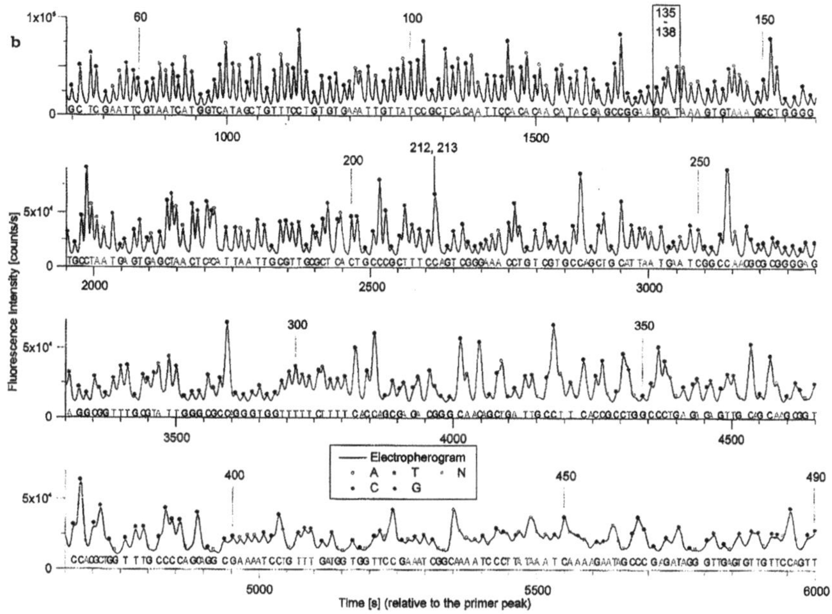

Figure 1.24. *(Continued)*

number of beads that possess no DNA molecule and therefore, the appropriate beads must be selected by staining the immobilized DNA with a fluorescent intercalating dye and using fluorescence to identify a bead with the appropriate number of DNAs attached.

The suspension of the bead containing the DNA molecule in the flow stream can be accomplished using an optical trapping technique. In this case, one or two tightly focused laser beams can be directed onto the bead. This generates an optical trap, which can hold the bead in the flow stream. The size of the trap is determined by the size of the focused laser beam(s) and is on the order of 5 μm. Since the trap is generated by a momentum exchange between the photons and the trapped bead, it does not depend on any type of electronic or vibrational transition. As such, the trapping laser can be a far red or IR laser so that it does not bleach the dyes incorporated into the target DNA molecule during polymerization.

The second technical challenge in this scheme is to produce a complement that contains dye-modified oligonucleotides. Since the detection is accomplished using fluorescence (typically visible fluorescence) single molecule detection, it is necessary to covalently label each dNTP with a probe and then more importantly, build a fluorescent complement of the target DNA using a polymerase enzyme. Incorporation of dye-modified dNTPs is a challenge, as it is with dye-labeled

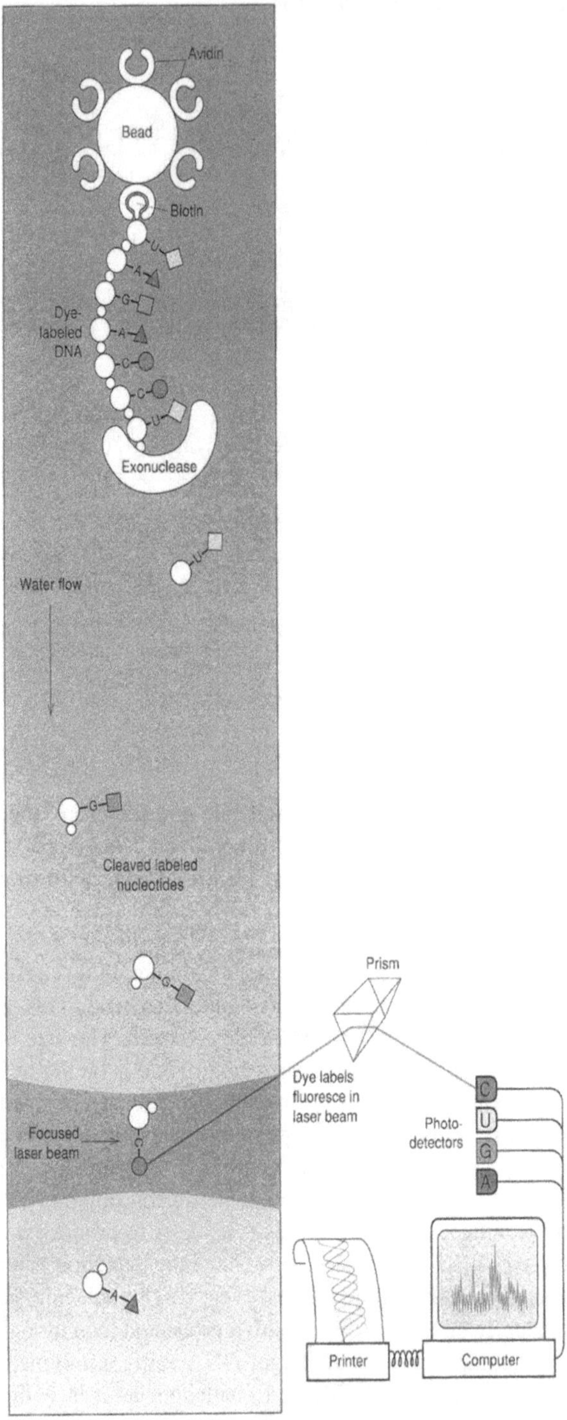
Avidin
Bead
Biotin
U
A
G
Dye-
labeled
DNA
A
C
C
U
Exonuclease
U
Water flow
G
Cleaved labeled
nucleotides
G
Prism
Dye labels
fluoresce in
laser beam
C
Focused
laser beam
Photo-
detectors
C
U
G
A
A
Printer
Computer

Figure 1.25.

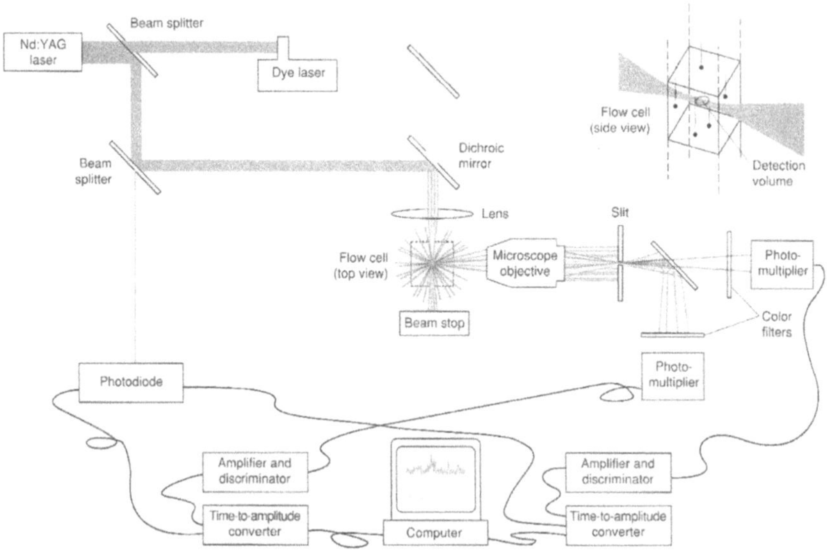

Figure 1.25. (A) Schematic diagram of single molecule sequencing. (B) Block diagram of a dual color single molecule detection apparatus. (C) Single molecule data of R6G and Texas Red using the apparatus described in (B). The dye concentration used in these experiments was set at 50 fM, which resulted in an arrival rate of ~1 molecule every 4 s. The dashed line represents the threshold, which defined the detection of a single molecule when the observed signal (photon burst) exceeded this level. (Adapted with permission from reference 60.) (*continued*)

ddNTPs, since the incorporation rate and fidelity of modified dNTPs by standard polymerase enzymes is not facile. Therefore, mutants will have to be prepared to accomplish this task. An alternative to using fluorescent dyes is to implement UV single molecule detection, which require unmodified dNTPs. While this strategy is much more forgiving in terms of the molecular biology, it places severe challenges on the detection phases of the technique since the nucleotide bases have absorption maxima around 260 nm and the fluorescence quantum yields of the bases are low (~10^{-4}) at room temperature in solution.

The final challenge in this approach is the ability to detect single dye molecules (nucleotides) in solution and additionally, identify the single molecules via spectral discrimination. Since the DNA molecule is composed of four different nucleotide bases, the bases as they are clipped by the exonuclease from the target DNA must be spectrally identified. While several researchers have demonstrated the ability to detect single molecules in flow streams,[57–59] the ability to color discriminate adds complexity to the instrumentation (multiple lasers, multiple detection channels) and requires careful selection of the dyes. The dyes must not only be well-resolved in terms of their emission maxima, but the photophysics of the dye must be conducive to single molecule detection, namely a high quantum

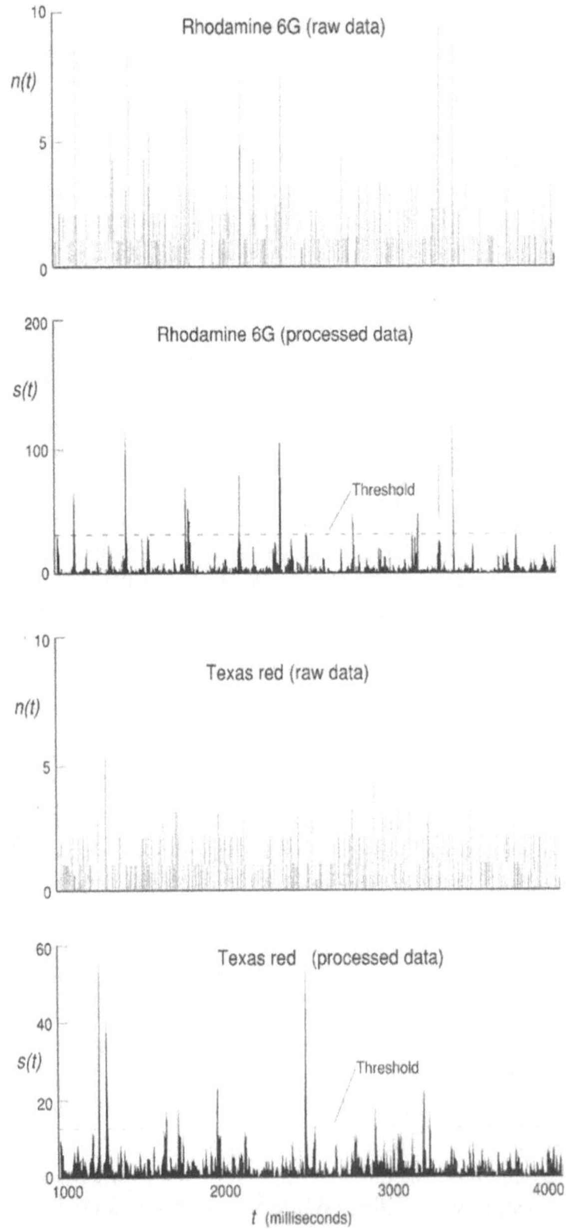

Figure 1.25. (*Continued*)

yield and favorable photochemical stability. An example of a dual color single molecule detection apparatus is shown in Figure 1.25.[60] It consists of a mode-locked Nd:YAG laser operating at 532 nm (second harmonic) and also, a synchronously pumped dye laser (λ_{em} = 585 nm). These wavelengths were chosen to match the absorption maxima of the two dyes selected for this experiment, R6G (λ_{ex} = 528 nm; λ_{em} = 555 nm) and Texas Red (λ_{ex} = 578 nm; λ_{em} = 605 nm). The fluorescence was processed on one of two photodetectors, which in this case consisted of microchannel plates (MCPs). The fluorescence from each dye was directed onto the appropriate MCP using a dichroic filter, with the emission further isolated from background photons and fluorescence from the other dye using a bandpass filter. The processing electronics consisted of conventional TCSPC electronics, which were used along with the pulsed lasers to allow implementation of time-gated detection. Time-gated detection reduces the amount of background scattered photons into the data stream, improving the signal-to-noise in the single molecule measurement. An example of the data output from this dual color single molecule detector is shown in Figure 1.25 (raw data and processed data). The raw data was filtered using a weight quadratic sum filter ($S(t)$) given by the following expression[57]

$$S(t) = \sum_{\tau=0}^{k-1} w(\tau)d(t + \tau)^2 \tag{1.6}$$

where k represents the time range covered by the molecular transit through the laser beam, $w(\tau)$ are weighting factors selected to best distinguish the signal from noise and $d(t)$ represents the raw data point at time (t). As can be seen from this data, large amplitude bursts of photons are clearly evident for both dyes. The detection efficiency was estimated to be 78% for R6G and 90% for Texas Red. In both cases, the error rates (defined as identifying a molecule when one was not present, false positive) was estimated to be <0.01 per second.

References

1. Watson, J. D., and Crick, F. H. C. Molecular structure of nucleic acid. A structure of deoxyribonucleic acid. *Nature 171*, 737, 1953.
2. Chen, E. Y., Schlessinger, D., and Kere, J. Ordered shotgun sequencing, a strategy for integrated mapping and sequencing of YAC clones. *Genomics 17*, 651–656, 1993.
3. Sanger, F., Nicklen, S., and Coulson, A. R. DNA sequencing with chain-terminating inhibitors. *Proc. Natl. Acad. Sci. USA 74*, 5463, 1977.
4. Maxam, A. M., and Gilbert, W. A new method for sequencing DNA. *Proc. Natl. Acad. Sci. USA 74*, 560, 1977.
5. Fung, E. N., and Yeung, E. S. High-speed DNA sequencing by using mixed poly(ethylene oxide) solutions in uncoated capillary columns. *Anal. Chem. 67*, 1913, 1995.

6. Smith, L. M., Saunders, J. Z., Kaiser, R. J., Hughes, P., Dodd, C. R., Connell, C. R., Heiner, C. Kent, S. B. H., and Hood, L. E. Fluorescence detection in automated DNA sequence analysis, *Nature 321*, 674, 1986.
7. Prober, J. M., Trainor, G. L., Dam, R. J., Hobbs, F. G., Robertson, C. W., Zagursky, R. J., Cocuzza, A. J., Jensen, M. A., and Baumeister, K. A system for rapid DNA sequencing with fluorescent chain-terminating dideoxynucleotides. *Science 238*, 336, 1987.
8. Ansorge, W., Sproat, B., Stegeman, J., Schwager, C., Zenke, M. Automated DNA sequencing: Ultrasensitive detection of bands during electrophoresis. *Nucleic Acids Res. 15*, 4593, 1987.
9. Metzker, M. L., Lu, J., and Gibbs, R. A. Electrophoretically uniform fluorescent dyes for automated DNA sequencing. *Science 271*, 1420, 1996.
10. Lee, L. G., Connell, C., Woo, S., Cheng, R., McArdle, B., Fuller, C., Halloran, N., and Wilson, R. DNA sequencing with dye-labeled terminators and T7 DNA polymerase: effects of dyes and dNTPs on incorporation of dye-terminators and probability analysis of termination fragments. *Nucleic Acids Res. 20*, 2471, 1992.
11. Tabor, S., and Richardson, C. C. A single residue in DNA polymerase from *E. coli* DNA polymerase I family is critical for distinguishing between deoxy- and dideoxynucleotides. *Proc. Natl. Acad. Sci. USA 92*, 6339, 1995.
12. Rosenblum, B. B., Lee, L. G., Spurgeon, S. L., Khan, S. H., Menchen, S. M., Heiner, C. R., and Chen, S. M. New dye-labeled terminators for improved DNA sequencing patterns. *Nucleic Acids Res. 25*, 4500, 1997.
13. Ju, J., Ruan, C., Fuller, C. W., Glazer, A. N., and Mathies, R. A. Fluorescence energy transfer dye-labeled primers for DNA sequencing and analysis. *Proc. Natl. Acad. Sci. USA 92*, 4347, 1995.
14. Ju, J. Y., Kheterpal, I., Scherer, J. R., Ruan, C. C., Fuller, C. W., Glazer, A. N., and Mathies, R. A. Design and synthesis of fluorescence energy-transfer dye-labeled primers and their application for DNA sequencing and analysis. *Anal. Biochem. 231*, 131–140, 1995.
15. Ju, J. Y., Glazer, A. N., and Mathies, R. A. Cassette labeling for facile construction of energy transfer fluorescent primers. *Nucleic Acids Res. 24*, 1144–1148, 1996.
16. Hung, S.-C., Mathies, R. A., and Glazer, A. N. Optimization of spectroscopic and electrophoretic properties of energy transfer primers. *Anal. Biochem. 252*, 78, 1997.
17. Lee, L. G., Spurgeon, S. L., Heiner, C. R., Benson, S. C., Rosenblum, B. B., Menchen, S. M., Graham, R. J., Constantinescu, A., Upadhya, K. G., and Cassel, J. M. New energy transfer dyes for DNA sequencing. *Nucleic Acids Res. 25*, 2816, 1997.
18. Hung, S.-C., Mathies, R. A., and Glazer, A. N. Comparison of fluorescence energy transfer primers with different donor-acceptor dye combinations. *Anal. Biochem. 255*, 32, 1998.
19. Williams, D. C., and Soper, S. A. 1995. Ultrasensitive near-IR fluorescence detection for capillary gel electrophoresis and DNA sequencing applications. *Anal. Chem. 67*, 3427–3432.
20. Middendorf, L. R., Bruce, J. C., Bruce, R. C., Eckles, R. D., Grone, D. L., Roemer, S. C., Sloiker, G. D., Steffens, D. L., Sutter, S. L., Brumbaugh, J. A., and Patonay, G. Continuous, on-line DNA sequencing using a versatile infrared laser scanner/electrophoresis apparatus. *Electrophoresis 13*, 487, 1992.
21. Shealy, D. B., Lipowska, M., Lipwoski, J., Narayanan, N., Sutter, S., Strekowski, L., and Patonay, G. Synthesis, chromatographic separation and characterization of near-infrared-labeled DNA oligomers for use in DNA sequencing. *Anal. Chem. 67*, 247, 1995.
22. Soper, S. A., and Mattingly, Q. Steady-state and picosecond laser fluorescence studies of non-radiative pathways in tricarbocyanine dyes: Implications to the design of near-IR fluorochromes with high fluorescence efficiencies. *J. Am. Chem. Soc. 116*, 3447, 1994.
23. Huang, X. C., Quesada, M. A., and Mathies, R. A. Capillary array electrophoresis using laser-excited confocal fluorescence detection. *Anal. Chem. 64*, 967, 1992.
24. Huang, X. C., Quesada, M. A., and Mathies, R. A. DNA sequencing using capillary array electrophoresis. *Anal. Chem. 64*, 2149, 1992.

25. Soper, S. A., Shera, E., Davis, L., Nutter, H., and Keller, R. The photophysical constants of several visible fluorescent dyes and their effects on ultrasensitive fluorescence detection. *Photochem. Photobiol. 57*, 972, 1993.
26. Takahashi, S., Murakami, K., Anazawa, T., and Kambara, H. Multiple sheath-flow gel capillary-array electrophoresis for multicolor fluorescent DNA detection. *Anal. Chem. 66*, 1021, 1994.
27. Kambara, H., and Takahashi, S. Multiple-sheathflow capillary array DNA analyzer. *Nature 361*, 565, 1993.
28. Swerdlow, H., Wu, S., Harke, H., and Dovichi, N. Capillary gel electrophoresis for DNA sequencing: Laser-induced fluorescence detection with the sheath flow cuvette. *J. Chromatogr. 516*, 61, 1990.
29. Swerdlow, H., Zhang, J. Z., Chen, D. Y., Harke, H. R., Grey, R., Wu, S., and Dovichi, N. J. Three DNA sequencing methods using capillary gel electrophoresis and laser-induced fluorescence. *Anal. Chem. 63*, 2835, 1991.
30. Ewing, B., Hillier, L., Wendl, M., Green, P. Base-calling of automated sequencer traces using *Phred*. I. Accuracy assesssment. *Genomics 8*, 175, 1998.
31. Tu, O., Knott, T., Marsh, M., Bechtol, K., Harris, D., Barker, D., and Bashkin, J. The influence of fluorescent dye structure on the electrophoretic mobility of end-labeled DNA. *Nucleic Acids Res. 26*, 2797, 1998.
32. Middendorf, L. R., Bruce, J. C., Bruce, R. C., Eckles, R. D., Grone, D. L., Roemer, S. C., Sloniker, G. D., Steffens, D. L., Sutter, S. L., Brumbaugh, J. A., Patonay, G. Continuous, on-line DNA sequencing using a versatile infrared laser scanner/electrophoresis apparatus. *Electrophoresis 13*, 487–494, 1992.
33. Ansorge, W., Zimmermann, C., Schwager, C., Stegemann, J., Erfle, H., and Voss, H. One label, one tube, Sanger DNA sequencing in one and two lanes on a gel. *Nucleic Acids Res. 18*, 3419, 1990.
34. Chen, D., Swerdlow, H. P., Harke, H. R., Zhang, J. Z., and Dovichi, N. J. Single-color laser-induced fluorescence detection and capillary gel electrophoresis for DNA sequencing. *Proc. Int. Soc. Opt. Eng. 1435*, 161, 1991.
35. Tabor, S., and Richardson, C. C. DNA sequence analysis with a modified bacteriophage T7 DNA polymerase. *Proc. Natl. Acad. Sci. USA 84*, 4767, 1987.
36. Tabor, S., and Richardson, C. C. DNA Sequence Analysis with a Modified Bacteriophage T7 DNA Polymerase. *J. Biol. Chem. 14*, 8322, 1990.
37. Chen, D. Y., Harke, H. R., and Dovichi, N. J. Two-label peak-height encoded DNA sequencing by capillary gel electrophoresis: Three examples. *Nucleic Acids Res. 20*, 4873, 1992.
38. Li, Q., and Yeung, E. Simple two-color base-calling schemes for DNA sequencing based on standard four-label Sanger chemistry. *Appl. Spectrosc. 49*, 1528, 1995.
39. Starke, H. R., Yan, J., Zhang, Z., Muhlegger, K., Effgen, K., and Dovichi, N. Internal fluorescence labeling with fluorescent deoxynucleotides in two-label peak-height encoded DNA sequencing by capillary electrophoresis. *Nucleic Acids Res. 22*, 3997, 1994.
40. Soper, S. A., Legendre, B. L., and Williams, D. C. On-line fluorescence lifetime determinations in capillary electrophoresis. *Anal. Chem. 67*, 4358, 1995.
41. Soper, S. A., and Legendre, B. L. Error analysis of simple algorithms for determining fluorescence lifetimes in ultradilute dye solutions. *Appl. Spectrosc. 48*, 400, 1994.
42. Köllner, M., Fischer, A., Arden-Jacob, J., Drexhage, K.-H., Müller, R., Seeger, S., and Wolfrum, J. Fluorescence pattern recognition for ultrasensitive molecule identification: comparison of experimental data and theoretical approximations. *Chem. Phys. Lett. 250*, 355, 1996.
43. Legendre, B. L., Williams, D. C., Soper, S. A., Erdmann, R., Ortmann, U., and Enderlein, J. An all solid-state near-infrared time-correlated single photon counting instrument for dynamic lifetime measurements in DNA sequencing applications. *Rev. Sci. Instrum. 67*, 3984, 1996.
44. Sauer, M., Zander, C., Müller, R., Ullrich, D., Drexhage, K. H., Kaul, S., Wolfrum, J. Detection

and identification of individual antigen molecules in human serum with pulsed semiconductor lasers. *Appl. Phys. B B65*, 427–431, 1997.

45. Sauer, M., Arden-Jacob, J., Drexhage, K. H., Marx, N. J., Karger, A. E., Lieberwirth, U., Muller, R., Neumann, M., Nord, S., Schulz, A., Seeger, S., Zander, C., and Wolfrum, J. On-line diode laser based time-resolved fluorescence detection of labeled oligonucleotides in capillary gel electrophoresis. *Biomed. Chromatogr. 11*, 81, 1997.
46. Lieberwirth, U., Arden-Jacob, J., Drexhage, K. H., Herten, D. P., Muller, R., Neumann, M., Schulz, A., Siebert, A., Siebert, S., Sagner, G., Klingel, S., Sauer, M., and Wolfrum, J. Multiplexed dye DNA sequencing in capillary gel electrophoresis by diode laser-based time-resolved fluorescence detection. *Anal. Chem. 70*, 4771, 1998.
47. Waddell, E., Stryjewski, W., and Soper, S. A. A fiber-optic-based multichannel time-correlated single photon-counting device with subnanosecond time resolution. *Rev. Sci. Instrum. 70*, 32, 1999.
48. Li, L. C., and McGowen, L. B. On-the-fly frequency-domain fluorescence lifetime detection in capillary electrophoresis. *Anal. Chem. 68*, 2737, 1996.
49. Li, L. C., He, H., Nunnally, B. K., and McGowen, L. B. On-the-fly fluorescence lifetime detection of labeled DNA primers. *J. Chromatogr. 695*, 85, 1997.
50. Nunnally, B. K., He, H., Li, L. C., Tucker, S. A., and McGowen, L. B. Characterization of visible dyes for four-decay fluorescence detection. *Anal. Chem. 69*, 2392, 1997.
51. He, H., Nunnally, B. K., Li, L. C., and McGowen, L. B. On-the-fly fluorescence lifetime detection of dye-labeled primers for multiplexed analysis. *Anal. Chem. 70*, 3413, 1998.
52. Li, L., and McGowen, L. B. Effects of gel material on fluorescence lifetime detection of dyes and dye-labeled DNA primers in capillary electrophoresis. *J. Chromatogr. 841*, 95, 1999.
53. Hall, P.; Sellinger, B. 1981. Better estimates of exponential decay parameters. *J. Phys. Chem. 85*, 2941, 1981.
54. Ballew, R. M., and Demas, J. N. 1989. An error analysis of the rapid lifetime determination method for the evaluation of single exponential decays. *Anal. Chem. 61*, 30, 1989.
55. Flanagan, J. H., Owens, C. V., Romero, S. E., Waddell, E., Kahn, S. H., Hammer, R. P., and Soper, S. A. 1998. Near-infrared heavy-atom-modified fluorescent dyes for base-calling in DNA-sequencing applications using temporal discrimination. *Anal. Chem. 70*, 2676, 1998.
56. Fairfield, E. R., Jett, J., Keller, R., Hahn, J., Krakowski, L., Marrone, B., Martin, J., Ratliff, R., Shera, E., and Soper, S. Rapid DNA sequencing based upon single molecule detection. *Gen. Anal. 8*, 1, 1991.
57. Shera, E. B., Seitzinger, N., Davis, L., Keller, R., and Soper, S. Detection of single fluorescent molecules. *Chem. Phys. Lett. 174*, 553, 1990.
58. Soper, S. A., Hahn, J., Nutter, H., Shera, E., Martin, J., Jett, J., and Keller, R. Single molecule detection of R-6G in ethanolic solutions utilizing CW excitation. *Anal. Chem. 63*, 432, 1991.
59. Soper, S. A., Mattingly, Q. L., and Vegunta, P. Photon burst detection of single near infrared fluorescent dye molecules. *Anal. Chem. 65*, 740, 1993.
60. Soper, S. A., Davis, L., and Shear, E. B. Detection and identification of single molecules in solution. *J. Opt. Soc. Am. B 9*, 1761, 1992.

2

Fluorescence in Nucleic Acid Hybridization Assays

Larry E. Morrison

2.1. Introduction

Fluorescence has progressively been finding a place in nucleic acid hybridization assays over the past two decades. At the beginning of the 1980s essentially all hybridization assays were detected using radioactive labels. Through the 1980s relatively few nonradioactive formats were investigated in much detail. This changed in the 1990s, and now at the end of the decade a large array of fluorescent hybridization assay formats have been reported. In this review, emphasis will be placed on describing the variety of assay approaches emerging in the 1990s, together with some of the key early developments from the previous decade. Discussion of both heterogeneous assay formats and homogeneous assay formats is included, as well as applications of these formats to amplified DNA and RNA assays. The discussion will be limited, however, to assays of nucleic acids *in vitro*, and will not include *in situ* hybridization or physical studies of the hybridization process. Particular attention will be placed on the full range of fluorescence characteristics which have been utilized in developing the various assay approaches, which includes fluorescence lifetimes, environmental sensitivity, quantum yields, fluorescence quenching, resonance energy transfer, excited state complex formation, and fluorescence polarization.

The purpose of hybridization assays is to identify or quantify the presence of particular nucleotide sequences within a specimen. These sequences may identify organisms contained in the specimen, such as in infectious disease testing, or they may identify particular genes, gene alterations, or transcription products. The primary reagents are nucleic acid probes, either DNA, RNA, or some synthetic analog of nucleic acid capable of specific base pairing and hybridization, such as peptide nucleic acid (PNA). In the case of fluorescent hybridization assays, these probes are modified to include a label that is fluorescent or that can bind fluores-

Larry E. Morrison • Vysis, Inc., Downers Grove, Illinois 60515-5400.

Topics in Fluorescence Spectroscopy, Volume 7: DNA Technology, edited by Joseph R. Lakowicz, Kluwer Academic / Plenum Publishers, New York, 2003

cent material in later steps. In some applications the probe is not required to hybridize with the analyte nucleic acid but may only need to differentiate double- from single-stranded nucleic acid, or nucleic acid from non-nucleic acid. Examples of these probes are intercalating dyes and dyes which bind to the major or minor grooves of the double helix. The analyte nucleic acid, or more specifically, a nucleotide sequence therein, is the target of the assay reagents. This target is either DNA or RNA that is released from cells within the specimen by lysis. It is the formation of a complex between the target and probe that is detected in hybridization assays, and indicates the presence of the target nucleic acid.

2.1.1. Heterogeneous Versus Homogeneous Hybridization Assays

Nucleic acid hybridization assays can be divided into the two broad categories of heterogeneous and homogeneous hybridization assays. Strictly speaking, heterogeneous assays are assays which require more than one phase, typically a solid phase and a liquid phase. The solid phase usually serves one of two purposes: (1) to immobilize probes specifically bound to the target nucleic acid, thereby providing a simple means for separating bound from unbound probes through washing of the solid phase, or (2) to fractionate the specimen such that different target nucleic acids are separated from one another and from other cellular components.

Homogeneous hybridization assays are assays that are performed in a single phase. Without the solid immobilization phase to facilitate physical separation of probe bound to target from probe free in the solution phase, the detectable characteristics of the probe label must change in response to hybridization, permitting measurement of hybridized probe in the presence of unbound probe. It is because of this requirement that fluorescence has become so popular in homogeneous assay development. No other class of labels has such a wealth of measurable properties that are strongly affected by various molecular interactions.

Heterogeneous fluorescence hybridization assays have advantages that include: (1) bound and unbound reagents can be physically separated so that intensity measurements can be used to quantify the amount of probe bound to target, and (2) hybridization can be driven to completion faster by increasing probe concentration. These advantages lead to the fact that lower amounts of target can be detected by heterogeneous assays relative to homogeneous assays. The disadvantages of heterogeneous fluorescence assays include: (1) surface hybridization rates are slower than solution hybridization rates (although this can be offset by advantage 2 above), (2) probe reagents can adsorb nonspecifically to the solid phase, increasing background fluorescence (probe adsorption increases with probe concentration providing a practical limitation to advantage 2 above), (3) heterogeneous assays require more steps including separation of phases and washing of the solid phase, and (4) the presence of the solid phase increases background light due to light scattering and fluorescent species contained within the solid phase.

Homogeneous fluorescence hybridization assays have advantages that include: (1) assays can be performed in a single step, (2) solution phase hybridization kinetics are faster that hybridizations onto surfaces, and (3) there is no solid phase that can scatter light, or contain endogenous luminescent compounds, or adsorb labeled reagents. The primary disadvantage of homogeneous fluorescence assays is that they are generally less sensitive than heterogeneous assays, as they are limited by the extent to which the fluorescence properties of bound probe is changed relative to the properties of unbound probe, as will be discussed later.

2.1.2. Amplified Hybridization Assays

Heterogeneous and homogeneous fluorescence hybridization assays use fluorescent probes to detect nucleic acid targets contained within the specimen. In order to increase sensitivity, the hybridization assays can also employ either target amplification or signal amplification. In target amplification the target nucleic acid sequence is copied many times. If only the original target is used as the template for making copies then the amplification is linear. If the copies can also serve as templates then the amplification is exponential. The best known target amplification method is the polymerase chain reaction (PCR) in which two primer DNA oligomers hybridize to the two complementary target strands, flanking the sequence to be amplified.[1] A DNA polymerase from a thermophilic organism is used to synthesize new DNA strands starting at the 3′-termini of the primers, making copies of the intervening sequence. After heating to denature the new stretches of double stranded DNA, the temperature is lowered and the primers anneal to the original target strands and the newly synthesized strands, and are extended again by the polymerase. Repeating this series of events provides a near doubling of the number of target strands on each cycle, easily providing millionfold and higher amplification levels. RNA targets can also be amplified by PCR after a preliminary copy of the RNA is made in DNA (cDNA) using the enzyme reverse transcriptase. Amplification of RNA targets by this method is called RT-PCR. Other forms of target amplification include the transcription based amplification system (TAS),[2] the self sustained sequence replication reaction (3SR),[3] and the strand displacement amplification (TDA).[4] 3SR and TDA have the advantage of being isothermal amplification reactions that do not require the temperature cycling of PCR.

Signal amplification does not rely on copying the target sequence, but instead increases assay sensitivity by either building probe structures containing large numbers of labels, or by enzymatically generating a large amount of detectable material. An example of building large labeled probe structures is the branched DNA (b-DNA) technique.[5] By using enzymatic probe labels, such as alkaline phosphatase (AP) or horseradish peroxidase (HRP), nonfluorescent compounds can be converted to fluorescent compounds, generating hundreds or thousands of fluorophores for each probe present. Other forms of signal amplification use polynucleotide probe labels which are amplified by polymerases or replicases in the reaction mixture. These include probes appended with a midivariant RNA sequence which can be exponentially amplified by the enzyme Q-beta replicase,[6] and so-called "padlock probes" which can be amplified by rolling circle amplification[7] using DNA polymerase.

2.2. Heterogeneous Hybridization Assays

2.2.1. Non-Amplified Heterogeneous Hybridization Assays

Most heterogeneous assay designs have the objective of immobilizing the target nucleic acid onto a solid support material. Any labeled probe that hybridizes to the target nucleic acid also becomes immobilized, and subsequent washing of the support material serves to remove labeled probe not hybridized to the target nucleic acid. Labeled probe detected on the support, or detected after release from the support, therefore, is a measure of the amount of target nucleic acid in the original specimen.

Three basic approaches to immobilizing target and labeled probe on the solid support are shown schematically in Figure 2.1. In the first approach (Figure 2.1A) the target is adsorbed onto the solid support. This is the manner in which DNA or RNA blots are prepared. The solid support is typically a thin membrane of nitrocellulose or nylon. The nucleic acid is heated to dissociate the complementary strands, referred to as denaturing the nucleic acid, and is then applied to the membrane. The membrane is allowed to dry, and is often baked in an oven to further adhere the nucleic acid. The membrane is then soaked in a solution containing labeled single-stranded probe with a nucleotide sequence complementary to the targeted sequence. After hybridization, the membrane is washed thoroughly to remove unhybridized probe, and the amount of label still bound to the membrane is measured.

In the hybridization step the buffer salt concentration and the temperature are selected to provide the most complete and rapid hybridization, the temperature generally set to between 15° and 20° C below the T_m of the probe-to-target hybridization. The T_m is the melting temperature, or more exactly, the temperature at which half of the probe nucleotides are hybridized with the target nucleotides,

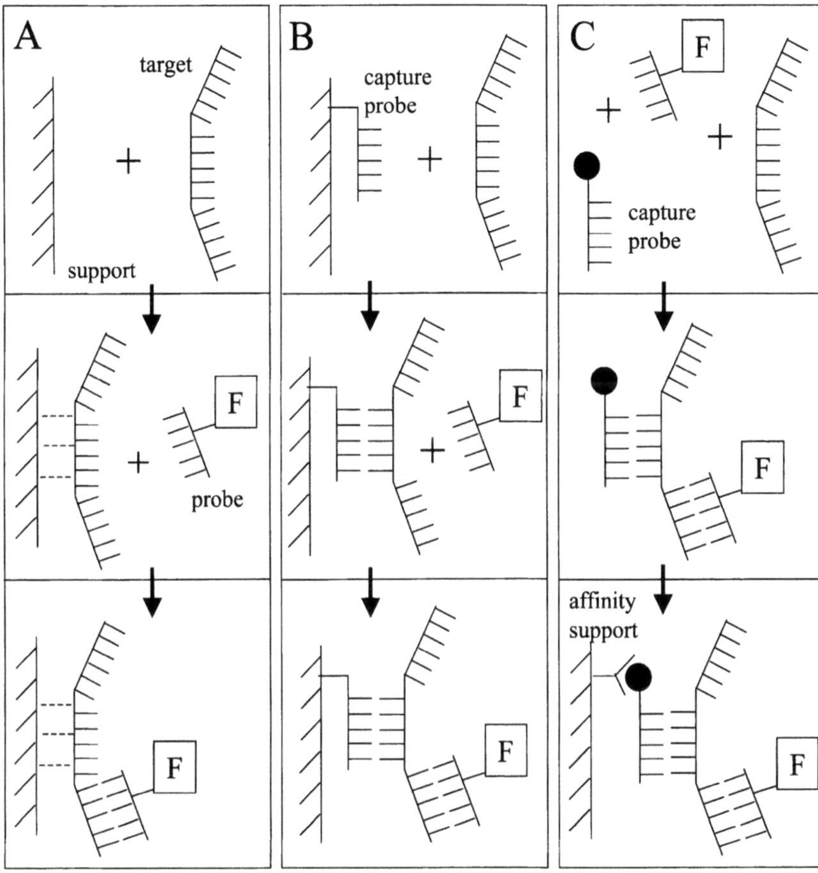

Figure 2.1. Schematic representations of 3 general heterogeneous assay formats. Refer to the text for a detailed description of each format. Nucleic acids are represented by comb structures in which the teeth of the combs represent individual nucleotides. The letter "F" represents the fluorescent probe label, solid circles represent an affinity-binding moiety, and Y-shaped structures represent the affinity binding partner, such as antibody or avidin. Solid support material is represented by vertical lines with appended diagonal lines.

for the particular buffer used (e.g., 50:50 formamide:water/2×SSC/pH 7.5, where 1×SSC = 0.15 M sodium chloride and 15 mM sodium citrate). Solvents such as formamide are often added to lower the T_m, thereby providing convenient assay temperatures, for example, 37°C for hybridization and 75°C for denaturation. Polymers such as dextran sulfate or polyethylene glycol are added to enhance the rate of hybridization, presumably by volume exclusion. Other materials may be added to the hybridization solution to reduce nonspecific adsorption of the labeled probe to the membrane (see reference 8 for a general review of solid phase hybridization conditions).

A second approach to heterogeneous assays (Figure 2.1B) attaches to the solid support a nucleic acid that contains a nucleotide sequence complementary to a region of sequence within the target nucleic acid. Instead of adsorbing the target nucleic acid nonspecifically to the support, the target is specifically captured from the solution by hybridization to the immobilized capture probe. Either before or after capture of the target, a labeled probe nucleic acid is hybridized to a different nucleotide sequence on the target nucleic acid, and the solid support is washed thoroughly to remove unbound labeled probe. Solid support materials in this "sandwich" assay format may be filters and membranes, as well as surfaces of microtiter wells, macro- or microscopic beads, and glass surfaces, which can be chemically derivatized to permit attachment of the capture probe. The sandwich format avoids the additional steps of purifying the specimen DNA and attaching it to the solid surface. Solid surfaces with immobilized capture probes can be made in quantity ahead of time and stored until needed in assays.

In a third heterogeneous assay approach (Figure 2.1C), also a sandwich format, all hybridizations take place in the solution phase. The hybridization step is followed by capture of the target-capture probe-detection probe complex onto the support material by an affinity reaction between a moiety attached to the capture probe and its binding partner attached to the solid support. Again, solid supports may be composed of a variety of materials which can be chemically modified to attach the binding partner or which strongly adsorb the binding partner, typically a protein such as antibody or avidin.

Long-lifetime fluorescent labels, such as the lanthanide ions, have been put to particularly good use in heterogeneous assays (for reviews on lanthanides in hybridization assays see references 9–11). Long lifetime emitters are valuable in heterogeneous assays because they allow rejection of background light due to scatter of the excitation light from the solid support surface and due to fluorescence of the support material itself. Since typical fluorescence occurs predominately from relatively short-lived emitters, in the range of a fraction of a nanosecond to 200 nanoseconds, the lanthanides, with emission lifetimes of hundreds of microseconds, can be easily isolated by time-resolved detection methods.

Lanthanide ions are coupled with nucleic acid probes via metal chelators such as diethylenetriamine pentaacetic acid (DTPA). With simple chelators, however, lanthanide ions are only weakly emissive, having small extinction coefficients for light absorption and having low quantum yields in aqueous media. For this reason, lanthanide ions are often released from the probe chelator with an "enhancing solution" following hybridization and washing to remove unbound probe. Chelators which have strongly absorbing substituents, and which can completely coordinate the ion and shield the lanthanide from water molecules, are included in the enhancing solution. The chelator chromophore absorbs the excitation light and efficiently transfers the excitation energy to the lanthanide ion, which fluoresces with high quantum yield.

The lanthanide and chelator can be coupled to the nucleic acid probe by either "indirect" attachment or "direct" attachment. In indirect attachment the nucleic acid probe is covalently coupled with a hapten or biotin, and the chelator is covalently attached to the affinity binding partner, either the antibody specific for the hapten or a biotin-binding protein, either avidin or streptavidin. Using the assay format shown in Figure 2.1A with indirect label attachment, 0.3 amol (8 pg) of adenovirus DNA blotted onto nitrocellulose filters was detected with hapten (sulfone and fluorene) labeled DNA probes.[12] Two layers of antibodies were used following hybridization, first sulfone or fluorene antibodies, followed by Eu^{3+}-labeled species-specific antibody against the first antibody. Europium fluorescence was measured after releasing the Eu^{3+} into the enhancing solution. In another example, adenovirus DNA was adhered to polystyrene microtiter wells, and probed with biotinylated DNA.[13] After incubation with Eu^{3+}-labeled streptavidin the Eu^{3+} was released and measured in enhancing solution to provide detection of 0.5 amol (10 pg) of the adenovirus target DNA. Hybridizations using chelators directly attached to the DNA probes were used to detect 1 amol[14] and 0.1 amol[15] of an HBV target sequence adhered to nitrocellulose filters and polystyrene microtiter wells, respectively.

A sandwich hybridization assay of the type shown in Figure 2.1B for adenovirus DNA, spiked into nasopharyngeal mucus, was shown to be 10-fold less sensitive than the assay in which the adenovirus DNA was directly blotted onto the nitrocellulose filter (approximately 3 amol).[12] However, purification of the specimens to remove interfering materials in the crude specimen matrix was not required, as would have been the case if the specimen DNA were directly blotted onto the nitrocellulose. Sandwich assays of this type have also been carried out on bead surfaces in which capture probe was synthesized directly on the bead surface by phosphoramidite chemistry. With synthetic oligomer targets representing the ΔF508 mutation in cystic fibrosis, and detection probe directly labeled with chelators containing absorbing chromophores that funnel energy into Eu^{3+}, two amol of target oligonucleotides could be detected on a 50 μm diameter microparticle (average over several beads).[16] By using a chelator with an attached chromophore, enhancing solution was not required for efficient detection of the target molecules. In a similar experiment, point mutations could be distinguished and 0.05 amol of targets were detected in 20 hour hybridizations.[17]

The large majority of unamplified heterogeneous assay formats utilize the long-lived lanthanide labels in order to reduce fluorescence background related to the use of solid support materials. However, the shorter-lived organic fluorophores have found application in fiber-optic-based sensors. The quartz fiber optics used in these applications exhibited low background fluorescence, and transmission of excitation light down the length of the fiber optic by total internal reflection avoided light scattering problems. In one fiber optic biosensor, a model capture probe (dT_{20}) was synthesized directly on the fiber optic surface via phosphoramidite chemistry. The immobilized probe was used to capture model target DNA or RNA (dA_{20} or rA_{20}) from the solution in which the fiber optic was immersed.[18] The intercalating dye, ethidium bromide, was added to indicate the formation of double-stranded DNA. An epifluorescence microscope was used to focus excitation light onto the end of the fiber optic within the collection angle of the fiber, while simultaneously collecting the fluorescence transmitted towards the objective due to total internal reflection down the length of the fiber optic. By this method, detectable signal was obtained in 86 μg/liter solutions of the target nucleic acid.

In another fiber optic application, a biotinylated capture probe (mixed based oligonucleotide) was bound to a fiber-optic surface coated with streptavidin, and unlabeled complementary oligonucleotide was detected by competitive hybridization with fluorescein labeled oligonucleotide of the same sequence.[19] The unlabeled target oligonucleotide was detectable at a concentration of 1.1 nM (132 pg).

By focussing the excitation light onto the end of an optical fiber, excitation light is efficiently carried down the length of the fiber by total internal reflection. The resulting evanescent wave penetrates the surrounding solution by only a fraction of a wavelength of the excitation light, permitting excitation of fluorescent labels bound to the surface via the immobilized capture probes, while excluding unbound labels in the bulk solution. Although not demonstrated in the above examples, this forms the basis of a nonseparation assay in which the test solution does not need to be washed from the surface of the fiber optic, For maximum sensitivity, however, the fiber would still require washing since a small amount of unbound detection probe would lie arbitrarily close enough to the fiber surface to be excited by the evanescent wave.

2.2.2. Amplified Heterogeneous Hybridization Assays

Various amplification strategies have been applied to heterogeneous hybridization assays to increase sensitivity. Although unamplified heterogeneous assays using lanthanide labels and time-resolved detection have shown sub-attomole detection levels, amplification can improve assay reproducibility by increasing the measurable signal above the variable levels of background fluorescence found in clinical specimens. Amplification also allows the more common organic fluorophores, such as fluorescein, to be used in highly sensitive assays. Depending upon the type and extent of amplification, single molecule detection is feasible when combined with even moderately sensitive detection systems.

2.2.2.1. Signal Amplification in Heterogeneous Hybridization Assays

An example of signal amplification in heterogeneous assays is the use of latex particles containing fluorophores entrapped during the polymerization process. Latex beads of 1.8 μm diameter and containing on the order of 100,000 fluorescent pyronine G molecules per bead, were conjugated to streptavidin and used to detect λ-phage DNA immobilized on nylon or cellophane membranes.[20] Denatured target DNA was adhered to the membranes, and allowed to hybridize with denatured biotinylated DNA probe. After washing, the membranes were incubated with the streptavidin coated beads, washed, and photographed under ultraviolet light. In this manner 0.3 pg of target, or about 0.01 amol, were detectable.

A protein complex containing a large number of chelated Eu^{3+} has also been used for signal amplification.[10] This streptavidin-based macromolecular complex (SMBC) is composed of thyroglobulin molecules conjugated to either the lanthanide chelator or chelator and streptavidin to give aggregates containing nearly 500 chelated Eu^{3+} each,[21] which can bind to biotinylated probes. The chelator in this complex was 4,7-bis (chlorosulfophenyl)-1,10-phenanthroline-2,9-dicarboxylic acid (BCPDA) which provided a chromophoric group to absorb light and efficiently transmit the energy to the Eu^{3+}. One application of this label complex was to detect DNA fragments on Southern blots.[21] The plasmid pBR328 was digested with restriction endonucleases, and the resulting DNA fragments were separated by agarose gel electrophoresis and transferred to a sheet of nitrocellulose. The blotted DNA was then hybridized with biotinylated probes (linearized pBR328) and incubated with the SBMC. A laser scanning system with a mechanical stage and time-resolved detection was used to detect bands containing as little as 4 ng of DNA fragments.

Instead of binding many fluorophores to probes, fluorophores can be created from nonfluorescent molecules via reactions catalyzed by enzyme labels associated with probes.[22] For this purpose, streptavidin or antibody conjugates of enzymes such as AP can be bound to biotin- or hapten-labeled DNA probes post hybridization. Alternatively, enzyme labels can be used to convert light-absorbing molecules into chelators that can transfer their absorbed energy to lanthanide ions following chelation.[23] A phosphorylated salicylic acid substrate was used for this purpose together with an AP-streptavidin conjugate.[24] Biotinylated probe was hybridized to pBR322 target DNA adhered to polystyrene microwells or nylon membranes. After washing, the wells or membranes were incubated with the enzyme conjugate, washed again, and incubated with the substrate to generate fluorescent terbium chelates. Southern blots were also detected by this enzyme-amplified lanthanide luminescence (EALL) after separating pBR322 restriction fragments on agarose gels and transferring the fragments to a nylon membrane. A more water soluble substrate (5-fluorosalicyl phosphate) was used in the microwell format while a more bulky and hydrophobic substrate (5-tert-octylsalicyl phosphate) was used for the blots where localization of the signal was important. Time-resolved detection of the long-lived terbium emission provided detection of less than 4 pg of target DNA.

A sandwich hybridization format using EALL with AP-anti-digoxigenin conjugate, and diflunisal phosphate to produce fluorescent terbium complexes, was demonstrated in streptavidin-coated microtiter wells.[25] Biotinylated capture probe was bound to the streptavidin-coated wells, and then incubated with the target DNA and detection probe labeled with the hapten. Following incubation with the enzyme conjugate and substrate, time-resolved measurement of terbium emission detected 0.2 fmol of the 200-base target DNA strand.

Two sequential rounds of signal amplification have been applied in a hybridization assay to further increase sensitivity.[26] The first round of amplification used the tyramine amplification system (TAS) and the second round used EALL with 5′-fluorosalicylic phosphate and Tb^{3+}. Biotinylated detection probe, hybridized to immobilized target, was incubated with a streptavidin-HRP conjugate and washed to remove the unbound conjugate. Addition of the TAS substrate, consisting of biotin covalently linked to tyramine, together with hydrogen peroxidase, led to the HRP-catalyzed oxidation of the tyramine moiety. This produced a highly reactive species that covalently attached to surrounding material on the surface of the immobilization support. The resultant multiplicity of immobilized biotin moieties then served to bind AP conjugates in a subsequent incubation step, and the AP was used in the EALL reaction to generate the fluorescent Tb^{3+} chelates. By combining TAS with EALL, a 10-fold enhancement in the signal-to-background ratio was achieved relative to EALL used alone.

2.2.2.2. Target Amplification in Heterogeneous Hybridization Assays

By far, PCR has been the preferred method of amplification for heterogeneous assays. The simplest way to couple PCR amplification with heterogeneous detection is to use the sandwich hybridization format to capture and detect the amplified target. No modifications to the PCR reagents or protocol are required. As an example, a biotinylated capture probe was used to collect a PCR-amplified region of HIV-1 DNA onto steptavidin-coated microtiter wells. Coupled with a Eu^{3+}-labeled detection probe, linear time-resolved fluorescence between 5 and 500 copies of original purified HIV-1 DNA was measured.[27] The same method was applied to detecting a specific point mutation for diagnosis of familial defective apolipoprotein B-100.[28] Sandwich hybridizations with unlabeled PCR products were also used in HLA typing.[29]

PCR assays can be performed with either or both of the primers labeled, using the 5′-terminus to attach substituents that facilitate capture or detection.

Biotinylated forward and reverse primers were used to immobilize resulting PCR products onto streptavidin-coated microtiter wells in an application which detected mitochondrial DNA point mutations to diagnose Leber hereditary optic neuroretinopathy.[30] After immobilization of the PCR products, the product strands were denatured with alkali, neutralized, and hybridized with allele specific detection probes labeled with chelated Eu^{3+}. Measurement of time-resolved fluorescence in enhancing solution identified the mutation in patient samples.

Instead of using labeled primers to facilitate immobilization, BCPDA-chelator-labeled primers have been used to provide time-resolved detection of products.[31] Agarose gel electrophoresis was used to separate the amplified product from unincorporated labeled primer. Eu^{3+} was introduced later by soaking the gel in $EuCl_3$, and the lanthanide fluorescence was measured with a scanning time-resolving fluorometer. A similar format was used with fluorescein-labeled primer to detect genomic ribosomal RNA targets of pathogenic bacteria.[32] By separating PCR products on a clear agarose gel, the light scattering and background fluorescence problems were greatly reduced relative to solid support materials such as nitrocellulose. This permitted the use of steady state fluorescence measurements, thereby allowing the more common shorter-lived organic fluorescent dyes to be used as labels. The amplified assay could detect a single lysed *Listeria* cell and could detect as few as 20 *Mycobacterium tuberculosis* cells spiked into sputum.

Inclusion of both a biotinylated primer for capture and a hapten-labeled reverse primer for detection was used for identifying HPV-16 sequences in cervical smears.[33] The hapten substituent was dinitrophenol (DNP) which was incubated with lanthanide-labeled anti-DNP following immobilization of the PCR products on streptavidin-coated microtiter wells. The lanthanide label consisted of Eu^{3+}-chelated by trisbipyridine cryptate [TBP(Eu^{3+})]. The TBP(Eu^{3+}) provided both strong chelation and light absorption, allowing the direct measurement of time-resolved europium emission without the need for enhancing solution. The fluorescence results in 14 clinical specimens correlated well with analyses using ^{125}I. This format, however, may be susceptible to false positives and high background since primer dimer formation, initiated by the partial hybridization of the forward primer to the reverse primer, would also give a positive signal.

Multiple PCR-amplified target sequences can be detected simultaneously using spectrally distinguishable fluorescent labels attached to primers or detection probes. Up to five different mutations and genes were detected simultaneously using the organic dyes 7-amino-4-methylcoumarin-3-acetic acid, 5′-carboxyfluorescein, 4′,5′-dichloro-2′,7′-dimethoxy-6-carboxyfluorescein, 6-carboxytetramethylrhodamine, and 6-carboxy-X-rhodamine attached to five different primer pairs.[34] The resulting amplified products could be detected in the same reaction mixture after removing the unincorporated primers by centrifugal filtration through a size selective membrane. Fluorescence in the resulting solution was recorded at 5 different pairs of excitation and emission wavelengths, each selective for a different fluorophore. Similarly, two different fluorophores attached to different primer pairs were used to detect homozygous and heterozygous alleles, as well as to detect translocations, and CMV infection.[34]

Multiplex assays were also achieved using biotinylated primers and capture of amplified products onto streptavidin coated supports. The captured DNA was denatured with alkali and incubated with multiple detection probes specific for different alleles and labeled with each of three different chelated lanthanide ions, Eu^{3+}, Tb^{3+}, and Sm^{3+}. After washing, the ions were released with enhancing solution and detected by time-resolved fluorescence at wavelengths selective for each ion. Three diabetes related HLA alleles were simultaneously detected with sensitivities of about 20 amol, 500 amol, and 500 amol for the Eu^{3+}, Tb^{3+}, and Sm^{3+} labels, respectively,[35] and cystic fibrosis mutations were detected similarly.[36] Combinations of the three lanthanide labels were attached to seven different allele-specific probes and used to type HPV infections.[37] Note that ambiguous results would be obtained if more than one of the seven different HPV organism types could be present within the same specimen.

Despite the high sensitivity PCR provides for detecting specific DNA sequences, PCR was not originally used for quantitative analysis. Variations to the PCR protocols, however, have lead to more quantitative assays. The use of internal standards and the reduction in the number of PCR cycles to keep the reaction in its exponential phase are such variations. Quantitative PCR using an internal standard and detected by EALL has been described.[38] The assay was found to be linear between 1000 and 200,000 molecules of target DNA. Quantitative PCR for determining chromosome aneuploidies has been implemented by a different strategy, however, in which chromosome-specific primer sets are used which encompass highly polymorphic small tandem repeat sequences.[39–41] When primers were labeled with 5-carboxyfluorescein, and the PCR products separated by electrophoresis on an automated DNA sequencer, fluorescence scans of the gel identified amplification products of different lengths corresponding to the different numbers of repeats. The high degree of polymorphism assures that most trisomic specimens will show three different length amplification products in 1:1:1 intensity ratios, or two products in a 1:2 intensity ratio. The few specimens showing a single peak were uninformative. The number of uninformative analyses could be reduced by amplifying two or three polymorphic loci per chromosome.[39,40] Alternatively, primers labeled with a different fluorophore and flanking a nonpolymorphic sequence could be included as a reference intensity for comparison to homozygous products.[40]

A variety of other approaches to fluorescence detection of PCR-generated products have also been tested. Flow cytometric detection of streptavidin-coated magnetic beads has been reported in which biotinylated PCR primers permitted collection of the amplification products after hybridization to digoxigenin-labeled detection probe.[42] Fluorescein-labeled anti-digoxigenin antibodies rendered the complexes fluorescent and magnetic collection simplified the washing procedures to remove unbound detection probes and antibodies. In another approach, PCR and primer extension with fluorescein-labeled primers detected point mutations by mobility shifts of the products separated by polyacrylamide gel electrophoresis.[43] The mobility shifts resulted from differential incorporation of a biotinylated nucleoside triphosphate, containing a long-linker arm, which was complementary to the nucleotide at the site of the mutation. Single base changes were also detected by endonuclease VII digestion at sites of mismatch created by heteroduplex formation between amplified reference and amplified sample DNA.[44] Fluorescein-labeled primers were used in the PCR amplifications, and gel electrophoresis in a DNA sequencer with automated fluorescence detection was used to distinguish the digested from undigested DNA fragments. PCR products have also been detected by the enzymatic ligation of two DNA probes that hybridize adjacent to one another on the amplified product strands.[45] When one of the two probes was labeled with a fluorescein derivative, the ligated product could be separated from unligated probes and detected using an automated DNA sequencer to perform the electrophoretic separation and fluorescence monitoring. This method proved particularly useful to detect point mutations at the 3′-terminal nucleotide of the 5′-probe. Detection of two different alleles was achieved simultaneously using two different 5′-probes labeled with different spectrally distinct fluorophores. Similar multiplex variations on this format have been reported,[46–48] as have related formats using lanthanide labels and timeresolved detection.[49,50]

Target amplification and signal amplification have been combined to detect RNA targets. Transcripts of the BCR-ABL translocation product were amplified using RT-PCR in which digoxigenin-dUTP was incorporated into the amplification products.[51] Biotinylated probe was bound to streptavidin-coated microtiter wells, and was used to capture the amplified products, which were in turn bound by AP-conjugated anti-digoxigenin. Detection using EALL and time-resolved fluorescence could identify BCR-ABL transcripts from one leukemia cell in a background of mRNA from 500,000 normal granulocytes. Similarly, prostate-specific antigen mRNA could be detected at a level of 160 molecules.[52]

Fluorescence hybridization assays using target amplification other than PCR have included use of 3SR and LCR. In LCR, two pairs of probes are used, the members of each pair hybridizing to contiguous regions of the same sequence, one pair to one strand of the target and the other pair to the complementary strand.[53] As such, the probes are complementary to each other, and more importantly, to the products of enzymatic ligation. For this purpose a thermostable DNA ligase is added to covalently join each pair of probes together when hybridized to the target strands. After denaturation, both the original target and the ligated pairs then serve to bring new pairs of probes together for ligation in a second round of target amplification, and so forth. In a variation on LCR, the probe pairs are separated from each other by several nucleotides when hybridized to the target strands and a DNA polymerase is used to incorporate nucleoside triphosphates to fill the "gaps," after which ligase joins the two probes. This Gap-LCR reduces the amount of ligation that occurs between probes in the absence of target. For heterogeneous detection of the amplified product, one member of each adjacent hybridizing probe pair was labeled with adamantane and the other member was labeled with carbazole.[54] The ligation products were captured onto microparticles coated with anti-carbazole, and after washing were detected with anti-adamantane conjugated to AP. The AP converted the substrate methylumbelliferone phosphate into a fluorescent product, thereby combining both target and signal amplification. DNA containing point mutations could be detected in a 10,000-fold excess of wild-type DNA.

3SR amplification uses reverse transcriptase to make copies of RNA targets, primed with a DNA oligonucleotide. Ribonuclease H degrades the RNA portion of the resulting RNA-DNA duplex leaving the newly synthesized DNA strand. The new DNA strand is primed by the reverse oligonucleotide primer and reverse transcriptase extends the primer to form double-stranded DNA. Both primers contain the T7 RNA polymerase promoter sequence so the resulting double-stranded DNA also contains this sequence and serves as a template for transcription by T7 RNA polymerase. The transcribed RNA is primed, extended, and the RNA degraded by ribonuclease H as the process begins to cycle isothermally. For heterogeneous sandwich format detection, the amplification product was captured onto polystyrene beads via an immobilized oligonucleotide complementary to a region within the product strands, and hybridized to a detection oligonucleotide labeled with a lanthanide chelator.[55] After incubation with Tb^{3+}, time-resolved fluorescence measurements could detect 100 molecules of purified *E. coli* ribosomal RNA, or one cell in a spiked sample.

2.3. Homogeneous Hybridization Assays

2.3.1. Non-Amplified Homogeneous Hybridization Assays

In homogeneous hybridization assays, a measurable property of the label must change upon hybridization of the probe to the target nucleic acid. This has led to a full exploitation of fluorescence properties in designing the variety of homogeneous assay formats, including the use of fluorescence quenching, resonance energy transfer, excimer formation, rotational diffusion, translational diffusion, and photon coincidence.

2.3.1.1. Dual Label Formats

The most common sequence-specific homogeneous formats have employed two labels, at least one of which is fluorescent. The two labels interact to varying degrees in a hybridization-dependent manner. In general, the assays are designed such that the two labels, and their appended nucleic acid probes, are free to diffuse in solution in one state and are bound close to one another in a second state. When bound near each other, the two labels can interact in a variety of ways. If the interactions result in radiationless loss of energy from an excited fluorophore, then a lowered light intensity is observed relative to the noninteracting state. The various radiationless processes compete with the emissive decay of the fluorophore as described in Equation 2.1:

$$\Phi = k_e/(k_e + k_{ic} + k_{isc} + k_q[Q] + k_{et}) \quad (2.1)$$

where Φ is the quantum yield for light emission (photons emitted per photons absorbed), k_e is the intrinsic emissive rate constant of the fluorophore, k_{ic} is the rate constant for internal conversion of energy from the excited state to the ground state, k_{isc} is the rate constant for intersystem crossing, k_q is the bimolecular rate constant for collisional quenching, $[Q]$ is the concentration of a quencher molecule, and k_{et} is the rate constant for resonance energy transfer. Internal conversion degrades the fluorophore excitation energy to heat via transitions through vibrational levels, while intersystem crossing usually creates an excited triplet state from the excited singlet state of the fluorophore, which at room temperature in solution typically leads to radiationless de-excitation instead of phosphorescence. Resonance energy transfer (also known as long-range, dipole-coupled, or Förster energy transfer) occurs between the energy donating fluorophore and an energy acceptor molecule which may be fluorescent or nonfluorescent. The rate constant for resonance energy transfer is proportional to the inverse sixth power of the distance separating two molecules (R, in cm), as shown in Equation 2.2:

$$k_{et} = 8.79 \times 10^{-25} \kappa^2 \Phi_D k_D \eta^{-4} R^{-6} \int_0^{\infty} F_D(\lambda) \varepsilon_A(\lambda) \lambda^4 d\lambda \quad (2.2)$$

where κ is an orientation factor ($\kappa^2 = 2/3$ for randomly oriented donor and acceptor molecules in solution), Φ_D is the donor fluorescence quantum yield, k_D is the first-order rate constant for de-excitation of the donor in the absence of energy transfer, η is the refractive index of the medium, $F_D(\lambda)$ is the normalized emission spectrum of the energy donor (integrated intensity = 1), and $\varepsilon_A(\lambda)$ is the spectrum of the acceptor's molar extinction coefficients for absorption, both expressed as functions of the wavelength, λ.[56] Efficient energy transfer requires overlap of the donor's emission spectrum and the acceptor's absorption spectrum, as indicated by the integral term, requiring transitions of comparable energy through which the energy transfer can occur.

In donor acceptor label pairs selected for good energy transfer, the values of R, at which energy transfer is as likely as de-excitation in the absence of energy transfer, are in the range of 30 to 100 angstroms. This may be compared to the 3.4 angstroms spacing between base pairs in the DNA double helix. Therefore, energy transfer can occur between label molecules that cannot contact one another, unlike collisional quenching which requires actual contact between labels. The length of the linkage between label molecules and DNA, the position of attachment, and interactions between the labels and DNA must also be considered when estimating the average separations distance between labels and whether labels may actually collide with one another.

Collisional quenching and resonance energy transfer both depend upon the presence of a second molecule and are the basis for several dual label interaction formats.[57,58] Figure 2.2 shows assay formats that have taken advantage of label interactions. The "adjacent probe" format, shown in Figure 2.2A, utilizes two nucleic acid probes, one labeled with a fluorophore (F) and the other labeled with a quencher or energy acceptor (Q), that are complementary to adjacent positions on the target nucleic acid.[59,60] The labels are placed near the 3′-terminus of one probe and the 5′-terminus of the other probe to assure close proximity of the two labels. In the case of collisional quenching, hybridization of the two probes to the target effectively confines the two labels to a very small volume, making $[Q]$ large enough that $k_q[Q]$ dominates the other terms in Equation 2.1, measurably decreasing the fluorescence quantum yield. In the case of resonance energy transfer, R becomes small enough that k_{et} dominates Equation 2.1 and quenching of the donor fluorescence is observed. If the acceptor label is also fluorescent, then characteristic emission from the acceptor can accompany donor quenching.

The adjacent label format was first demonstrated using a chemiluminescent isoluminol label attached to the 5′-terminus of one probe, serving as the energy

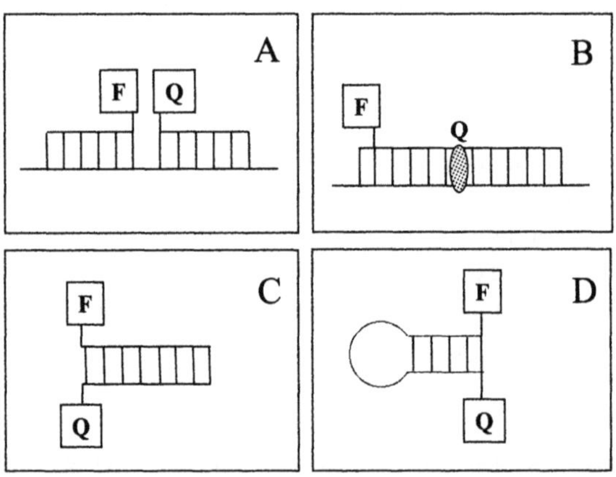

Figure 2.2. Dual label probe formats used in homogeneous hybridization assays. Refer to the text for a detailed description of each format. Vertical lines represent complementary base pairs joined to the sugar/phosphate backbones (horizontal lines) of polynucleotide strands. "F" represents the primary or energy donor fluorescent label, and "Q" represents the quencher or energy acceptor label.

donor, and tetramethylrhodamine serving as the fluorescent acceptor label on the 3′-terminus of the other probe.[60] In the presence of target polynucleotide the rhodamine fluorescence was found to increase 24%. Similarly, using fluorescein and tetramethylrhodamine labels, a 71% quenching of the fluorescein emission was observed in the presence of target.[61] A variation on the adjacent probe format used two triple helix-forming probes hybridizing to adjacent positions on a double-stranded DNA target.[62]

It may be noted that the response of quenching and energy transfer to target concentration in the adjacent probe format is maximal when the target concentration equals that of the probes.[58] At higher concentrations the interaction decreases since the adjacent probes become hybridized to different target strands. This response characteristic can lead to ambiguity in assessing target concentration unless two or more target dilutions are tested in order to determine on which side of the maximum the target concentration lays.

A second interacting label format, shown in Figure 2.2B, uses one labeled probe and a double-helix-specific nucleic acid binding dye. The dye is added at a concentration that assures a dye molecule will be bound within several base pairs of the probe label when the probe is hybridized to the target. This approach was demonstrated with an energy donating isoluminol-labeled probe and the intercalator, ethidium bromide.[60] Correspondence between absorbance hypochromicity measurements at 260 nm and ethidium bromide fluorescence proved that the energy transfer from the isoluminol to the intercalator was a direct result of the probe hybridizing to the target. In a similar experiment, the binding dye, acridine orange, served as the energy donor to a tetramethylrhodamine probe label, and melting curves were again used to verify the sensitivity of the energy transfer to hybridization.[61]

Figure 2.2C shows the "complementary probe pair" format in which two labeled probes have complementary sequences.[63] Following denaturation of target and probes, the complementary probes strands compete for hybridization to each other and to the target strands. At high target concentrations the probes are hybridized primarily to target strands and the labels are prevented from interacting. At low target concentrations the probes are primarily hybridized to each other, thereby allowing their labels to interact. The competitive assay format was demonstrated with a variety of labels on the 3′-terminus of one probe and the 5′-terminus of the other probe.[57,64] This labeling arrangement allows the two labels to collide. Both energy transfer and collisional quenching were observed, depending upon the label combination and the manner in which the labels were attached to the DNA probes, although collisional quenching was more common.[57,58] Work with adjacent probes has indicated that the two labels should be separated by a several nucleotide distance in order to favor the energy transfer mechanism.[65]

Unlike the adjacent probes format, in the complementary probe pair format quenching and energy transfer decrease steadily as target concentration increases, with a midpoint in the response occurring at a target concentration equal to the probe concentration. Measured fluorescence values were fit well by the predicted relationship in Equation 2.3:

$$[T1 \cdot P2]/[P1 \cdot P2]_0 = [T1 \cdot T2]_0/([P1 \cdot P2]_0 + [T1 \cdot T2]_0) \quad (2.3)$$

where $[P1 \cdot P2]_0$ and $[T1 \cdot T2]_0$ are the initial concentrations of complementary probes and target prior to denaturation, respectively, and $[T1 \cdot P2]$ is the concentration of probe-target hybrids following competitive hybridization.[58,64] Plasmids containing an *E. coli* enterotoxin gene target could be distinguished from background fluorescence at concentrations down to 4 pM when 5 probe pairs complementary to different portions of the target were each present at a concentration of 20 pM.[64]

Energy transfer between 5′- and 3′-labels was also demonstrated using lumazine donor labels and a ruthenium II metal complex as the energy acceptor label.[66] Ru II emission increased 110% when the probes were allowed to hybridize to one another. Sixty-three percent quenching of a 5′-fluorescein label on one probe by a 5′-tetramethylrhodamine label on a complementary probe was also reported.[61] It should be noted, though, that use of two 5′-labels restricts the assay to fairly short probes in order to obtain effective label interaction.

A variation on the complementary probe pair is the so called "molecular beacon" in which one end of the complementary probe pair has been joined together to form a hairpin, as shown in Figure 2.2D.[67] The region of the hairpin complementary to the target nucleic acid forms the loop and is accessible for hybridization. The stem and loop size and base compositions are selected so that the hairpin structure predominates in the absence of target, but in the presence of target hybridization of the loop region to target predominates. In the closed hairpin, the 3′- and 5′-terminal labels interact to quench fluorescence, as in complementary probe pairs. Hybridization to target serves to open the hairpin structure and separate the terminal labels. The rapid intramolecular rate of hairpin closure might appear to offer advantages over two probe formats. At the concentrations beacons are generally used (e.g., 0.3 μM), however, the rate of linear probe-to-probe hybridization is also fast (e.g., 2×10^7 M^{-1} s^{-1} at 37°C)[58] and would be essentially complete in several seconds. Probe-to-target hybridization rates should be similar for both hairpin and complementary linear probes at the same concentration. The added structure of the hairpin probe, however, has been shown to be important in detecting point mutations.[68,69] While hairpin probes and linear probe pairs showed similar melting temperatures with perfectly base-paired targets, the hairpin probe displayed a lower melting temperature with targets containing a single base mismatch, relative to linear probes.

Assays based on fluorescence quenching can produce a false positive result if the specimen contains sufficiently high levels of an endogenous quencher species.

Assays based on energy transfer can identify the presence of endogenous quenchers if both donor and acceptor fluorescence are measured. This is because quenching of the donor fluorescence must be accompanied by a corresponding increase in acceptor fluorescence if the intended energy transfer reaction is the only process occurring. Unfortunately, energy transfer is monitored often by measuring the donor fluorescence alone, either because the acceptor label is nonfluorescent, or because the acceptor fluorescence resulting from energy transfer is small compared to background acceptor fluorescence resulting from direct absorption of the excitation light. A method for discriminating against this background acceptor fluorescence was demonstrated in an immunoassay format, and combined the use of a long-lifetime donor fluorophore with a short-lifetime acceptor fluorophore.[70] In this situation, the decay of excited acceptor molecules following a short pulse of excitation light is given by Equation 2.4:

$$A^* = k_{et}D_0^*(k_A - k_D - k_{et})^{-1}[\exp\{-(k_D + k_{et})t\} - \exp\{-k_At\}] + A_0^*\exp\{-k_At\} \tag{2.4}$$

where A^* is the number of excited acceptor fluorophores present at time, t, following the light pulse, A_0^* and D_0^* are the numbers of excited acceptors and donors initially formed ($t = 0$), k_A and k_D are the respective rate constants of acceptor and donor de-excitation in the absence of energy transfer, and $\exp\{x\} \equiv e^x$. The terms in this equation have been grouped to reflect excited acceptor fluorophores formed by direct absorption of the excitation light pulse, in the last term, and the excited acceptor fluorophores formed by energy transfer, in the preceding term. Following the excitation pulse, excited acceptor molecules produced by absorption of the excitation light decay rapidly and the last term becomes small compared to the preceding term. New excited acceptor fluorophores are generated from the longer-lived donor fluorophores, providing the appearance of long-lived acceptor fluorescence. By gating the fluorescence detector, emission from the acceptor fluorophores at short times following the excitation pulse can be ignored, and only the longer-lived acceptor fluorescence is integrated, thereby providing the needed discrimination against the background acceptor fluorescence. A variation on this time-resolved energy transfer method uses a long-lived donor fluorophore, with a low quantum yield, and a short-lived acceptor fluorophore, with a high quantum yield, such that energy transfer produces an increase in total light emission, in addition to allowing discrimination against directly excited acceptor fluorescence.[71]

Interactions between labels other than quenching and energy transfer have been proposed as the basis of homogeneous assays. Excimers, complexes formed between an excited state molecule and a ground state molecule of the same compound, were applied to the adjacent probe format using 3′-pyrene and 5′-pyrene-labeled probes.[72,73] In the absence of the target nucleic acid, the two probes exhibited fluorescence characteristic of the pyrene monomer (3 narrow emission peaks between 370 and 430 nm) since few pyrene labels would collide within the lifetime of the pyrene excited state. In the presence of target, the pyrene labels were brought together, and excimer fluorescence was observed (broad emission centered near 490 nm). Exciplexes, formed between an excited state molecule and ground state molecule of two different compounds, could similarly form the basis of an assay.

Another form of label interaction was demonstrated using a lanthanide label and time-resolved fluorescence,[74] again in the adjacent probe format. A strong metal chelator was used as the 5′-label of one probe, and a weaker but light absorbing chelator was used as the 3′-label of the second probe. The strong chelator held the Tb^{3+} ion tightly in single- and double-stranded forms, while the weaker light-absorbing chelator only bound the Tb^{3+} when the two were brought together by the presence of the target nucleic acid. Once complexed, light absorbed by the one chelator could efficiently excite the Tb^{3+} producing the characteristic lanthanide emission.

A dual label assay format that does not require label interactions has been demonstrated using the coincidence of photons detected from the two different probe labels.[75] Since interaction is not utilized, the two labeled probes are not restricted to hybridizing at adjacent positions on the target strand, but can hybridize at any two positions on the same strand. A flow system, similar to those used for single molecule detection, was modified with two detection channels for coincidence detection of the two different probe labels. Pulsed laser light (70 ps pulse width at 82 MHz) was focussed to a 10 μm spot on a flow cell through which the assay solution was pumped at a rate of 87 μm/s. The probe concentration was such that there was a very low probability that two probes free in solution would traverse the 10 μm excitation region simultaneously. Under these conditions, only target molecules would be detected by the coincidence of photons emitted from the two different fluorophores, as they passed through the light beam together. Using rhodamine 6G and Bodipy-TR labels on probes specific for a *B. thuringiensis* toxin trans gene, one transgene per maize genome (3 billion base pairs) could be detected.

2.3.1.2. Single Label Formats

Several single label homogeneous assay formats have been described in the literature. Without the presence of a second label, single label assays rely solely on the ability of the target DNA to substantially alter a property of the label fluorescence. Some fluorescent labels have been shown to exhibit large fluorescence quantum yield changes upon hybridization to target nucleic acid. Pyrene labels have exhibited both fluorescence increases and decreases depending upon where the label is attached and the length and flexibility of the linker arm.[76] In some cases this may be due to intercalation of the pyrene into double-stranded nucleic acids. In other cases it may be due to interaction with specific nucleic acid bases, for example, through electron transfer via the pyrene excited state, or due to changes in the polarity of the pyrene environment. Two studies reported quenching of pyrene fluorescence when a pyrene labeled probe was hybridized to target polynucleotide. In one of the studies, pyrene label fluorescence was reduced 75% by probe-to-target hybridization.[77] In the other study, pyrene fluorescence was reduced 81% to 99% by hybridization to target, depending upon the particular probe sequence and length.[58] In this latter work, the pyrene label was particularly sensitive to quenching by thymidine bases, in either the probe strand or target strand. To take advantage of this, probes should be selected to be as short as possible with adenosine nucleotides located near the position of pyrene attachment in order to maximize the pyrene quantum yield of the single-stranded probe and maximize quenching in the hybridized state. Quenching of fluorescein emission upon hybridization of labeled probes to complementary DNA has also been reported.[78,79] Similarly, an intercalator dye, oxazole yellow, which shows enhanced fluorescence efficiency upon intercalation, has been used as a covalently attached probe label to provide enhanced fluorescence intensity in response to hybridization.[80]

Another fluorescence property that can respond to hybridization is fluorescence polarization. In a fluorescence polarization experiment, polarized light is used to preferentially excite molecules that have their absorption transition moments aligned with the electric vector of the polarized excitation light. If the molecules are rotating slowly with respect to the lifetime of their excited states, then fluorescence is expected to be polarized also. Polarization, p, is defined in Equation 2.5:

$$p = (I_{par} - I_{perp})/(I_{par} + I_{perp}) \tag{2.5}$$

where I_{par} is the fluorescence intensity measured at 90° to the excitation light path through a polarizer oriented parallel to the polarization of the excitation light, and I_{perp} is the fluorescence intensity measured through the polarizer rotated 90° relative to the first measurement. Rotation of the molecule within the lifetime of the molecule's excited state tends to depolarize the emission. This is because the angular positions of the transition moments at the time of emission become rotated relative to their positions at the time of light absorption. The faster the rotation of the molecule, the lower the polarization of the emitted light, until $I_{par} = I_{perp}$, at which point the fluorescence is completely depolarized and $p = 0$. Larger molecules are expected to rotate slower than smaller molecules, and the relationship between the effective molecular volume, V, and the fluorescence polarization is described by Equation 2.6:

$$1/p - 1/3 = (1/p_0 - 1/3)[1 + (RT\tau_0)/(\eta V)] \tag{2.6}$$

where R is the gas constant, T is the absolute temperature, η is the solvent viscosity, τ_0 is the lifetime of the excited state, and p_0 is the limiting value of the label's fluorescence polarization, determined in the absence of rotation.[81] The value of p_0 depends upon the angle between the absorption and emission transition moments and has a maximum value of 0.5 for colinear transition moments, and $-1/3$ for perpendicular transition moments.[82]

Hybridization of probe to target can be viewed as increasing V, since the entire hybridized complex is larger than the probe alone and, therefore, should rotate at a slower rate. The freedom of the label to rotate about its linkage to the probe, however, can reduce the effect of the slower overall rotation of the complex. Therefore, more rigid linkages to the probe, or other restrictions to the label motion, can accentuate the change in label rotational diffusion resulting from hybridization.

The effect of probe-to-target hybridization on fluorescence polarization has been demonstrated using fluorescein labels attached to DNA oligomers.[83,84] In one study p increased from about 0.06 for probe alone to 0.09 in the presence of target DNA.[84] It may be noted that these p values correspond to ratios of the two polarization components, I_{par}/I_{perp}, of 1.13 and 1.20, respectively, which would require a high degree of instrument precision and low levels of background interference to accurately differentiate at low probe concentrations. With 1 nM probe concentration, a detection limit of about 0.1 nM target was estimated. Increased polarization has been obtained by incorporating a binding site into the probe sequence for the double-strand-specific DNA binding protein, Eco R1.[83] Polarization of the probe fluorescence went from about 0.054 in the absence of target, to 0.082 in the presence of target, and to 0.105 in the presence of target and protein (p values converted from anisotropy values in original article).

Probe translational diffusion rates are also sensitive to molecular size and this sensitivity has been combined with autocorrelation measurements to detect the presence of nucleic acid targets.[85] Using focussed laser light on a droplet of sample, translational diffusion rates of 0.166 and 0.181 ms were measured for Bodipy and tetramethylrhodamine labeled DNA oligomers (18 nucleotide length), respectively. These diffusion rates increased to 2.94 and 2.86 ms, respectively, in the presence of M13mp18(+) strand DNA (7.5 kb). Because of the extremely small volume interrogated by the focused laser beam, single molecule detection is possible, however, it should be noted that this corresponds to nanomolar target concentrations. As with other microvolume techniques, the target must be concentrated into the microvolumes in order to fully realize the assay potential.

Homogeneous assays are generally less sensitive than heterogeneous assays. This is because homogeneous assays require a detectable change in a fluorescence property to signal the presence of the target nucleic acid. This change is never 100% complete. For example, in the complementary probe format, excellent quenching levels of 90 to 99% have been achieved in fluorescence quenching assays, however, 10 to 1% of the unmodified signal still remains. At 1 nM probe concentration, in the absence of target, the unmodified signal would be equivalent to 100 to 10 pM target concentration. Reducing the probe concentration increases the assay sensitivity, but also decreases the rate of hybridization. Assays of several minutes at 1 nM probe concentration become hours at 10 pM probe concentration. Practically speaking, homogeneous assays can provide femtomole or higher levels of target sequence detection, assuming assay volumes of 0.1 to 1 ml. Of course, very small assay volumes can be used to improve the detection limits, if efficient methods of sample concentration are available. For clinical relevance in infectious disease testing, however, attomole detection levels are desirable. In light of these facts, it has been reasoned that homogeneous detection methods are best applied to the detection of amplified polynucleotide targets.[57,64]

2.3.2. Amplified Homogeneous Hybridization Assays

In general, target amplification reactions are homogeneous by nature and coupling with a homogeneous hybridization system for detecting the amplified target can provide a completely homogeneous amplified hybridization assay. The higher sensitivity of heterogeneous detection systems for target amplification is not often needed since target amplification can routinely produce million- to 100 millionfold amplification of target sequences. Signal amplification is not addressed here since it is by nature typically a heterogeneous process.

2.3.2.1. Dual Interacting Label Formats

Sequence-specific detection of amplified PCR products in a homogeneous format was first performed with fluorescence quenching in complementary probe pairs.[64] After completion of 25 cycles of PCR, the probe pair, 5′-labeled with fluorescein and 3′-labeled with quencher (pyrenebutyrate), was mixed with the PCR solution, the solution heated to denature and then cooled to permit competitive hybridization. Fluorescence intensity versus the input number of λ phage DNA target molecules showed a detection level below 100,000 molecules, in this early example.

End point detection of PCR products, as was done in the competitive probe assay and in heterogeneous detection assays, has the drawbacks that extra sample manipulation is required and contamination of other reactions with the amplified PCR product is possible. An assay format developed specifically for detecting PCR products can potentially circumvent this problem by utilizing the 5′-to-3′ exonuclease activity of DNA polymerase, as shown in Figure 2.3A.[86] A dual labeled probe, complementary to the amplified sequence and internal to the

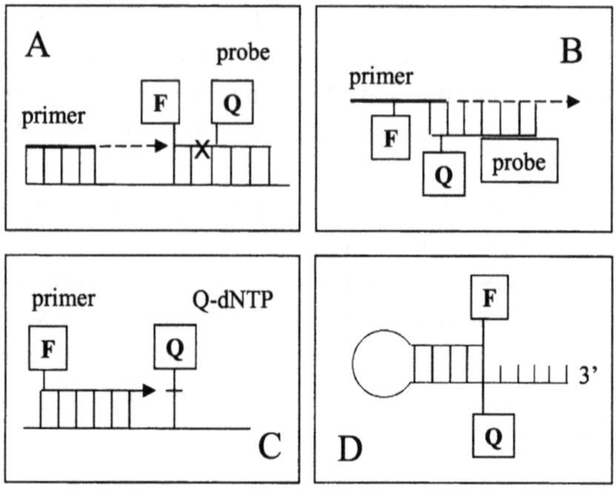

Figure 2.3. Dual label formats developed for use in amplified homogeneous hybridization assays using DNA polymerases. Refer to the text and the caption to Figure 2.2 for a detailed description of the assay formats.

priming sites, was added to the amplification mixture prior to temperature cycling. One label was attached to the 5′-terminus of the probe and a second label was attached several nucleotides internal to the 5′-terminus. During the priming step, primers and probe both hybridized to targets strands. When primer extension began, the polymerase would extend the primer until it reached the probe. At this point the polymerase would digest the probe using its 5′-to-3′ exonuclease activity, thereby separating the two labels and relieving quenching of one label by the other. By using two probes, each with a spectrally distinct fluorescing label, wild type target or target containing a point mutation could be detected in the same amplification reaction. Fluorescence could be measured at the end of the reaction, or in later work, could be monitored continuously throughout PCR using a thermocycling instrument with fluorescence excitation and detection capability. In either situation, it is possible to measure fluorescence without opening the reaction tube, thereby preventing possible contamination of other samples with the amplified product.

The 5′-to-3′ nuclease format has been used in a variety of assays. Examples of the 5′-to-3′ nuclease format utilizing endpoint detection of fluorescence quenching include sequence-specific priming for HLA typing,[87] quantitative analysis of c-erbB-2 oncogene amplification,[88] detecting a single-base polymorphism in *Orthopoxvirus*,[89] and detecting a number of human papillomavirus virus genotypes in cervical specimens.[90] Measurement of fluorescence at each amplification cycle has been used in quantitative PCR to quantify myc, ccndl, and erbB2 gene amplification in breast tumors,[91] and has been used in RT-PCR to quantify the cystic fibrosis transmembrane receptor (CFTR) mRNA in support of a gene therapy project.[92] With fluorescence measured at each cycle, quantitative PCR can be based on the cycle number at which fluorescence first starts to rise above background fluorescence (kinetic PCR), before the PCR reagents become limiting. Standard curves can be generated from plotting known amounts of input target versus the cycle number at which the fluorescence rise is first detected. DNA regions of known copy number can also be used as internal standards.

The molecular beacon hairpin probe was initially developed around PCR, although it has also been applied to nonamplified assays, such as targeting 16S ribosomal RNA from ruminal bacteria.[93] In PCR applications, fluorescence quenching is generally measured at each cycle. Examples of hairpin probes used in PCR assays include detecting drug resistance in *Mycobacterium tuberculosis*,[94] and the simultaneous detection of four different pathogenic retroviruses.[95] In the latter study four different hairpin probes were labeled with four spectrally distinguishable fluorophores: fluorescein, tetrachloro-6-carboxyfluorescein, tetramethylrhodamine, and carboxyrhodamine 6G. The quencher label on all hairpin probes was the nonfluorescent 4-(4′-dimethylaminophenylazo)benzoic acid. Ten retroviral genomes could be detected and 10 copies of one retrovirus could be detected in the presence of 100,000 copies of a second retrovirus.

The 5′-to-3′ nuclease assay and the hairpin probe assay formats showed that PCR could proceed effectively in the presence of hybridization probes. In fact, the older adjacent probe assay format has since been used to continuously monitor amplification during PCR.[96] Likewise, it should be possible to use the complementary probe pairs in the same manner.

Two other assay formats have been developed in conjunction with primer extension by DNA polymerase that occurs during PCR or subsequent to PCR. In one format, shown in Figure 2.3B, a primer labeled with Cy5 near the 3′ terminus was used together with a 3′-fluorescein-labeled oligomer that was complementary to the sequence of the extended primer adjacent to the primer sequence.[97] Hybridization placed the two labels near one another, similar to the adjacent probe format, and energy transfer from fluorescein to Cy5 was indicative of primer extension. If the oligomer was complementary to a sequence that might contain a single base change, a fluorescence melting curve served to differentiate the target containing the mismatch from the perfectly matched target. In this manner, patients homozygous and heterozygous for factor V Leiden were genotyped. Similarly, patient samples were genotyped for the C677T base substitution in the methylenetetrahydrofolate reductase gene.[98]

In the second assay format, depicted in Figure 2.3C, a fluorescein-labeled primer was complementary to a sequence adjacent to the location of a single nucleotide polymorphism on an amplified target strand.[99] Primer extension reactions run in different tubes, each labeled with a different 6-carboxy-x-rhodamine-labeled dideoxynucleoside triphosphate, incorporated the energy acceptor label only if the target contained the complementary nucleotide. Energy transfer from fluorescein to rhodamine indicated which polymorphism was present. This assay format was used to detect mutations in the CTFR gene, the human leukocyte antigen H gene, and the tyrosine kinase receptor protooncogene.[100]

Another approach to detecting amplified PCR products is to use dual interacting labels on PCR primers. One such format used a primer strand that could form a hairpin structure, shown in Figure 2.3D, placing two labels next to one another, similar to the dual-labeled hairpin probe, resulting in quenching of the fluorescence of one label by the other.[101] The priming sequence extended beyond the stem of the hairpin on the 3′-portion of the primer. The priming region of the probe hybridized to the target and was extended by DNA polymerase. In the next cycle of PCR, the opposite end of the newly synthesized fragment was primed and extended by polymerase, eventually reaching the incorporated hairpin primer, displacing the stem structure and opening up the hairpin. Primer incorporation into product is signaled, therefore, by increased fluorescence. A related primer format used two complementary oligomers, instead of a single hairpin-forming strand, much like the complementary probe pair format.[102] Again the priming region extended beyond the complementary probe, and primer extension eventually resulted in displacement of the complementary labeled strand, thereby decreasing energy transfer between fluorescein and Texas Red labels on the complementary oligomers.

It should be noted that both of these labeled primer approaches only measure incorporation of primer into a double-stranded product. The amplified sequence between the primers is not being interrogated. As such, the labeled primer approaches may be susceptible to a higher level of false positive results, since nonspecific priming of the specimen DNA, or primer-to-primer hybridization, will result in primer incorporation that is not indicative of the intended target sequence. An approach which combines a hairpin primer with interrogation of the amplified sequence has been described.[103] As in the molecular beacon, fluorophore and quencher labels on opposite sides of the stem structure interact to reduce fluorescence in the absence of target. In this embodiment of the hairpin primer, a polymerase-blocking linkage between the primer sequence region and the hairpin stem region prevents opening of the hairpin structure due to second strand synthesis. The loop region of the hairpin primer is selected to be complementary to a region of the extended hairpin primer, just to the 3′-side of the priming sequence. Thus, after synthesis of the new strand, and denaturation, the loop region undergoes intramolecular hybridization to the newly synthesized region, forming a larger hairpin loop structure and opening the original stem region. This separates the fluorophore and quencher labels, thereby increasing fluorescence specifically in response to the amplified sequence.

2.3.2.2. *Single Label Formats*

Isothermal SDA reactions have been developed using fluorescence polarization of single labels to detect the amplified target sequences. In SDA, copies of target strands are continuously displaced from their templates by DNA polymerase extension of restriction enzyme-nicked sites.[4] Fluorophore-labeled oligomers are added which are complementary to the amplified strands. As the number of amplified strands increases, an increasing fraction of the fluorescent oligomers hybridize to the larger targets, resulting in measurable increases in the polarization of the label fluorescence, due the slower rotation of the target-probe complex. Binding of protein to the target-probe complex, via introduction of an EcoR1 restriction site and addition of a cleavage defective EcoR1 restriction enzyme, was found to increase polarization further.[104] Examples of SDA amplification monitored by increases in fluorescence polarization include detection of as few as 10 genomes of *Mycobacterium tuberculosis*,[105,106] and detection of *Chlamydia trachomatis* DNA.[107]

Fluorescence polarization has also been applied to monitoring PCR amplification products.[108] Fluorophore-labeled oligomers complementary to the amplified products were included in PCR solutions. As with SDA, increases in the amplified product resulted in increased polarization of the label fluorescence. This method was applied to genotyping with respect to the ΔF508 mutation in cystic fibrosis.

A single label format using fluorescence correlation analysis has also been reported for detecting amplification in PCR.[109] A fluorophore-labeled DNA oligomer that hybridized to the amplified product between the primer sequences was included in the PCR solution at a fairly low concentration (nanomolar). When the amplified product built up to a sufficient concentration, hybridization of the labeled probe to amplified target became appreciable, and the DNA polymerase also extended the labeled probe to form a higher molecular weight labeled product. The presence of amplified target was then detected as the accumulation of the slower diffusing fluorescent strands. In this manner *Mycobacterium tuberculosis* genomic DNA was detected down to 10 target molecules.

2.3.2.3. *Nucleic Acid Binding Dyes*

Nucleic acid binding dyes include dyes which bind to the major or minor grooves of helical nucleic acids or intercalate between the nucleotide bases. Many of these binding dyes show large increases in their fluorescence quantum yields upon binding to nucleic acids, such as the double-strand-specific intercalator dye, ethidium bromide. Dyes such as ethidium bromide have been used to measure nucleic acid concentrations for decades, but although they often have some base composition specificities, they have not been useful for identifying the presence of specific nucleic acid sequences. In PCR amplification, however, sequence specificity is provided by the oligonucleotide primers, and production of double-stranded product can be monitored by double-strand specific DNA binding dyes.

By including ethidium bromide in a PCR reaction and continuously monitoring fluorescence, the increase in amplified PCR product could be detected as a function of PCR cycle number[110] The cycle at which the intercalator fluorescence became detectable above background fluorescence correlated with the input number of HIV template molecules over six orders of magnitude down to 100 target strands. This not only provided a kinetic method of quantitative PCR, but also provided a sealed system which minimized potential contamination problems from release of amplified PCR products. Using the double-strand-specific dye SYBR Green I, a human β-globin gene fragment could also be detected in a range between one billion and 100 input target molecules.[96] Similarly, hepatitis c viral RNA was measured using RT-PCR in the presence of the intercalator, YO-PRO-1.[111] In a different approach, PCR was performed with primers that incorporated a T7 polymerase promoter into the amplified product, and the product was in turn transcribed, producing a double-stranded RNA product detected by propidium iodide intercalation.[112] PCR using competitive PNA primers and SYBR Green I to detect point mutations (PNA clamping) was reported in the identification of hereditary hemochromatosis.[113]

One major weakness in the use of double-strand-specific nucleic acid binding dyes to detect PCR amplification products is that unintended amplifications of nontarget sequences can occur, and double-stranded primer dimers often result from partial hybridization of primers to one another. The binding dyes would not discriminate between these nontarget products producing false positive results. One approach to circumventing this problem was to perform DNA melting curves, monitoring the dye fluorescence, to differentiate the amplified target sequences by their characteristic melting temperatures.[114] This also formed the basis of a multiplex PCR where multiple amplified products could be simultaneously distinguished by their individual melting temperatures.

2.4. Conclusions

Fluorescence has provided a large variety of DNA hybridization assay formats, both heterogeneous and homogeneous. Heterogeneous formats offer high sensitivity, with time-resolved lanthanide fluorescence providing detection down to 0.01 amol, or about 6000 molecules of target nucleic acid. With amplification, the detection limit drops to the tens or hundreds of molecules, or even the single molecule level. For many applications amplification would not be necessary. The variety of environment-sensitive fluorescence properties available makes fluorescence particularly valuable to homogeneous assays where measurable properties must change in response to hybridization. Homogeneous assays conserve time and labor, but suffer in sensitivity due to fundamental constraints on probe concentration. Homogeneous assay detection levels are closer to the fmol or 0.1 fmol level. Since homogeneous assay detection levels are limited by probe concentration, sensitivity would increase tremendously if sample analytes could be concentrated into microvolumes, thereby providing amol or lower detection levels. Combining hybridization assays with amplification, however, provides 10 molecule or lower detection levels to even homogeneous assays. Since target amplification methods tend to be homogeneous, combination with a homogeneous detection system provides a completely homogeneous high sensitivity assay that can be performed in sealed vessels, thereby reducing the risk of sample cross contamination. Given the wide range of assay formats already demonstrated in fluorescence approaches to hybridization assays, it should be relatively easy to tailor either a heterogeneous or homogeneous assay to any particular need.

References

1. Mullis, K. B., and Faloona, F. A. Specific synthesis of DNA in vitro via a polymerase-catalyzed chain reaction. *Methods Enzymol. 155*, 335–350, 1987.
2. Kwoh, D. Y., Davis, G. R., Whitfield, K. M., Chappelle, H. L., DiMichele, L. J., and Gingeras, T. R. Transcription-based amplification system and detection of amplified human immunodeficiency virus type 1 with a bead-based sandwich hybridization format. *Proc. Natl. Acad. Sci. USA 86*, 1173–1177, 1989.
3. Guatelli, J. C., Whitfield, K. M., Kwoh, D. Y., Barringer, K. J., Richman, D. D., and Gingeras, T. R. Isothermal, in vitro amplification of nucleic acids by a multienzyme reaction modeled after retroviral replication [published erratum appears in *Proc. Natl. Acad. Sci. USA 87*, 7797, 1990]. *Proc. Natl. Acad. Sci. USA 87*, 1874–1878, 1990.
4. Walker, G. T., Fraiser, M. S., Schram, J. L., Little, M. C., Nadeau, J. G., and Malinowski, D. P. Strand displacement amplification—An isothermal, in vitro DNA amplification technique. *Nucleic Acids Res. 20*, 1691–1696, 1992.
5. Urdea, M. S., Kolberg, J., Clyne, J., Running, J. A., Besemer, D., Warner, B., and Sanchez-Pescador, R. Application of a rapid non-radioisotopic nucleic acid analysis system to the detection of sexually transmitted disease-causing organisms and their associated antimicrobial resistances. *Clin. Chem. 35*, 1571–1575, 1989.
6. Lomeli, H., Tyagi, S., Pritchard, C. G., Lizardi, P. M., and Kramer, F. R. Quantitative assays based on the use of replicatable hybridization probes. *Clin. Chem. 35*, 1826–1831, 1989.
7. Lizardi, P. M., Huang, X., Zhu, Z., Bray-Ward, P., Thomas, D. C., and Ward, D. C. Mutation detection and single-molecule counting using isothermal rolling-circle amplification. *Nature Genet. 19*, 225–232, 1998.
8. Meinkoth, J., and Wahl, G. Hybridization of nucleic acids immobilized on solid supports. *Anal. Biochem. 138*, 267–284, 1984.
9. Diamandis, E. P., and Christopoulos, T. K. Europium chelate labels in time-resolved fluorescence immunoassays and DNA hybridization assays. *Anal. Chem. 62*, 1149A–1157A, 1990.

10. Diamandis, E. P. Time-resolved fluorometry in nucleic acid hybridization and western blotting techniques. *Electrophoresis 14*, 866–875, 1993.
11. Dickson, E. F., Pollak, A., and Diamandis, E. P. Time-resolved detection of lanthanide luminescence for ultrasensitive bioanalytical assays. *J Photochem. Photobiol. B 27*, 3–19, 1995.
12. Syvanen, A. C., Tchen, P., Ranki, M., and Soderlund, H. Time-resolved fluorometry: A sensitive method to quantify DNA-hybrids. *Nucleic Acids Res. 14*, 1017–1028, 1986.
13. Dahlen, P. Detection of biotinylated DNA probes by using Eu-labeled streptavidin and time-resolved fluorometry. *Anal. Biochem. 164*, 78–83, 1987.
14. Oser, A., Roth, W. K., and Valet, G. Sensitive non-radioactive dot-blot hybridization using DNA probes labelled with chelate group substituted psoralen and quantitative detection by europium ion fluorescence. *Nucleic Acids Res. 16*, 1181–1196, 1988.
15. Oser, A., and Valet, G. Improved detection by time-resolved fluorometry of specific DNA immobilized in microtiter wells with europium/metal-chelator labelled DNA probes. *Nucleic Acids Res. 16*, 8178, 1988.
16. Lovgren, T., Heinonen, P., Lehtinen, P., Hakala, H., Heinola, J., Harju, R., Takalo, H., Mukkala, V. M., Schmid, R., Lonnberg, H., Pettersson, K., and Iitia, A. Sensitive bioaffinity assays with individual microparticles and time-resolved fluorometry. *Clin. Chem. 43*, 1937–1943, 1997.
17. Hakala, H., Heinonen, P., Iitia, A., and Lonnberg, H. Detection of oligonucleotide hybridization on a single microparticle by time-resolved fluorometry: Hybridization assays on polymer particles obtained by direct solid phase assembly of the oligonucleotide probes. *Bioconjug. Chem. 8*, 378–384, 1997.
18. Piunno, P. A., Krull, U. J., Hudson, R. H., Damha, M. J., and Cohen, H. Fiber-optic DNA sensor for fluorometric nucleic acid determination. *Anal. Chem. 67*, 2635–2643, 1995.
19. Abel, A. P., Weller, M. G., Duveneck, G. L., Ehrat, M., and Widmer, H. M. Fiber-optic evanescent wave biosensor for the detection of oligonucleotides. *Anal. Chem. 68*, 2905–2912, 1996.
20. Vener, T. I., Turchinsky, M. F., Knorre, V. D., Lukin Yu, V., Shcherbo, S. N., Zubov, V. P., and Sverdlov, E. D. A novel approach to nonradioactive hybridization assay of nucleic acids using stained latex particles. *Anal. Biochem. 198*, 308–311, 1991.
21. Christopoulos, T. K., Diamandis, E. P., and Wilson, G. Quantification of nucleic acids on nitrocellulose membranes with time-resolved fluorometry. *Nucleic Acids Res. 19*, 6015–6019, 1991.
22. Mansfield, E. S., Worley, J. M., McKenzie, S. E., Surrey, S., Rappaport, E., and Fortina, P. Nucleic acid detection using non-radioactive labelling methods. *Mol. Cell. Probes 9*, 145–156, 1995.
23. Evangelista, R. A., Pollak, A., and Templeton, E. F. Enzyme-amplified lanthanide luminescence for enzyme detection in bioanalytical assays. *Anal. Biochem. 197*, 213–224, 1991.
24. Templeton, E. F., Wong, H. E., Evangelista, R. A., Granger, T., and Pollak, A. Time-resolved fluorescence detection of enzyme-amplified lanthanide luminescence for nucleic acid hybridization assays. *Clin. Chem. 37*, 1506–1512, 1991.
25. Chiu, N. H., Christopoulos, T. K., and Peltier, J. Sandwich-type deoxyribonucleic acid hybridization assays based on enzyme amplified time-resolved fluorometry. *Analyst 123*, 1315–1319, 1998.
26. Ioannou, P. C., and Christopoulos, T. K. Two-round enzymatic amplification combined with time-resolved fluorometry of $Tb3+$ chelates for enhanced sensitivity in DNA hybridization assays. *Anal. Chem. 70*, 698–702, 1998.
27. Dahlen, P. O., Iitia, A. J., Skagius, G., Frostell, A., Nunn, M. F., and Kwiatkowski, M. Detection of human immunodeficiency virus type 1 by using the polymerase chain reaction and a time-resolved fluorescence-based hybridization assay. *J. Clin. Microbiol. 29*, 798–804, 1991.
28. Eggertsen, G., Eriksson, M., Wiklund, O., Iitia, A., Olofsson, S. O., Angelin, B., and Berglund, L. Time-resolved fluorometry in the genetic diagnosis of familial defective apolipoprotein B-100. *J. Lipid Res. 35*, 1505–1508, 1994.

29. Sjoroos, M., Ilonen, J., Reijonen, H., and Lovgren, T. Time-resolved fluorometry based sandwich hybridisation assay for HLA-DQA1 typing. *Dis. Markers 14*, 9–19, 1998.
30. Huoponen, K., Juvonen, V., Iitia, A., Dahlen, P., Siitari, H., Aula, P., Nikoskelainen, E., and Savontaus, M. L. Time-resolved fluorometry in the diagnosis of Leber hereditary optic neuroretinopathy. *Hum. Mutat. 3*, 29–36, 1994.
31. Chan, A., Diamandis, E. P., and Krajden, M. Quantification of polymerase chain reaction products in agarose gels with a fluorescent europium chelate as label and time-resolved fluorescence spectroscopy. *Anal. Chem. 65*, 158–163, 1993.
32. Maher, M., Dowdall, D., Glennon, M., Walshe, S., Cormican, M., Wiesner, P., Gannon, F., and Smith, T. The sensitive detection of fluorescently labelled PCR products using an automated detection system. *Mol. Cell. Probes 9*, 265–276, 1995.
33. Lopez, E., Chypre, C., Alpha, B., and Mathis, G. Europium(III) trisbipyridine cryptate label for time-resolved fluorescence detection of polymerase chain reaction products fixed on a solid support. *Clin. Chem. 39*, 196–201, 1993.
34. Chehab, F. F., and Kan, Y. W. Detection of specific DNA sequences by fluorescence amplification: A color complementation assay. *Proc. Natl. Acad. Sci. USA 86*, 9178–9182, 1989.
35. Sjoroos, M., Iitia, A., Ilonen, J., Reijonen, H., and Lovgren, T. Triple-label hybridization assay for type-1 diabetes-related HLA alleles. *Biotechniques 18*, 870–877, 1995.
36. Heinonen, P., Iitia, A., Torresani, T., and Lovgren, T. Simple triple-label detection of seven cystic fibrosis mutations by time-resolved fluorometry. *Clin. Chem. 43*, 1142–1150, 1997.
37. Samiotaki, M., Kwiatkowski, M., Ylitalo, N., and Landegren, U. Seven-color time-resolved fluorescence hybridization analysis of human papilloma virus types. *Anal. Biochem. 253*, 156–161, 1997.
38. Bortolin, S., Christopoulos, T. K., and Verhaegen, M. Quantitative polymerase chain reaction using a recombinant DNA internal standard and time-resolved fluorometry. *Anal. Chem. 68*, 834–840, 1996.
39. Mansfield, E. S. Diagnosis of Down syndrome and other aneuploidies using quantitative polymerase chain reaction and small tandem repeat polymorphisms. *Hum. Mol. Genet. 2*, 43–50, 1993.
40. Adinolfi, M., Pertl, B., and Sherlock, J. Rapid detection of aneuploidies by microsatellite and the quantitative fluorescent polymerase chain reaction. *Prenat. Diagn. 17*, 1299–1311, 1997.
41. Verma, L., Macdonald, F., Leedham, P., McConachie, M., Dhanjal, S., and Hulten, M. Rapid and simple prenatal DNA diagnosis of Down's syndrome. *Lancet 352*, 9–12, 1998.
42. Vlieger, A. M., Medenblik, A. M., van Gijlswijk, R. P., Tanke, H. J., van der Ploeg, M., Gratama, J. W., and Raap, A. K. Quantitation of polymerase chain reaction products by hybridization-based assays with fluorescent, colorimetric, or chemiluminescent detection. *Anal. Biochem. 205*, 1–7, 1992.
43. Livak, K. J., Hobbs, F. W., and Zagursky, R. J. Detection of single base differences using biotinylated nucleotides with very long linker arms. *Nucleic Acids Res. 20*, 4831–4837, 1992.
44. Del Tito, B. J., Jr., Poff, H. E., 3rd, Novotny, M. A., Cartledge, D. M., Walker, R. I., 2nd, Earl, C. D., and Bailey, A. L. Automated fluorescent analysis procedure for enzymatic mutation detection. *Clin. Chem. 44*, 731–739, 1998.
45. Landegren, U., Kaiser, R., Sanders, J., and Hood, L. A ligase-mediated gene detection technique. *Science 241*, 1077–1080, 1988.
46. Grossman, P. D., Bloch, W., Brinson, E., Chang, C. C., Eggerding, F. A., Fung, S., Iovannisci, D. M., Woo, S., Winn-Deen, E. S., and Iovannisci, D. A. High-density multiplex detection of nucleic acid sequences: Oligonucleotide ligation assay and sequence-coded separation [published erratum appears in *Nucleic Acids Res. 26*, 5539, 1998]. *Nucleic Acids Res. 22*, 4527–4534, 1994.
47. Eggerding, F. A. A one-step coupled amplification and oligonucleotide ligation procedure for multiplex genetic typing. *PCR Methods Applic. 4*, 337–345, 1995.

48. Eggerding, F. A., Iovannisci, D. M., Brinson, E., Grossman, P., and Winn-Deen, E. S. Fluorescence-based oligonucleotide ligation assay for analysis of cystic fibrosis transmembrane conductance regulator gene mutations. *Hum. Mutat. 5*, 153–165, 1995.
49. Samiotaki, M., Kwiatkowski, M., Parik, J., and Landegren, U. Dual-color detection of DNA sequence variants by ligase-mediated analysis. *Genomics 20*, 238–242, 1994.
50. Kwiatkowski, M., Samiotaki, M., Lamminmaki, U., Mukkala, V. M., and Landegren, U. Solid-phase synthesis of chelate-labelled oligonucleotides: Application in triple-color ligase-mediated gene analysis. *Nucleic Acids Res. 22*, 2604–2611, 1994.
51. Bortolin, S., and Christopoulos, T. K. Detection of BCR-ABL transcripts from the Philadelphia translocation by hybridization in microtiter wells and time-resolved immunofluorometry. *Clin. Chem. 41*, 693–699, 1995.
52. Galvan, B., Christopoulos, T. K., and Diamandis, E. P. Detection of prostate-specific antigen mRNA by reverse transcription polymerase chain reaction and time-resolved fluorometry. *Clin. Chem. 41*, 1705–1709, 1995.
53. Wu, D. Y., and Wallace, R. B. The ligation amplification reaction (LAR)—Amplification of specific DNA sequences using sequential rounds of template-dependent ligation. *Genomics 4*, 560–569, 1989.
54. Abravaya, K., Carrino, J. J., Muldoon, S., and Lee, H. H. Detection of point mutations with a modified ligase chain reaction (Gap-LCR). *Nucleic Acids Res. 23*, 675–682, 1995.
55. Bush, C. E., Vanden Brink, K. M., Sherman, D. G., Peterson, W. R., Beninsig, L. A., and Godsey, J. H. Detection of *Escherichia coli* rRNA using target amplification and time-resolved fluorescence detection. *Mol. Cell. Probes 5*, 467–472, 1991.
56. Forster, T. Transfer mechanisms of electronic excitation. *Disc. Faraday Soc. 27*, 7–17, 1959.
57. Morrison, L. E. in *Nonisotopic Probing, Blotting, and Sequencing*, Kricka, L. J., ed. San Diego, CA: Academic Press, 1995, pp. 429–471.
58. Morrison, L. Homogeneous detection of specific DNA sequences by fluorescence quenching and energy transfer. *J. Fluores. 9*, 187–196, 1999.
59. Heller, M. J., Morrison, L. E., Prevatt, W. D., and Akin, C. Light-emitting polynucleotide hybridization diagnostic method. European patent application 070 685, 1983.
60. Heller, M. J., and Morrison, L. E. in *Rapid Detection and Identification of Infectious Agents*, Kingsbury, D. T., and Falkow, S., eds. Orlando, FL: Academic Press, 1985, pp. 245–256.
61. Cardullo, R. A., Agrawal, S., Flores, C., Zamecnik, P. C., and Wolf, D. E. Detection of nucleic acid hybridization by nonradiative fluorescence resonance energy transfer. *Proc. Natl. Acad. Sci. USA 85*, 8790–8794, 1988.
62. Mergny, J. L., Garestier, T., Rougee, M., Lebedev, A. V., Chassignol, M., Thuong, N. T., and Helene, C. Fluorescence energy transfer between two triple helix-forming oligonucleotides bound to duplex DNA. *Biochemistry 33*, 15321–15328, 1994.
63. Morrison, L. E. Competitive homogeneous assay. European patent application 232 967, 1987; United States patent 5,928,862, 1999.
64. Morrison, L. E., Halder, T. C., and Stols, L. M. Solution-phase detection of polynucleotides using interacting fluorescent labels and competitive hybridization. *Anal. Biochem. 183*, 231–244, 1989.
65. Heller, M. J., and Jablonski, E. J. Fluorescent Stokes shift probes for polynucleotide hybridization assays. European patent application 229 943, 1987.
66. Bannwarth, W., and Muller, F. Energy transfer from a lumazine (= pteridine-2,4(1H,3H)-dione) chromophore to bathophenanthroline-ruthenium(II) ucomplexes during hybridization processes of DNA. *Helv. Chim. Acta 74*, 2000–2008, 1991.
67. Tyagi, S., and Kramer, F. R. Molecular beacons: Probes that fluoresce upon hybridization. *Nature Biotechnol. 14*, 303–308, 1996.
68. Tyagi, S., Bratu, D. P., and Kramer, F. R. Multicolor molecular beacons for allele discrimination. *Nature Biotechnol. 16*, 49–53, 1998.

69. Bonnet, G., Tyagi, S., Libchaber, A., and Kramer, F. R. Thermodynamic basis of the enhanced specificity of structured DNA probes. *Proc. Natl. Acad. Sci. USA 96*, 6171–6176, 1999.
70. Morrison, L. E. Time-resolved detection of energy transfer: Theory and application to immunoassays. *Anal. Biochem. 174*, 101–120, 1988.
71. Mathis, G. Rare earth cryptates and homogeneous fluoroimmunoassays with human sera. *Clin. Chem. 39*, 1953–1959, 1993.
72. Ebata, K., Masuko, M., Ohtani, H., and Kashiwasake-Jibu, M. Nucleic acid hybridization accompanied with excimer formation from two pyrene-labeled probes. *Photochem. Photobiol. 62*, 836–839, 1995.
73. Paris, P. L., Langenhan, J. M., and Kool, E. T. Probing DNA sequences in solution with a monomer-excimer fluorescence color change. *Nucleic Acids Res. 26*, 3789–3793, 1998.
74. Oser, A., and Valet, G. Nonradioactive assay of DNA hybridization by DNA-template-mediated formation of a ternary TbIII complex in pure liquid phase. *Chem. Int. Ed. Engl. 29*, 1167–1169, 1990.
75. Castro, A., and Williams, J. G. Single-molecule detection of specific nucleic acid sequences in unamplified genomic DNA. *Anal. Chem. 69*, 3915–3920, 1997.
76. Kierzek, R., Li, Y., Turner, D. H., and Bevilacqua, P. C. 5′-Amino pyrene provides a sensitive, nonperturbing fluorescent probe of RNA secondary and tertiary structure formation. *J. Am. Chem. Soc. 115*, 4985–4992, 1993.
77. Yguerabide, J., Talavera, E., Alvarez, J. M., and Afkir, M. Pyrene-labeled DNA probes for homogeneous detection of complementary DNA sequences: Poly(C) model system. *Anal. Biochem. 241*, 238–247, 1996.
78. Lee, S. P., Porter, D., Chirikjian, J. G., Knutson, J. R., and Han, M. K. A fluorometric assay for DNA cleavage reactions characterized with BamHI restriction endonuclease. *Anal. Biochem. 220*, 377–383, 1994.
79. Talavera, E., Alvarez-Pez, J., Ballesteros, L., and Bermejo, R. Fluorescein-labeled DNA probes for homogeneous hybridization assays: Application to DNA *E. coli* renaturation. *Appl. Spectrosc. 51*, 401–406, 1997.
80. Ishiguro, T., Saitoh, J., Yawata, H., Otsuka, M., Inoue, T., and Sugiura, Y. Fluorescence detection of specific sequence of nucleic acids by oxazole yellow-linked oligonucleotides. Homogeneous quantitative monitoring of in vitro transcription. *Nucleic Acids Res. 24*, 4992–4997, 1996.
81. Perrin, M. F. Polarisation de la lumiere de fluorescence. Vie moyenne des molecules dans l'etat excite. *J. Phys. Radium 7*, 390–401, 1926.
82. Weber, G. in *Fluorescence and Phosphorescence Analysis*, Hercules, D. M., ed., New York: Interscience, 1966, pp. 217–240.
83. Kumke, M. U., Li, G., McGown, L. B., Walker, G. T., and Linn, C. P. Hybridization of fluorescein-labeled DNA oligomers detected by fluorescence anisotropy with protein binding enhancement. *Anal. Chem. 67*, 3945–3951, 1995.
84. Tsuruoka, M., Yano, K., Ikebukuro, K., Nakayama, H., Masuda, Y., and Karube, I. Optimization of the rate of DNA hybridization and rapid detection of methicillin resistant *Staphylococcus aureus* DNA using fluorescence polarization. *J. Biotechnol. 48*, 201–208, 1996.
85. Kinjo, M., and Rigler, R. Ultrasensitive hybridization analysis using fluorescence correlation spectroscopy. *Nucleic Acids Res. 23*, 1795–1799, 1995.
86. Lee, L. G., Connell, C. R., and Bloch, W. Allelic discrimination by nick-translation PCR with fluorogenic probes. *Nucleic Acids Res. 21*, 3761–3766, 1993.
87. Faas, S. J., Menon, R., Braun, E. R., Rudert, W. A., and Trucco, M. Sequence-specific priming and exonuclease-released fluorescence detection of HLA-DQB1 alleles. *Tissue Antigens 48*, 97–112, 1996.
88. Gelmini, S., Orlando, C., Sestini, R., Vona, G., Pinzani, P., Ruocco, L., and Pazzagli, M. Quantitative polymerase chain reaction-based homogeneous assay with fluorogenic probes to measure c-erbB-2 oncogene amplification. *Clin. Chem. 43*, 752–758, 1997.

89. Ibrahim, M. S., Esposito, J. J., Jahrling, P. B., and Lofts, R. S. The potential of 5′ nuclease PCR for detecting a single-base polymorphism in Orthopoxvirus. *Mol. Cell. Probes 11*, 143–147, 1997.
90. Swan, D. C., Tucker, R. A., Holloway, B. P., and Icenogle, J. P. A sensitive, type-specific, fluorogenic probe assay for detection of human papillomavirus DNA. *J. Clin. Microbiol. 35*, 886–891, 1997.
91. Bieche, I., Olivi, M., Champeme, M. H., Vidaud, D., Lidereau, R., and Vidaud, M. Novel approach to quantitative polymerase chain reaction using real-time detection: Application to the detection of gene amplification in breast cancer. *Int. J. Cancer 78*, 661–666, 1998.
92. Gibson, U. E., Heid, C. A., and Williams, P. M. A novel method for real time quantitative RT-PCR. *Genome Res. 6*, 995–1001, 1996.
93. Schofield, P., Pell, A. N., and Krause, D. O. Molecular beacons: Trial of a fluorescence-based solution hybridization technique for ecological studies with ruminal bacteria. *Appl. Environ. Microbiol. 63*, 1143–1147, 1997.
94. Piatek, A. S., Tyagi, S., Pol, A. C., Telenti, A., Miller, L. P., Kramer, F. R., and Alland, D. Molecular beacon sequence analysis for detecting drug resistance in Mycobacterium tuberculosis [see comments]. *Nature Biotechnol. 16*, 359–363, 1998.
95. Vet, J. A., Majithia, A. R., Marras, S. A., Tyagi, S., Dube, S., Poiesz, B. J., and Kramer, F. R. Multiplex detection of four pathogenic retroviruses using molecular beacons. *Proc. Natl. Acad. Sci. USA 96*, 6394–6399, 1999.
96. Wittwer, C. T., Herrmann, M. G., Moss, A. A., and Rasmussen, R. P. Continuous fluorescence monitoring of rapid cycle DNA amplification. *Biotechniques 22*, 130–131, 134–138, 1997.
97. Lay, M. J., and Wittwer, C. T. Real-time fluorescence genotyping of factor V Leiden during rapid-cycle PCR. *Clin. Chem. 43*, 2262–2267, 1997.
98. Bernard, P. S., Lay, M. J., and Wittwer, C. T. Integrated amplification and detection of the C677T point mutation in the methylenetetrahydrofolate reductase gene by fluorescence resonance energy transfer and probe melting curves. *Anal. Biochem. 255*, 101–107, 1998.
99. Chen, X., and Kwok, P. Y. Template-directed dye-terminator incorporation (TDI) assay: A homogeneous DNA diagnostic method based on fluorescence resonance energy transfer. *Nucleic Acids Res. 25*, 347–353, 1997.
100. Chen, X., Zehnbauer, B., Gnirke, A., and Kwok, P. Y. Fluorescence energy transfer detection as a homogeneous DNA diagnostic method. *Proc. Natl. Acad. Sci. USA 94*, 10756–10761, 1997.
101. Nazarenko, I. A., Bhatnagar, S. K., and Hohman, R. J. A closed tube format for amplification and detection of DNA based on energy transfer. *Nucleic Acids Res. 25*, 2516–2521, 1997.
102. Chiang, P. W., Song, W. J., Wu, K. Y., Korenberg, J. R., Fogel, E. J., Van Keuren, M. L., Lashkari, D., and Kurnit, D. M. Use of a fluorescent-PCR reaction to detect genomic sequence copy number and transcriptional abundance. *Genome Res. 6*, 1013–1026, 1996.
103. Whitcombe, D., Theaker, J., Guy, S. P., Brown, T., and Little, S. Detection of PCR products using self-probing amplicons and fluorescence. *Nature Biotechnol. 17*, 804–807, 1999.
104. Walker, G. T., Linn, C. P., and Nadeau, J. G. DNA detection by strand displacement amplification and fluorescence polarization with signal enhancement using a DNA binding protein. *Nucleic Acids Res. 24*, 348–353, 1996.
105. Walker, G. T., and Linn, C. P. Detection of Mycobacterium tuberculosis DNA with thermophilic strand displacement amplification and fluorescence polarization. *Clin. Chem. 42*, 1604–1608, 1996.
106. Walker, G. T., Nadeau, J. G., Linn, C. P., Devlin, R. F., and Dandliker, W. B. Strand displacement amplification (SDA) and transient-state fluorescence polarization detection of Mycobacterium tuberculosis DNA [see comments]. *Clin. Chem. 42*, 9–13, 1996.
107. Spears, P. A., Linn, C. P., Woodard, D. L., and Walker, G. T. Simultaneous strand displacement amplification and fluorescence polarization detection of Chlamydia trachomatis DNA. *Anal. Biochem. 247*, 130–137, 1997.

108. Gibson, N. J., Gillard, H. L., Whitcombe, D., Ferrie, R. M., Newton, C. R., and Little, S. A homogeneous method for genotyping with fluorescence polarization. *Clin. Chem. 43*, 1336–1341, 1997.
109. Walter, N. G., Schwille, P., and Eigen, M. Fluorescence correlation analysis of probe diffusion simplifies quantitative pathogen detection by PCR. *Proc. Natl. Acad. Sci. USA 93*, 12805–12810, 1996.
110. Higuchi, R., Fockler, C., Dollinger, G., and Watson, R. Kinetic PCR analysis: Real-time monitoring of DNA amplification reactions. *Biotechnology (New York) 11*, 1026–1030, 1993.
111. Ishiguro, T., Saitoh, J., Yawata, H., Yamagishi, H., Iwasaki, S., and Mitoma, Y. Homogeneous quantitative assay of hepatitis C virus RNA by polymerase chain reaction in the presence of a fluorescent intercalater. *Anal. Biochem. 229*, 207–213, 1995.
112. Livache, T., Fouque, B., and Teoule, R. Detection of HIV1 DNA in biological samples by an homogeneous assay: Fluorescence measurement of double-stranded RNA synthesized from amplified DNA. *Anal. Biochem. 217*, 248–254, 1994.
113. Kyger, E. M., Krevolin, M. D., and Powell, M. J. Detection of the hereditary hemochromatosis gene mutation by real-time fluorescence polymerase chain reaction and peptide nucleic acid clamping. *Anal. Biochem. 260*, 142–148, 1998.
114. Ririe, K. M., Rasmussen, R. P., and Wittwer, C. T. Product differentiation by analysis of DNA melting curves during the polymerase chain reaction. *Anal. Biochem. 245*, 154–160, 1997.

3

Energy Transfer Fluorescent Labels for DNA Sequencing and Analysis

Jin Xie, Su-Chun Hung, Alexander N. Glazer, and Richard A. Mathies

3.1. Introduction

Fluorescence resonance energy transfer (FRET) is a photophysical process where the excited-state energy of a donor fluorophore is transferred to an acceptor fluorophore through a resonance dipole–dipole interaction between the donor and the acceptor. This form of sensitized fluorescence has been widely exploited as a "spectroscopic ruler" to determine proximity relationships between donor and acceptor labels bound to biological structures.[1] FRET has been exploited for a variety of labeling applications in polymer science, biochemistry, and structural biology. In particular, the concept of FRET has been invaluable in the development of a variety of new analytical methods, instruments, and reagents for high-speed and high-throughput analyses of nucleic acids which has been driven by the Human Genome Project.[2] A central challenge in this project has been the enhancement of DNA sequencing capabilities to the gigabase-per-year level. Two key technologies have contributed to the recent improvements that have made it feasible to advance the target date for sequencing the human genome: capillary array electrophoresis[3] and energy-transfer (ET) sequencing reagents.[4,5] By exploiting the unique spectroscopic properties of FRET, Ju and coworkers[6] developed this new class of high-performance fluorescent labeling reagents, thereby defining a new labeling paradigm for DNA sequencing and analysis. Their work was the first to exploit energy transfer concepts in the design of fluorescent labeled primers to overcome the limitations imposed by the use of single dye-labeled primers and terminators.[7,8] Since the initial conception of ET primers, several alternative ET primers as well as ET terminators have been developed, including cyanine-donor based ET primers,[9] dipyrrometheneboron (BODIPY) based ET

Jin Xie, Su-Chun Hung, Alexander N. Glazer • Department of Molecular and Cell Biology, University of California, Berkeley, California 94720-3200. *Richard A. Mathies* • Department of Chemistry, University of California, Berkeley, California, 94720-1460.

Topics in Fluorescence Spectroscopy, Volume 7: DNA Technology, edited by Joseph R. Lakowicz, Kluwer Academic / Plenum Publishers, New York, 2003

primers,[10] DYEnamic™ ET primers and terminators by Amersham Biosciences,[5,11] and BigDye™ primers and terminators by PE Applied Biosystems.[12,13]

This chapter will present an overview of recent advances in the design, synthesis, and application of ET fluorescent primers and terminators for DNA analyses, including sequencing, fragment sizing, and short tandem repeat (STR)-based cancer diagnosis. Brief introductions of the design and synthesis of universal ET cassettes for facile construction of ET-labeled primers, as well as the synthesis and potential applications of biotinylated ET cassettes for bioanalysis will also be discussed. Finally, a new homogeneous DNA diagnostic assay based on template-directed primer extension detected by FRET will also be presented.

3.2. Theory of Fluorescence Resonance Energy Transfer

FRET occurs when an excited donor moiety transfers energy nonradiatively to an acceptor moiety by long-range resonance coupling between the transition dipoles of the donor and the acceptor. No photons are "passed" between the donor and the acceptor. Each dye moiety acts like a classical dipole antenna and the energy is directly transferred. The fluorescence intensity of the donor emission is reduced dramatically due to the energy transfer; the acceptor is promoted to its excited state and emits sensitized emission.

According to the Förster theory,[14] the rate of energy transfer (k_T) from the excited singlet state of a donor to an acceptor is described by the following equation.

$$k_T = (8.71 \times 10^{23}) J \kappa^2 k_F n^{-4} R^{-6} \qquad (\text{in s}^{-1}) \qquad (3.1)$$

Here, R is the distance between the centers of the donor and the acceptor, and the energy transfer rate is dependent on $1/R^6$.[15] κ^2 is the orientation factor for dipole–dipole interaction, the average value of κ^2 is 2/3 for a random orientation between the chromophores. The refractive index, n, of the medium between the chromophores is generally taken as 1.33 in aqueous solution. k_F is the rate constant for fluorescence emission of the donor, and J is the spectral overlap integral, which is proportional to the overlap in the emission spectrum of the donor and the absorption spectrum of the acceptor.

If the donor of the donor-acceptor pair is excited, three processes can occur simultaneously: energy transfer from the donor to the acceptor (k_T), donor fluorescence emission (k_F), and nonradiative decay (k_R) as indicated in the scheme below, where D is the donor, and A is the acceptor chromophore. In order to have efficient energy transfer, k_T needs to be much faster than the rates of other processes that compete for the excited state of the donor. The power of FRET for the development of new fluorescent reagents arises in large part from the ability to tune independently the distance between the donor and the acceptor, and the absorption and emission properties of the FRET-coupled dyes so that k_T can be optimized.

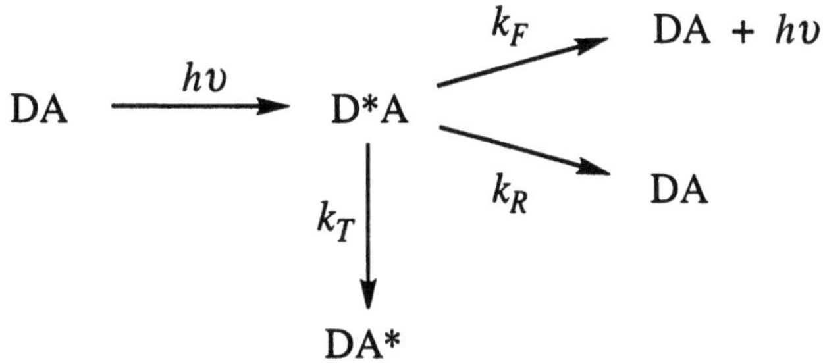

3.3. Sanger Dideoxy Chain-Termination Sequencing

The Sanger dideoxy chain-termination method[16] is the most popular method used in large-scale DNA sequencing. Sanger sequencing methodology can employ fluorescence detection by labeling either primers[7] or terminators.[8] Fluorescence detection is attractive because it eliminates the use of radioactive reagents and it can be readily automated. Figure 3.1 presents the strategy of the Sanger dideoxy chain-termination method for DNA sequencing using dye-labeled primers. Four sequencing reactions are set up, one for each base. Each reaction mixture contains: DNA template; one of the four dye-labeled primers; thermostable DNA polymerase; four deoxyribonucleoside triphosphates (dNTPs); and one of the four 2′,3′-dideoxynucleoside triphosphates (ddNTPs). The synthesis starts at a specific priming site by using a complementary dye-labeled primer. Extension of the new complementary DNA chain ends with the incorporation of a ddNTP terminator, since it lacks the 3′-hydroxyl terminus needed to form the next phosphodiester bond. By adjusting the ratio of ddNTP to dNTP, a random distribution of DNA fragments of various lengths can be produced. At this point, the fragments can either be pooled and sequenced, or they can be carried through another 20–30 cycles of thermal denaturation, annealing, and extension/termination to increase signal levels. Four sets of chain-terminated reaction mixtures are then combined and electrophoresed together through a sieving matrix. The DNA fragments terminating with the four ddNTPs are identified by their four distinct fluorescence emissions. The order of these fluorescence signals directly gives the base sequence.

3.4. Energy-Transfer Fluorescent Primers

For high-throughput and high-accuracy DNA sequencing, an "ideal" set of four fluorophore labels would need to embody the following properties: (1) the dyes should have similar, high molar absorbances at a common excitation wavelength; (2) they should exhibit strong and well-separated fluorescence emissions

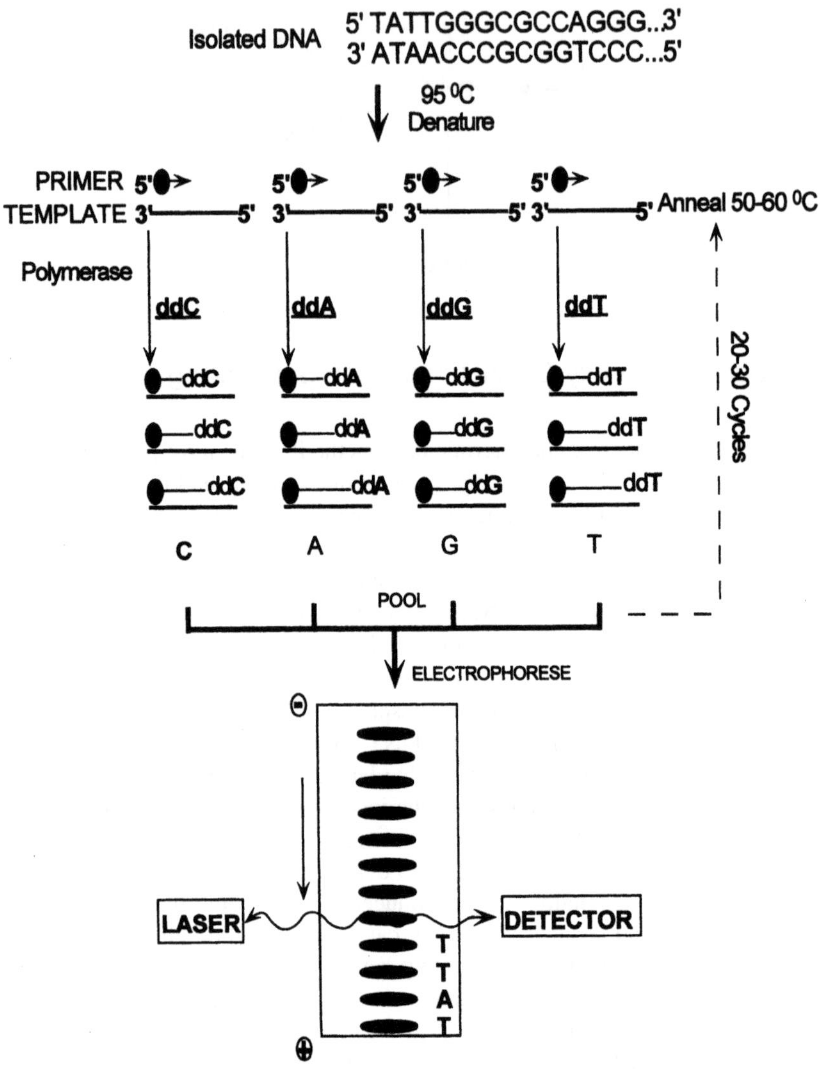

Figure 3.1. The Sanger dideoxy chain termination method and cycle sequencing procedures.

with high fluorescence quantum yields; and (3) ideally, there should be minimal differential mobility shift of the DNA sequencing fragments resulting from the four dye labels.

Fluorescein and rhodamine derivatives (Figure 3.2) are the dyes traditionally used for single-dye labeled primers: FAM (or F), 5-carboxyfluorescein; JOE (or J),

Structure	Dye	Solvent	A_{max}	ε_{max}	E_{max}
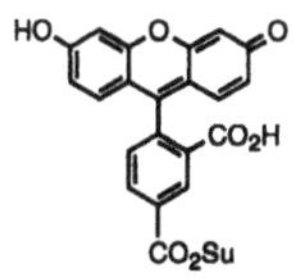	5-FAM	1xTBE 7M Urea	500 nm	8.0×10^4	525 nm
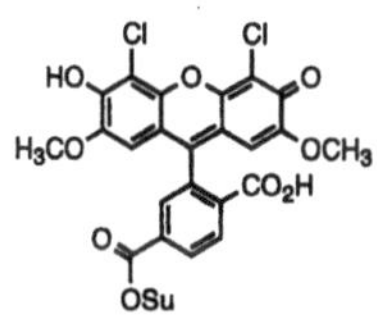	6-JOE	1xTBE 7M Urea	525 nm	7.1×10^4	555 nm
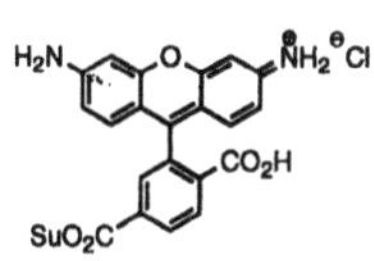	6-R110	1xTBE 7M Urea	498 nm	7.8×10^4	530 nm
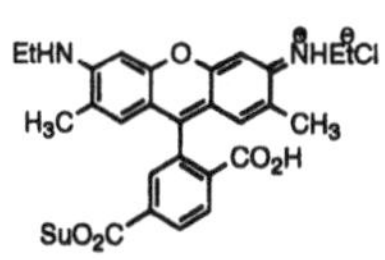	6-R6G	1xTBE 7M Urea	533 nm	9.8×10^4	555 nm
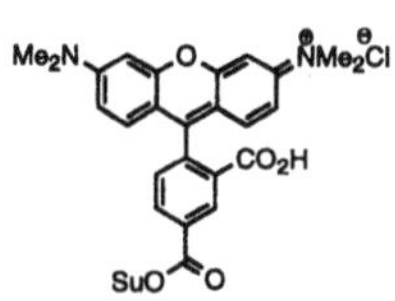	5-TAMRA	1xTBE 7M Urea	555 nm	9.0×10^4	580 nm
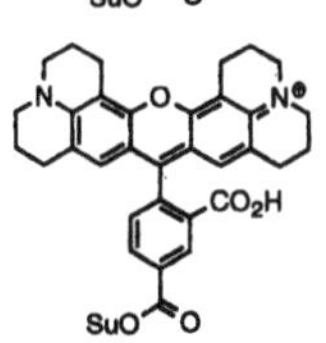	5-ROX	1xTBE 7M Urea	585 nm	8.2×10^4	605 nm
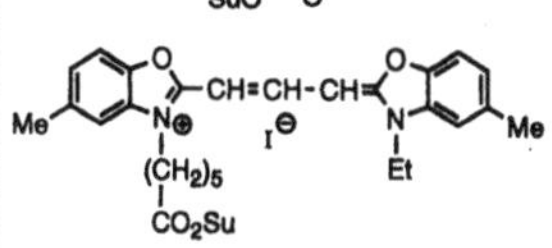	CYA	1xTBE 7M Urea	491 nm	1.5×10^5	508 nm

Figure 3.2. Spectroscopic properties of fluorescent dyes used for sequencing.

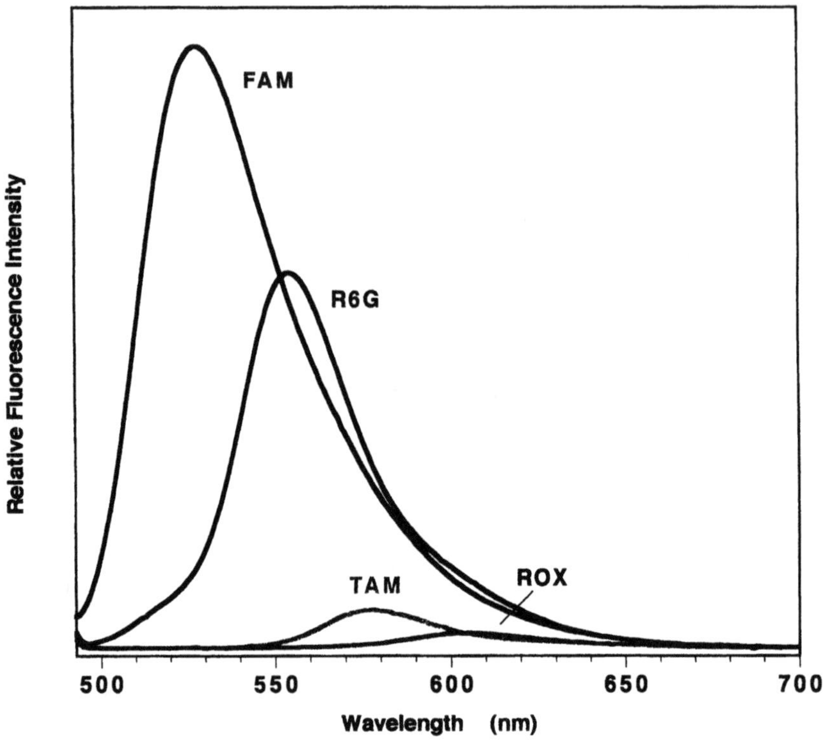

Figure 3.3. Relative fluorescence emission spectra of FAM, R6G, TAMRA, and ROX. Emission spectra were determined with equimolar solutions of dyes in 1 × TBE containing 7 M urea with excitation at 488 nm.

2′,7′-dimethoxy-4′,5′-dichloro-6-carboxyfluorescein; TAMRA (or T) N,N,N′,N′-tetramethyl-6-carboxyrhodamine; and ROX (or R), 6-carboxy-X-rhodamine. Figure 3.3 presents the emission spectra of equimolar solutions of these four dyes following the excitation at 488 nm. 6-carboxyrhodamine 6G (R6G or G) can be used in place of JOE because it exhibits preferable spectroscopic and electrophoretic properties.[17] Even though these four dyes have high extinction coefficients, and fluorescence quantum yields with well-separated emissions (wavelengths 525, 555, 580, and 605 nm for FAM, R6G, TAMRA, and ROX, respectively), they poorly satisfy the two spectroscopic criteria mentioned above. The four dyes cannot be excited efficiently at a common wavelength. When excited at 488 nm, TAMRA and ROX exhibit very weak emission; the relative emission intensity of ROX at its maximum wavelength is about 27-fold weaker than that of FAM. Many researchers have been forced to employ dual wavelength excitation systems (488 and 514 nm) to compensate this deficiency.[18,19] Nevertheless, TAMRA and ROX are still excited much less efficiently at either wavelength than FAM or R6G because of their very weak molar absorbances at 488 and 514 nm. Moreover, fluorescein and rhodamine derivatives have different chemical structures with different charges, which make it difficult to match the electrophoretic mobilities of the DNA sequencing fragments.[7,20]

The development of ET fluorescent primers[5,6,17,21] successfully addressed all the shortcomings imposed by the single dye-labeled primers. A schematic representation of a set of ET primers is presented in Figure 3.4. Each primer can be assembled by employing standard solid-phase phosphoramidite chemistry on a DNA synthesizer.[6,17,21] Thus, the synthesis of ET primers is simple and practical. The ET primers carry a FAM at the 5′-end as a common energy donor and a fluorescein or a rhodamine derivative (F, G, T, and R) as an acceptor.[17] The acceptor is attached to a modified thymidine residue (T*) within the primer sequence. The nomenclature used for these ET primers is D-N-A, where D is the donor, A is the acceptor, and N is the number of nucleotides between D and A. A primer with FAM at the 5′-end and the acceptor ROX on a T* ten nucleobases away is thus designated as F10R.

This set of ET primers produces nearly ideal labels for detection. First, the acceptor fluorophore with poor absorbance at the excitation wavelength can still be efficiently excited to give enhanced fluorescence emission through ET from the donor to the acceptor. Second, FRET can result in higher signal-to-noise by providing a large Stokes-shift of the emission maximum of acceptors. Third, the presence of the common donor fluorophore allows the use of a single laser line at 488 nm to efficiently excite all four acceptor fluorophores, which simplifies the instrumentation and makes multiplex detection convenient. Since the efficiency of ET depends on the distance between the donor and the acceptor, the efficiency can be easily adjusted through changes in the nucleotide spacing N between the donor and the acceptor.

Systematic studies have shown that there is a complex dependence of energy transfer efficiency, fluorescence quantum yield, and electrophoretic mobility on the number of nucleotides between D and A.[5,22] It is difficult to define *a priori* the distance between the donor and the acceptor in ET primers, because both dyes are linked to the oligonucleotide primer through flexible linkers, and because the conformation of the oligonucleotide also plays a role in determining the mean distance between the donor and the acceptor. Thus, the separation distance must be optimized experimentally by adjusting N. If the two fluorophores within the ET primer are too close to each other, the acceptor emission will be quenched through direct interaction of the dye moieties. On the other hand, if they are too far away the acceptor emission will decrease because of the inefficient energy transfer. Ju *et al.*[21] and Hung *et al.*[22] have systematically examined the fluorescence and electrophoretic properties of a library of up to 56 ET primers with different donor-

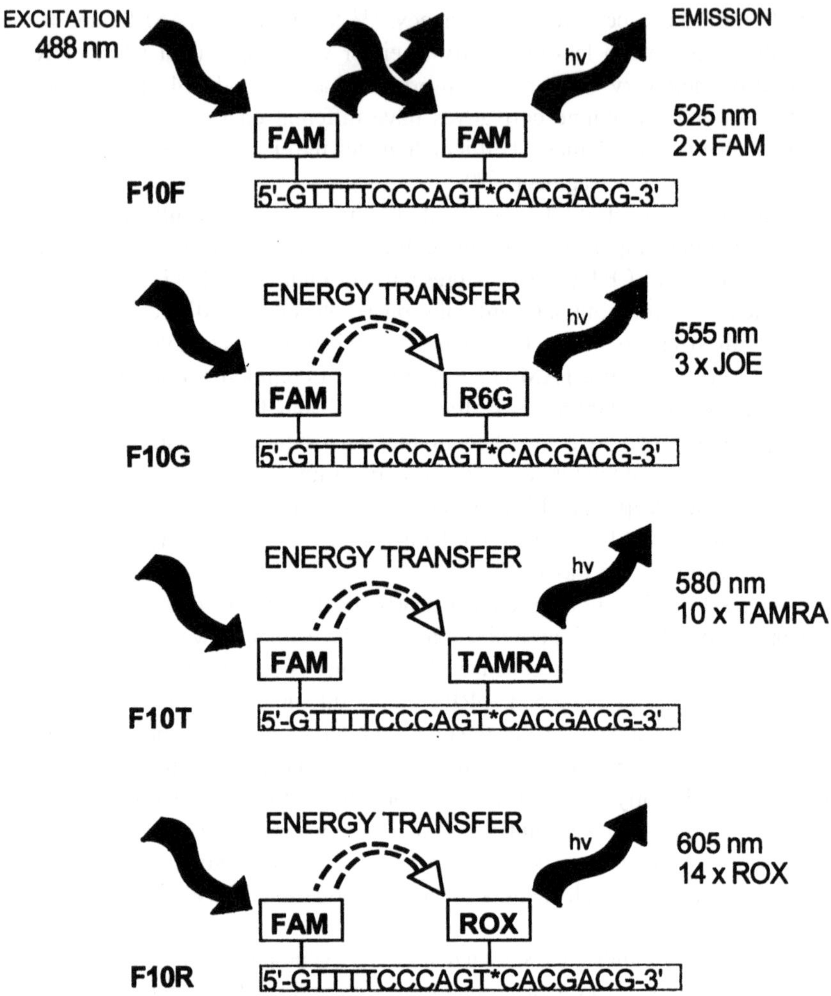

Figure 3.4. Schematic representation of the structure and spectroscopic properties of ET primers.

acceptor nucleotide spacing ranging from 1 to 12. Their studies show that in F-N-A series the strongest acceptor fluorescence intensity is obtained when the nucleotide spacing N is 8–10. When excited at 488 nm, the fluorescence intensity of the acceptor emission from this set of ET primers (F10F, F10G, F10T and F10R) is from 2- to 14-fold higher than that of the corresponding single dye-labeled primers. This set also generates extended DNA fragments with substantially similar electrophoretic mobility.[17] As a result, the four ET primers enable high-quality, four-color DNA sequencing on capillary electrophoresis DNA sequencers using a single laser line at 488 nm for excitation.[17] Furthermore, much less DNA template is used in sample preparation (8-fold in this example). These ET primer constructs are commercially available as DYEnamic™ ET primers from Amersham Biosciences. A typical run with the F-10 ET primers on a commercial capillary array electrophoresis sequencer (MegaBACE 1000 Sequencer, Molecular Dynamics, Sunnyvale, CA) provides DNA sequence with 99–100% accuracy in the first 800 bases and extended to over 1000 bases (Figure 3.5).

The efficiency of ET primer technology for DNA sequencing has been further enhanced by the use of 3-(ε-carboxypentyl)-3′-ethyl-5,5′-dimethyloxacarbocyanine (CYA or C; ε = 142,000 M^{-1} cm^{-1} at 488 nm) rather than FAM (ε = 60,000 M^{-1} cm^{-1} at 488 nm) as a common energy donor.[9] A set of four C-10 ET primers (C10F, C10G, C10T, and C10R) were synthesized and compared with the F-10 ET primer set. For example, Figure 3.6 shows, as anticipated from the comparison of the extinction coefficients of the two donors, that the donor absorbance of C10R is twice that of F10R. Moreover, the acceptor emission intensity of C10R is nearly double that of F10R. Since the fluorescence quantum yield of CYA is about 10-fold lower than that of FAM, a very weak donor emission leakage is also observed for C10R. This result shows that energy transfer competes successfully with radiative and other nonradiative de-excitation pathways from the CYA singlet excited state. This study has expanded the range of dyes that can be used as donors in the construction of ET coupled systems by showing that a dye with low fluorescence quantum yield but high absorption cross section can be exploited as an effective donor in ET constructs. Also, the resulting ET primers will provide higher spectral purity due to low residual donor emission. With 488-nm excitation, the fluorescence emission intensity of C10F, C10G, C10T, and C10R is from 1.4- to 24-fold stronger than that of the corresponding primers labeled only with the single acceptor dye, and from 0.8- to 1.7-fold stronger than that of the corresponding ET primers with FAM as the donor. As expected, much less crosstalk among the four detection channels was observed for the CYA-based ET primers (Figure 3.7).

Another ET primer set (C10R110, C10G, C10T, and C10R; where R110 is 5:6-carboxyrhodamine-110), where only rhodamine dyes are used as acceptors, provides the best match in the electrophoretic mobility coupled with intense acceptor emission. This set also gives outstanding results in cycle sequencing of double-stranded DNA templates with long read-lengths and high base-calling accuracy. This set of ET primers is by far the best for DNA sequencing.[23] The unique properties of this set of labels have been exploited to successfully perform high speed DNA sequencing on microfabricated capillary electrophoresis channels. Liu *et al.*[24] have demonstrated that high-speed DNA sequencing can be achieved in only 20 min with 99.4% accuracy to 500 base pairs. These results support the feasibility of performing high-speed genomic sequencing with microfabricated electrophoretic devices in conjunction with cyanine-based ET primers.

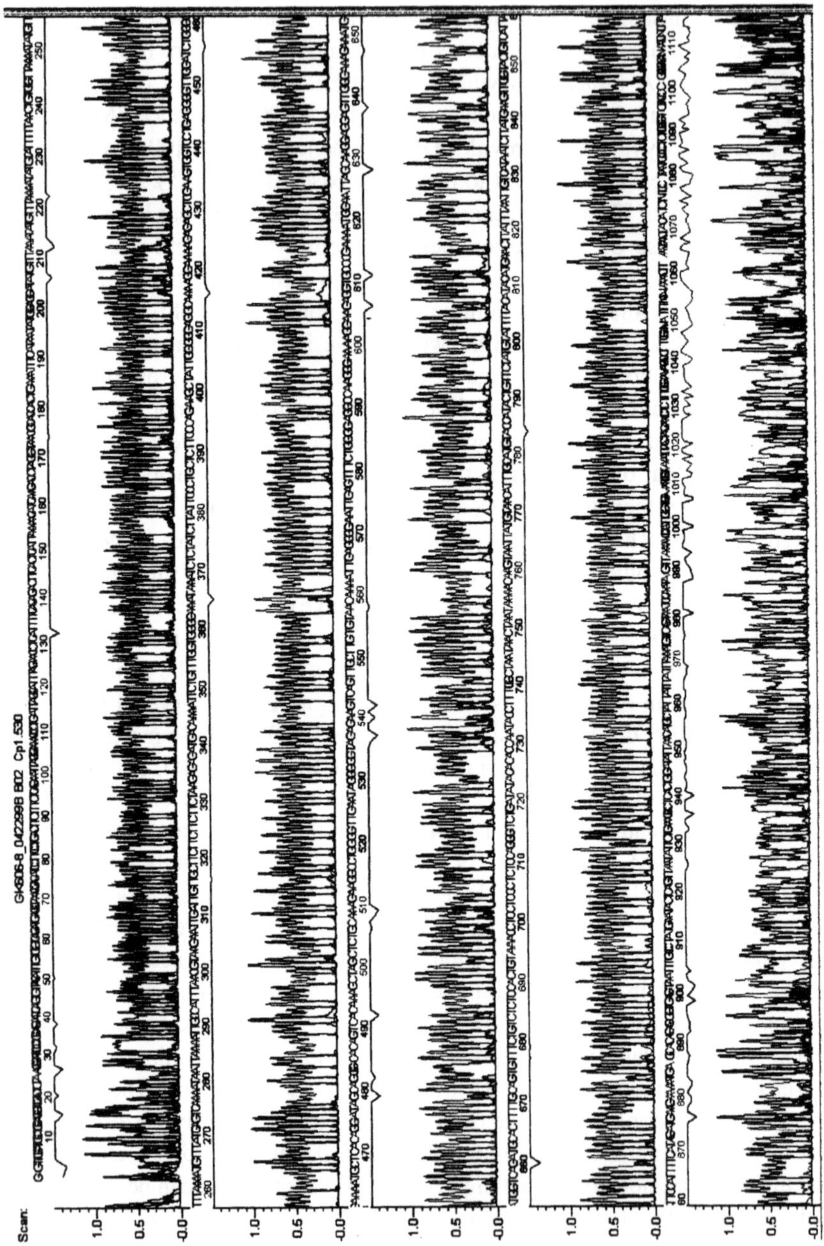

Figure 3.5. A 5.5-hour sequencing run on a MegaBACE 1000 Sequence. The M13 clone was sequenced with Amersham's DYEnamic ET primer sequencing kit. (Data kindly provided by Amersham Pharmacia Biotech.)

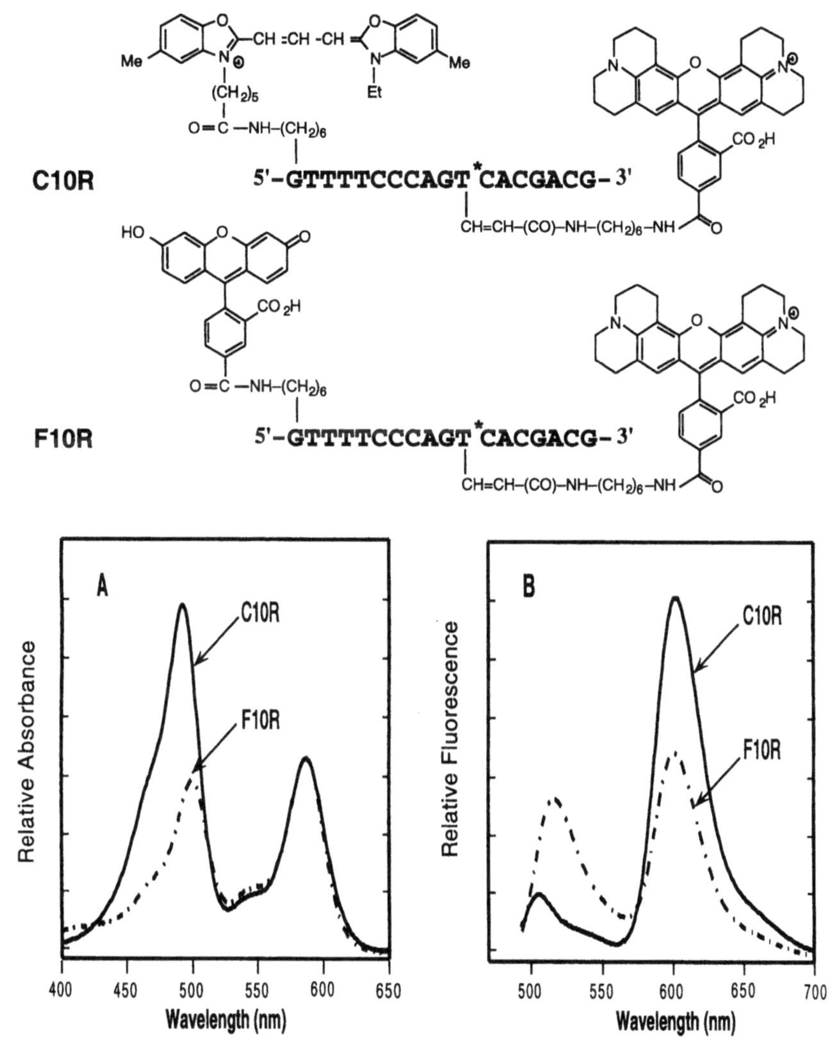

Figure 3.6. Comparison of the absorbance and emission spectra of ET primers based on F10R and C10R.

Based on the spectroscopic properties of 56 ET primers having different donors, acceptors, or nucleotide spacings between donor and acceptor, Hung *et al.*[22] found that alternative spacing combinations are valuable in multiplex genetic analyses, such as forensic identification and short tandem repeat (STR) analysis.[25] They developed new optimized C-4 or C-6 ET primer sets (C4G/C4R, C6G/C6R)

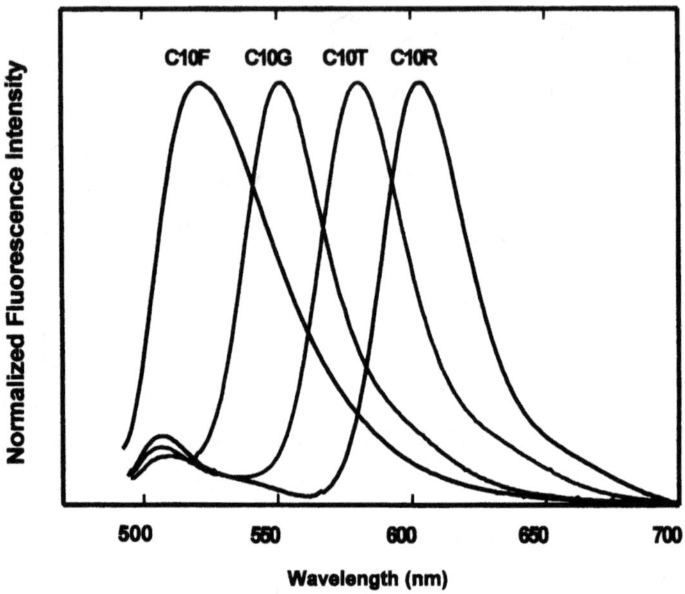

Figure 3.7. Normalized fluorescence emission spectra of the four CYA ET primers.

that offer excellent acceptor emission intensities and negligible donor emission. An excellent match of the electrophoretic mobility of single-base extension DNA fragments is also obtained with these two primer sets. Since these two primer sets allow precise quantitation of the ratio of signals from DNA fragments labeled with one or another of two different primers, these two primer sets have been applied to bladder cancer diagnosis based on two-color electrophoretic analyses of PCR-amplified STRs.[25]

3.5. ET Cassette for Construction of Energy-Transfer Fluorescent Primers

In 1996, Ju and coworkers developed a general method to tag primers of any sequence with ET coupled fluorophores using a universal ET cassette composed of six 1′,2′-dideoxyribose phosphate monomers (S_6) as a spacer.[26] The use of the ET cassette is advantageous in several respects: the spacer maintains the natural nucleic acid phosphate functionality to avoid possible anomalies in electrophoretic mobility; the spacer will not hybridize to the template, thus avoiding false priming. The major advantages offered in this method are the elimination of the constraints imposed by the requirement of complementarity to a particular target DNA sequence, and by the spacing between a 5′-terminal base and a modified thymidine residue within the primer sequence.[26] As presented in Figure 3.8A, the designed primer containing a six ribose (S_6 = R-R-R-R-R-R-) spacer and a modified base T* can be incorporated by conventional synthesis at the 5′-end of the oligonucleotide primer of interest. The donor is attached to the 5′ side of the ribose spacer and the acceptor to the T*. The resulting four-color primers (FS_6F, FS_6J, FS_6T, and FS_6R) display well-separated acceptor emission spectra with 2- to 12-fold enhancement in fluorescence intensity relative to that of the corresponding single dye-labeled primers. The only drawback of this approach is that the preparation of this ET cassette requires automated synthesis of the cassette on the 5′ end of the desired primer and then post-conjugation of dyes to the cassette.

Recently, a new ET cassette technology was developed by Berti *et al.*[27] that can be used to label any PCR, sequencing, or other primer of interest. The new ET cassette consists of a donor (3′ end) and an acceptor (5′ end) dye pair separated by a 1′,2′-dideoxyribose chain (Figure 3.8B). This new cassette also bears a key component, a pyridyldisulfide group, that can be quantitatively coupled to a thiol-activated target through a disulfide exchange reaction (Figure 3.8C). An example of the two-step ET cassette primer conjugation reaction is shown is Figure 3.8D. In order to optimize the energy transfer process, the length of the sugar chain in the new cassette designs is selected to be seven or eight.[22] Berti *et al.*[28] and Medintz *et al.*[29] have demonstrated the use of the new ET cassettes for four-color primer sequencing, fragment sizing, and high-performance STR analysis with high accuracy and performance.

Other ET cassettes containing a succinimidyl ester for primer conjugation were also developed. Lee and coworkers have synthesized an alternative set of ET primers[12] based on the same concept discussed above. This so-called BigDye™ cassette is not a single dye molecule, but a compound with two dye chromophores linked together in a donor–acceptor relationship. A fluorescein derivative (4′-aminomethyl-5-carboxyfluorescein) is chosen as the donor, and a set of four 4,7-dichloro-substituted rhodamine dyes (d-rhodamine) are used as the acceptors. As shown in Figure 3.9, a 4-aminomethylbenzoic acid backbone or linker is used to conjugate both the donor and the acceptor and hold them at a particular distance to facilitate ET. The entire dye complex is attached via an N-hydroxysuccinimide ester to the 5′-end of the primer. Their results show that BigDye ET primers give a 2-fold increase in signal over the single dye-labeled primers, and the ET efficiency is only 60% with the 4-aminomethylbenzoic acid linker.[12] Furthermore, the acceptors in BigDye™ ET cassettes have both lower extinction coefficients and relative quantum yields than the corresponding d-rhodamine dye alone. All these observations are consistent with the systematic results reported by Ju and Hung.[5,22] The use of the short spacer in these constructs increases self-quenching, and degrades the overall signal strength compared with ET primers that use the optimal donor-acceptor spacing.

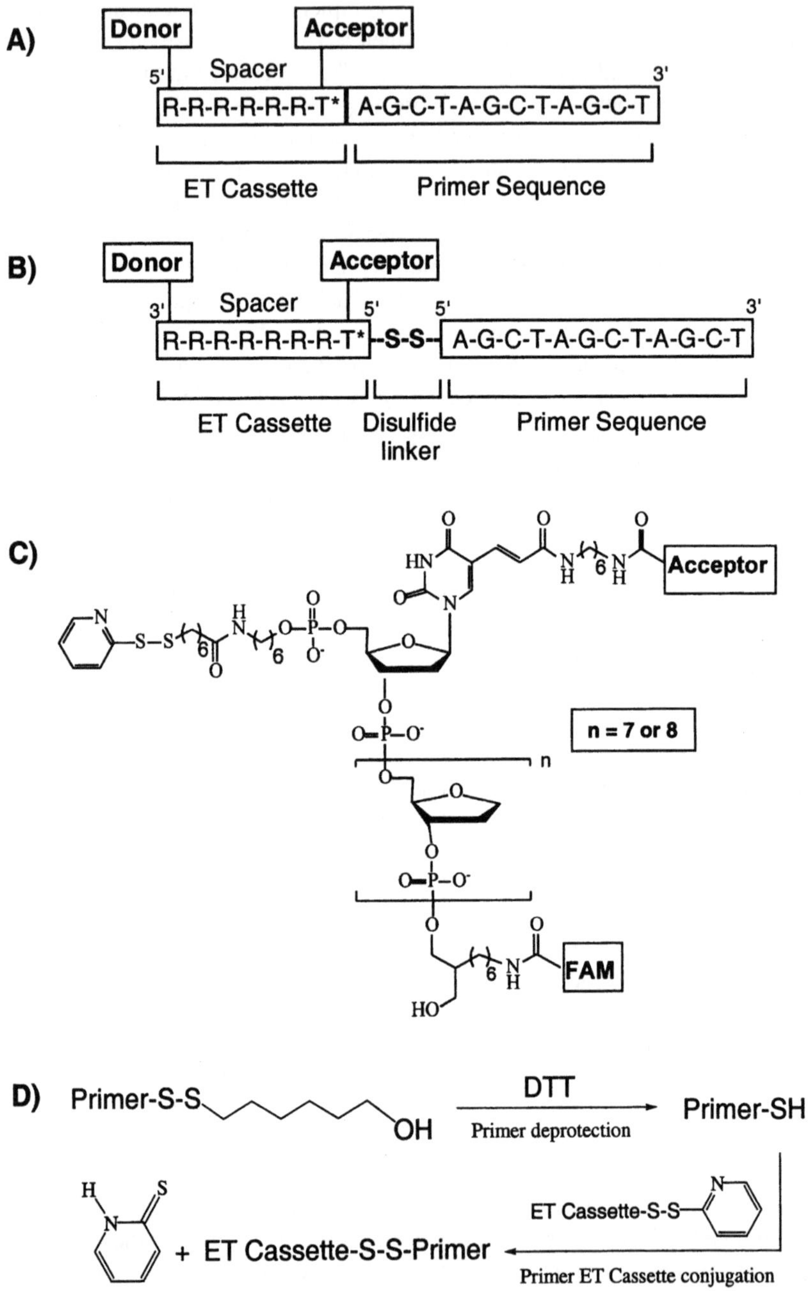

Figure 3.8. A) ET cassette labeled primer. R = 1′,2′-dideoxyribose monomer. B) New ET cassette labeled primer exploiting disulfide linkage chemistry. C) Structure of the new reactive ET cassette. D) Schematic of the two-step ET cassette primer conjugation.

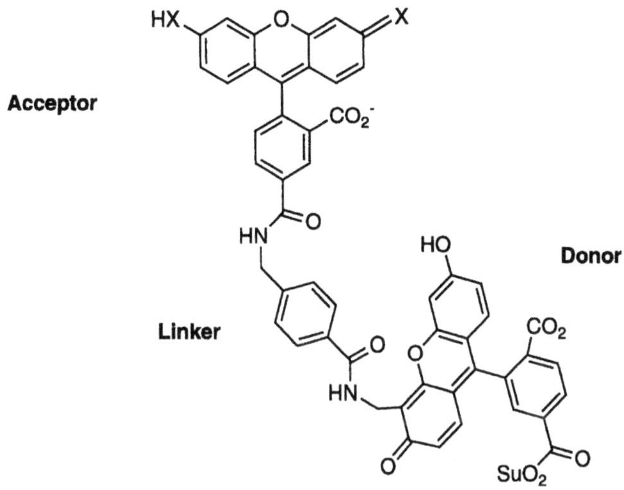

Figure 3.9. Chemical structure of the BigDye ET labeling cassette.

3.6. Energy-Transfer Fluorescent Terminators

Dye-labeled terminator sequencing has several advantages over sequencing with labeled primers.[30] The sequencing reactions are more simply performed because the four base-specific terminations can be carried out in one tube. Only a single extension reaction is required for each template, and the synthesis of a labeled primer is unnecessary. Thus the "primer walking strategy" can be applied to sequence a clone completely or to close gaps. A second significant advantage is that false terminations, in which DNA fragments are terminated by a deoxynucleotide rather than a dideoxynucleotide, are not observed as these products are unlabeled. The major disadvantage of dye-labeled terminator chemistry is that with every polymerase the intensity pattern has been found to be less uniform than for dye-labeled primers. The presence of very small or very large peaks can result in errors in automated base-calling. Nevertheless, the application of ET labeling technology to dye-terminator chemistry would be very valuable for all the reasons discussed above.

Prober and coworkers[8] first demonstrated that the dye-labeled alkynylamino-dideoxynucleoside triphosphates can be used as terminators for DNA sequencing. Thermo Sequenase[31] has also facilitated the automation of dye-terminator cycle sequencing reactions, because of its greater acceptance of dye-labeled terminators compared to other DNA polymerases. In order to obtain higher peak intensities and hence more accurate base-calling compared with single dye-labeled terminators, Amersham Biosciences has developed the DYEnamic™ ET terminator kit,[32] and PE Applied Biosystems has developed the BigDye™ terminator kit,[13] both of which are based on the ET-labeling concept. The DYEnamic™ ET terminator kit[33] features novel ET terminators and a new DNA polymerase (Thermo Sequenase™ II DNA polymerase). Fluorescein is chosen as the donor and the acceptor dyes are standard rhodamine dyes (R110, R6G, TAMRA, and ROX). Thermo Sequenase™ II DNA polymerase has advantageous biochemical properties including tolerance to high salt concentration, efficient utilization of dITP, high processivity, and excellent performance on GC-rich templates. Most importantly, it performs well with extension steps as short as 15 seconds at 70°C[33] while most of the other polymerases call for a 60°C, 4 minutes extension step. The alternative BigDye cassette is attached via an NHS ester through a propargylamino or propargylethoxyamino linker to dideoxynucleoside triphosphates. The BigDye™ terminator set has mobility shifts within the half-base requirement for minimal artifacts, and produces improved signal-to-noise ratios and more even peak heights compared to the single dye-labeled terminators.[13] Both terminator sets make the Sanger dideoxy chain-termination method more efficient and robust. However, the signal strength available from the DYEnamic™ ET terminator constructs is significantly greater than that of the BigDye terminators because there is less self-quenching.

3.7. Short Tandem Repeat (STR) Analysis with Capillary Array Electrophoresis and ET Labels

Microsatellite DNA sequences containing di-, tri-, tetra-, and pentanucleotide repeats are characterized by a high degree of genetic polymorphism.[34–36] Because of their high abundance and relative ease of detection following amplification by the polymerase chain reaction (PCR), STRs have found widespread use in gene mapping,[37,38] especially as distinctive markers in health care diagnostics.[39,40] The development of sensitive, rapid, and accurate methods and apparatus for high-throughput STR analysis will be critical for exploiting microsatellite variation in cancer screening. Two-color sizing of multiplexed STRs on replaceable hydroxyethyl cellulose (HEC) sieving matrices using ET primers gives reproducible results with single base resolution.[41,42] Wang and coworkers have also demonstrated the application of capillary array electrophoresis (CAE) and ET primers in a two-color assay for rapid detection of chromosome alterations that are used in the diagnosis of bladder cancer.[25] Rapid separations ($\leq$35 min) are achieved on capillary arrays using replaceable separation matrices, and the allelic ratios are quantitatively determined with a precision of $\pm$10%. With this precision, a variation of 20% is considered diagnostically significant.[25]

In such a two-color STR assay, PCR amplicons from normal DNA are labeled with one fluorescent ET primer, PCR amplicons from urine or tumor DNA

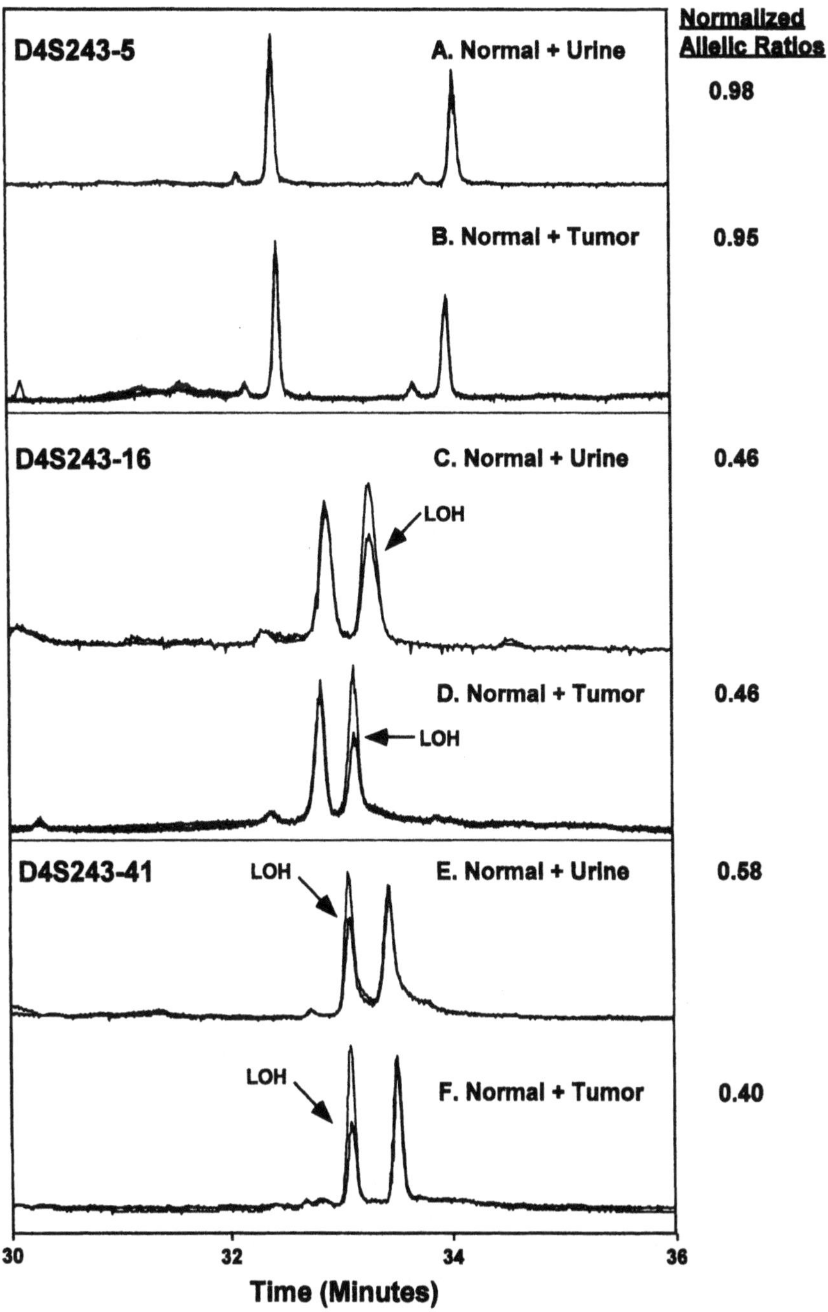

Figure 3.10. Two-color LOH analyses with CAE using ET primers.

are labeled with a second distinctive fluorescent ET primer, and both samples are co-injected into the same capillary for separation. Using the C6G/C6R and C4G/C4R ET primer pairs, labeled normal and urine or tumor PCR amplicons can be excited at a common laser wavelength and the intense and nonoverlapping emissions can be detected. These two ET primer sets are advantageous for multicolor fragment sizing, because of the minimal crosstalk between the donor and the acceptor, and the uniform mobility shift of the PCR products. Figure 3.10 presents two-color electropherograms for samples from different bladder cancer patients amplified at the D4S243 loci. A size change in one of the alleles detected in the urine and/or tumor compared to the normal blood sample is used as an indication of the extension or deletion of the microsatellite marker, and an allelic ratio intensity change indicates the loss of heterozygosity (LOH). The normal sample is detected in the green channel while the urine or tumor samples are detected in the red channel. The top set of traces acts as a control since this particular marker is not altered in this patient. The near unity allelic ratios provide a measure of the precision. Figure 3.10C presents analyses of the cells in the urine from a patient where there is a significant LOH in the heavy D4S243 allele. The results are consistent with the findings of Steiner *et al.*[43] based on the assays performed with radioisotope labeling. This method provides a significant improvement in the speed, ease, and precision of STR analyses.

3.8. Biotinylated Energy Transfer Cassettes

The biotin-streptavidin system is very attractive for *in vitro* labeling of biotargets, because biotin has an extraordinarily high affinity for streptavidin with a reported dissociation constant of ~10^{-15} M.[44] For example, a streptavidin-phycoerythrin-Cy5 conjugate has been used to stain biotinylated targets hybridized to high density oligonucleotide arrays in two-color mutational assays.[45] The very strong binding affinity of the biotin-streptavidin system has also been exploited for DNA purification and DNA sequencing.[46–48]

In labeling applications, it would be useful to develop ET tags relevant to the streptavidin-biotin system for the advantages noted above in connection with sequencing. Most importantly, FRET can make the streptavidin-biotin system robust for multicolor spectral analysis while allowing the use of a single line laser excitation source. In 1999, Xie *et al.*[49] described the synthesis, characterization, and application of biotinylated ET cassettes and fluorescent ET-labeled streptavidins. The biotinylated ET cassette is made by using an oligonucleotide as a "scaffold," with the donor (CYA)-acceptor ($\mathbf{T^{ROX}}$) fluorophore pairs separated by 10 intervening nucleobases: 5′-CYA-NNNNNNNNN$\mathbf{T^{ROX}}$NN$\mathbf{T^{B}}$NNNNNNN-3′. Biotin-labeled thymidine ($\mathbf{T^{B}}$) is introduced by the use of biotin-dT phosphoramidite at different locations in the oligonucleotides. In this example, $\mathbf{T^{B}}$ is introduced two bases 3′ to the position of the base $\mathbf{T^{ROX}}$. A systematic study indicates that the spectroscopic properties of these cassettes are not sensitive to the location of biotin as long as $\mathbf{T^{B}}$ is not placed between the donor and the acceptor. Biotinylated ET cassettes with an acceptor-biotin spacing of three nucleotides (C10R**3**B) offer excellent acceptor emission intensity coupled with negligible donor emission. With 488-nm excitation, the fluorescence emission intensity of C10R**3**B is 18-fold stronger than that of the corresponding oligonucleotide labeled with ROX alone.

Xie *et al.* have also developed a synthetic method that can be used to prepare sets of fluorescent ET dye-labeled streptavidins for multiplex detection. ET-labeled streptavidins can be prepared either by the covalent attachment of an ET cassette to streptavidin through a thioether bond or by the noncovalent streptavidin-biotin interaction between the ET cassette and streptavidin (Figure 3.11). For the noncovalent approach, streptavidin is first mixed with two or three molar equivalents of the biotinylated ET cassette. Since the bound streptavidin still has additional biotin binding sites available, the fluorescent ET label can still interact with a biotinylated target to provide detection. By using this sandwich detection format, up to three ET fluorescent dye pairs can be attached to streptavidin to increase the assay sensitivity.

The biotinylated ET cassettes and fluorescent ET-labeled streptavidin reagents have a broad range of applications. The fluorescent ET dye-labeled streptavidins have been used for cell-sorting analysis, which can directly detect any specific cell surface antigen by using a biotinylated antibody. The fluorescent ET dye-labeled streptavidins demonstrated high assay sensitivity. The ET dye-labeled streptavidins should also find application in mutational analysis using oligonucleotide arrays. The ET fluorescent streptavidins are advantageous compared to phycoerythrin-conjugates in that they can provide more evenly matched signal intensities while exhibiting strong and distinct fluorescent emissions in two-color assays, and maintaining compatibility with 488-nm excitation.

3.9. Template-Directed Dye-Terminator Incorporation

Chen and Kwok[50–52] have developed a homogeneous assay that allows the rapid analysis of single nucleotide polymorphisms using FRET. Their approach combines the specificity of enzymatic discrimination between the two alleles of a single nucleotide polymorphism in a template-directed primer extension reaction and the sensitivity of FRET. This method eliminates the need for product separation and the use of radioactivity. The DNA fragment containing a mutation site is first amplified from genomic DNA by PCR followed by enzymatic degradation of

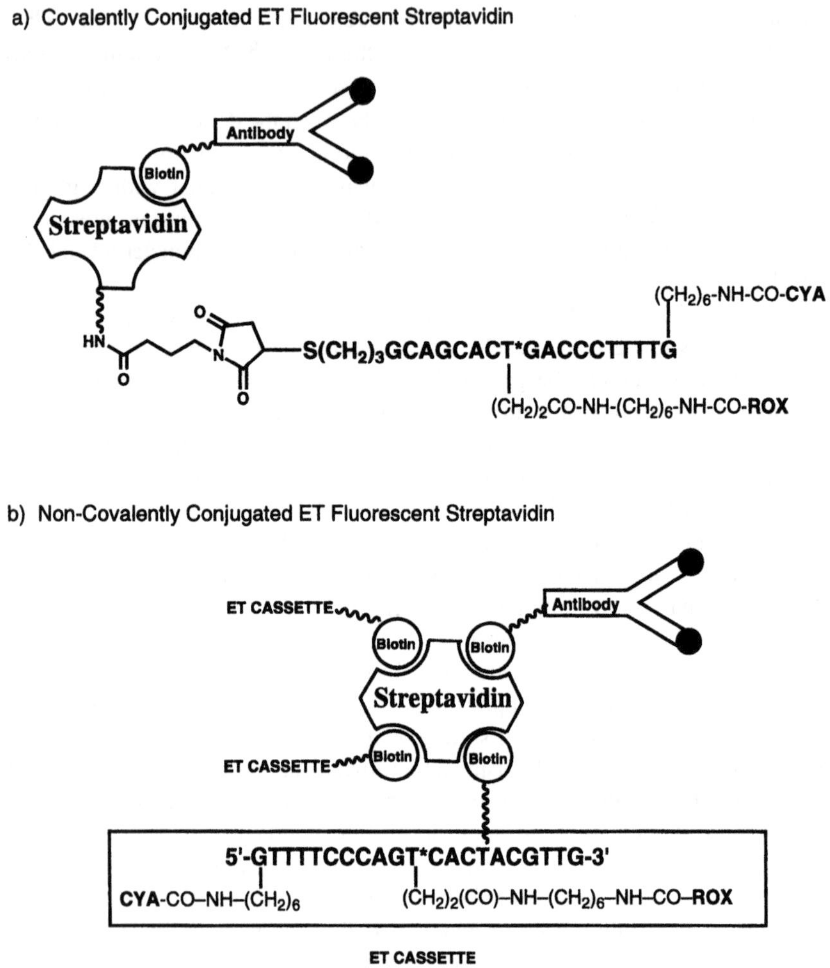

Figure 3.11. Chemical structures of ET-labeled streptavidins.

excess primers and deoxyribonucleoside triphosphates before the primer extension reaction is performed. Then, amplified DNA fragments are incubated with a 5′-fluorescein-labeled primer, which is designed to hybridize to the DNA template adjacent to the polymorphic site. Two dye-labeled terminators (ROX-ddNTP and TAMRA-ddNTP) are used, one for each allele. The dye-labeled primer is extended one base by the incorporation of a complementary dye-terminator specific to the allele present on the template. FRET occurs when the dye-labeled ddNTP is incorporated into the primer. Thus, the acceptor allelic status can be easily determined by monitoring which dye-labeled ddNTP is incorporated at the mutation site. The method has proven to be highly sensitive, specific, and suitable for automated genotyping of large numbers of samples.[50] This homogenous DNA diagnostic method has been successfully used to detect mutations in the cystic fibrosis transmembrane conductance regulator gene and in the human leukocyte antigen H gene.[51]

3.10. Conclusions

Fluorescence resonance energy transfer technology has done much to enhance DNA sequencing capabilities and the Human Genome Project. The enhanced emission intensities and improved spectral purity provided by ET primers and terminators have led to a major improvement in the performance of automated DNA sequencers. Fluorescent ET dye-labels should be equally valuable for the analysis of PCR fragments, restriction fragments, single nucleotide polymorphisms, and microsatellites in research, pharmaceutical, and health care diagnostic applications.

Acknowledgments

The authors are indebted to the members of the UC Berkeley High Sensitivity DNA Analysis Project whose work is in part reviewed here. We also thank Mark Lewis of Molecular Dynamics for providing CAE sequencing data. These studies were supported by the Director, Office of Energy Research, Office of Biological and Environmental Research of the U.S. Department of Energy under contract DE-FG-9161125. Additional support from Amersham Biosciences is gratefully acknowledged.

References

1. Stryer, L. *Annu. Rev. Biochem. 47*, 819–846, 1978.
2. Collins, F., Patrinos, A., Jordan, E., Chakravarti, A., Gesteland, R., Walters, L., and the members of the DOE and NIH planning groups. New goals for the U.S. Human Genome Project: 1998–2003. *Science 282*, 682–689, 1998.
3. Kheterpal, I., and Mathies, R. A. *Anal. Chem. 71*, 31A–37A, 1999.
4. Glazer, A. N., and Mathies, R. A. *Curr. Opin. Biotechnol. 8*, 94–102, 1997.
5. Ju, J., Glazer, A. N., and Mathies, R. A. *Nature Med. 2*, 246–249, 1996.
6. Ju, J., Ruan, C., Fuller, C. W., Glazer, A. N., and Mathies, R. A. *Proc. Natl. Acad. Sci. USA 92*, 4347–4351, 1995.

7. Smith, L. M., Sanders, J. Z., Kaiser, R. J., Hughes, P., Dodd, C., Connell, C. R., Heiner, C., Kent, S. B. H., and Hood, L. E. *Nature 321*, 674–679, 1986.
8. Prober, J. M., Trainor, G. L., Dam, R. J., Hobbs, F. W., Robertson, C. W., Zagursky, R. J., Cocuzza, A. J., Jensen, M. A., and Baumeister, K. *Science 238*, 336–341, 1987.
9. Hung, S.-C., Ju, J., Mathies, R. A., and Glazer, A. N. *Anal. Biochem. 243*, 15–27, 1996.
10. Metzker, M. L., Lu, J., and Gibbs, R. A. *Science 271*, 1420–1422, 1996.
11. *Amersham Life Science News*, 1–7, 1999.
12. Lee, L. G., Spurgeon, S. L., Heiner, C. R., Benson, S. C., Rosenblum, B. B., Menchen, S. M., Graham, R. J., Constantinescu, A., Upadhya, K. G., and Cassel, J. M. *Nucleic Acids Res. 25*, 2816–2822, 1997.
13. Rosenblum, B. B., Lee, L. G., Spurgeon, S. L., Khan, S. H., Menchen, S. M., Heiner, C. R., and Chen, S. M. *Nucleic Acids Res. 25*, 4500–4504, 1997.
14. Förster, T. *Delocalized Excitation and Excitation Transfer*. New York; Academic Press, 1965.
15. Stryer, L., and Haugland, R. *Proc. Natl. Acad. Sci. USA 74*, 5463–5467, 1967.
16. Sanger, F., Nicklen, S., and Coulson, A. R. *Proc. Natl. Acad. Sci. USA 74*, 5463–5467, 1977.
17. Hung, S.-C., Ju, J., Mathies, R. A., and Glazer, A. N. *Anal. Biochem. 238*, 165–170, 1996.
18. Carson, S., Cohen, A. S., Belenkii, A., Ruiz-Martinez, M. C., Berka, J., and Karger, B. L. *Anal. Chem. 65*, 3219–3226, 1993.
19. Takahashi, S., Murakami, K., Anazawa, T., and Kambara, H. *Anal. Chem. 66*, 1021–1026, 1994.
20. Tu, O., Knott, T., Marsh, M., Bechtol, K., Harris, D., Barker, D., and Bashkin, J. *Nucleic Acids Res. 26*, 2797–2802, 1998.
21. Ju, J., Kheterpal, I., Scherer, J. R., Ruan, C., Fuller, C. W., Glazer, A. N., and Mathies, R. A. *Anal. Biochem. 231*, 131–140, 1995.
22. Hung, S.-C., Mathies, R. A., and Glazer, A. N. *Anal. Biochem. 252*, 78–88, 1997.
23. Hung, S.-C., Mathies, R. A., and Glazer, A. N. *Anal. Biochem. 255*, 32–38, 1998.
24. Liu, S., Shi, Y., Ja, W. W., and Mathies, R. A. *Anal. Chem. 71*, 566–573, 1999.
25. Wang, Y., Hung, S. C., Linn, J. F., Steiner, G., Glazer, A. N., Sidransky, D., and Mathies, R. A. *Electrophoresis 18*, 1742–1749, 1997.
26. Ju, J., Glazer, A. N., and Mathies, R. A. *Nucleic Acids Res. 24*, 1144–1148, 1996.
27. Berti, L., Xie, J., Medintz, I. L., Glazer, A. N., Mathies, R. A. *Anal. Biochem 292*, 188–197, 2001.
28. Berti, L., Medintz, I. L., Tom, J., Mathies, R. A. *Bioconj. Chem. 12*, 493–500, 2001.
29. Medintz, I. L., Berti, L., Emrich, C. A., Tom, J., Scherer, J. R., Mathies, R. A. *Clin. Chem. 47*, 1614–1621, 2001.
30. Rosenthal, A., and Charnock-Jones, D. *Linear Amplification Sequencing with Dye Terminators*, Totowa, NJ: Humana Press, 1993.
31. Reeve, M., and Fuller, C. *Nature 376*, 796–797, 1995.
32. BioDirectory '99 from Amersham Pharmacia Biotech.
33. DYEnamic™ ET terminators cycle sequencing premix kit from Amersham Pharmacia Biotech.
34. Charlesworth, B., Sniegowski, P., and Stephan, W. *Nature 371*, 215–220, 1994.
35. Reed, P. W., Davies, J. L., Copeman, J. B., Bennett, S. T., Palmer, S. M., Pritchard, L. E., Gough, S. C. L., Kawaguchi, Y., Cordell, H. J., Balfour, K. M., Jenkins, S. C., Powell, E. E., Vignal, A., and Todd, J. A. *Nature Genet. 7*, 390–395, 1994.
36. Oda, S., Oki, E., Maehara, Y., and Sugimachi, K. *Nucleic Acids Res. 25*, 3415–3420, 1997.
37. Gyapay, G., Morissette, J., Vignal, A., Dib, C., Fizames, C., Millasseau, P., Marc, S., Bernardi, G., Lathrop, M., and Weissenbach, J. *Nature Genet. 7*, 246–339, 1994.
38. Dib, C., Faure, S., Fizames, C., Samson, D., Drouot, N, Vignal, A., Millasseau, P., Marc, S., Hazan, J., Seboun, E., Lathrop, M., Gyapay, G., Morissette, J., and Weissenbach, J. *Nature 380*, 152–154, 1996.
39. Hammond, H. A., Jin, L., Zhong, Y., Caskey, C. T., and Chakraborty, R. *Am. J. Hum. Genet. 55*, 175–189, 1994.

40. Mansfield, E. S., Vainer, M., Harris, D. W., Gasparini, P., Estivill, X., Surrey, S., and Fortina, P. *J. Chromatogr. A 781*, 295–305, 1997.
41. Wang, Y., Ju, J., Carpenter, B. A., Atherton, J. M., Sensabaugh, G. F., and Mathies, R. A. *Anal. Chem. 67*, 1197–1203, 1995.
42. Wang, Y., Wallin, J. M., Ju, J., Sensabaugh, G. F., and Mathies, R. A. *Electrophoresis 17*, 1485–1490, 1996.
43. Steiner, G., Schoenberg, M. P., Linn, J. F., Mao, L., and Sidransky, D. *Nature Med. 3*, 621–624, 1997.
44. Green, N. M. *Methods Enzymol. 184*, 51–67, 1980.
45. Hacia, J. G., Brody, L. C., Chee, M. S., Fodor, S. P. A., and Collins, F. S. *Nature Genet. 14*, 441–447, 1996.
46. Hawkins, T. L., O'Connor-Morin, T., Roy, A., and Santillan, C. *Nucleic Acids Res. 22*, 4543–4544, 1994.
47. Holmberg, A., Fry, G., and Uhlen, M. Automatic preparation of DNA templates for sequencing on the ABI Catalyst robotic workstation, in *Automated DNA Sequencing and Analysis*, Adams, M. D., Fields, C., Venter, J. C. eds. San Diego, CA: Academic Press, 1994, pp. 139–145.
48. Ju, J., Yan, H., Zaro, M., Doctolero, M., Goralski, T., Konrad, K., Lachenmeier, E., and Cathcart, R. *Ninth International Genome Sequencing and Analysis Conference*, Hilton Head, SC, 1997. Meeting Abstract 223.
49. Xie, J., Mathies, R. A., and Glazer, A. N. *DOE Human Genome Program Contractor-Grantee Wokshop VII*, Oakland, CA, 1999. Meeting Abstract 19.
50. Chen, X., and Kwok, P. Y. *Nucleic Acids Res. 25*, 347–353, 1997.
51. Chen, X., Zehnbauer, B., Gnirke, A., and Kwok, P. K. *Proc. Natl. Acad. Sci. USA 94*, 10756–10761, 1997.
52. Chen, X., Livak, K., and Kwok, P. K. *Genome Res. 8*, 549–556, 1998.

4

On-the-Fly Fluorescence Lifetime Detection in Capillary Electrophoresis for DNA Analysis

Linda B. McGown

4.1. Introduction

Fluorescence detection for capillary electrophoresis (CE) has achieved widespread use in the past decade for applications such as DNA sequencing and fingerprinting. Fluorescent labels are a relatively safe alternative to the traditional radioactive labels, providing similar detection limits, high sensitivity, and ease of automation and on-line detection.[1,2] Moreover, the selectivity of fluorescence detection makes it suitable for multiplex detection, i.e., determination of multiple analytes in a single lane or capillary in gel electrophoresis. Multiplex methods for detection not only improve the throughput by a factor equal to the number of multiplexed analytes but also eliminate the problem of capillary-to-capillary misalignment.[3]

Multiplex methods generally have relied on the use of dye labels of different emission and/or excitation spectra to distinguish among analytes. The dyes may be attached to the analytes using covalent linkers or by physical association. In relying on spectral (color) discrimination in multiplex schemes, it is necessary to achieve wavelength selectivity in the fluorescence emission spectrum using one or more excitation wavelengths. Detection using only filter selection of color may be inadequate to resolve overlapping peaks, while the use of array detection to collect a spectrum in order to better identify and resolve colors can decrease sensitivity because of dispersion. Moreover, since multiplex detection relies on spectral resolution of the different dye labels, broadly overlapping emission spectra may result in crosstalk between detection channels. Dyes with narrow emission spectra have been used to increase spectral resolution,[4,5] but some overlap is inevitable due to the continuous nature of the spectra.

Linda B. McGown • Department of Chemistry, Duke University, Durham, North Carolina 27708.

Topics in Fluorescence Spectroscopy, Volume 7: DNA Technology, edited by Joseph R. Lakowicz, Kluwer Academic / Plenum Publishers, New York, 2003

Thus, it has been desirable to explore new approaches to fluorescence detection that have the potential to increase accuracy and resolution in on-the-fly, multiplex detection. One such approach is to use fluorescence lifetime instead of color to discriminate among the labeled analytes. Fluorescence lifetime is a discrete function generally governed by the simple, exact relationships of first-order kinetics for each decay component, and is therefore easier to model and resolve than spectral color.[6]

Fluorescence lifetime detection can be achieved using either time-domain or frequency-domain techniques.[7] The time domain uses pulsed excitation to generate a decay curve, from which the fluorescence lifetime profile is extracted. The frequency domain uses continuous (cw) excitation that is intensity modulated at a high (MHz–GHz) frequency, producing a time-dependent intensity signal that is comprised of a sinusoidal (a.c.) component superimposed on a time-independent (d.c.) intensity component. The fluorescence emission is phased-shifted and demodulated relative to the exciting light and the fluorescence lifetime can be calculated independently from the phase-shift ϕ and the extent of demodulation of the excitation beam, expressed as the modulation ratio m.

In its earliest application to HPLC, on-the-fly frequency-domain fluorescence lifetime detection was limited by available instrumentation that could only collect data at one modulation frequency per chromatogram and therefore required multiple injections of the sample.[8–11] The resulting chromatograms had to be perfectly overlaid and the data combined at each point along the chromatographic peaks in order to determine fluorescence lifetimes and intensity at each of the points.

The potential utility of fluorescence lifetime detection was greatly increased by the introduction of a multiharmonic Fourier transform phase-modulation fluorometer (MHF).[12] In the MHF, the intensity of the exciting light, provided by a continuous wave laser, is electrooptically modulated by a Pockels cell at as many as 75 frequencies in the MHz range, simultaneously. The resulting emission is modulated at the same multiple frequencies and is detected using cross-correlation electronics.[13] The data is then digitized and Fourier transformed into the resultant phase and modulation information used to determine fluorescence lifetime.

The introduction of the MHF revolutionized on-the-fly fluorescence lifetime detection in CE.[14,15] An entire frequency response that contains sufficient information to analyze multiexponential decays can be acquired on the order of milliseconds, providing many points per electrophoretic peak. This is a major advantage over time-domain detection, which generally must assume monoexponential decay for on-the-fly detection even when the actual decay is more complex.[16,17] Fluorescence intensity is simultaneously derived from the same measurement to produce a lifetime-intensity electropherogram in no more time than is required for an intensity-based electropherogram alone.

Measuring Fluorescence Lifetime On-the-Fly in CE

In the frequency-domain, or phase-modulation, fluorescence lifetime measurement,[7] the excitation beam is intensity-modulated and the resulting a.c. component of the fluorescence emission is phase-shifted and demodulated relative to the a.c. component of the excitation beam. The phase shift (ϕ) and demodulation factor (m) can be used to calculate the fluorescence lifetime as follows:

$$\tau_p = \omega^{-1} \tan \phi \tag{4.1}$$

$$\tau_m = \omega^{-1}[((1/m^2) - 1)]^{1/2} \tag{4.2}$$

where τ_p and τ_m are the lifetimes calculated from phase and demodulation, respectively, and ω is the angular modulation frequency ($\omega = 2\pi F$, where F is the linear modulation frequency). For a single, exponential decay,

$$\tau_p = \tau_m = \tau \tag{4.3}$$

where τ is the true fluorescence lifetime. However, if the decay law is more complex, then the observed lifetimes τ_p and τ_m are only apparent values and measurements must be made at multiple modulation frequencies in order to resolve the lifetimes and fractional intensity contributions of the individual lifetime components.

The first phase-modulation fluorescence lifetime instruments allowed measurements of phase and modulation at only one modulation frequency at a time. Measurements had to be made sequentially at a series of frequencies in order to obtain the equivalent of an entire decay curve. This was changed with the introduction of the multiharmonic Fourier transform instrument (MHF) for fluorescence lifetime measurements.[15] The MHF generates a base frequency (F) which is then used by a comb generator to produce several harmonics (F, 2F, 3F, ..., nF). The sum of these harmonics is used by a Pockels cell to modulate the intensity of the exciting laser beam. A cross-correlation frequency, f, is combined with the base frequency (F + f, 2F + 2f, 3F + 3f, ..., nF + nf) to gate the PMTs, resulting in low frequency (f, 2f, 3f, ..., nf) signals which contain all of the high frequency information. Conversion to low frequency allows the phase and modulation data to be easily detected by commercial electronics. In the MHF, the cross-correlation frequency determines the rate at which the lifetime measurements are made. For example, a cross-correlation frequency of 10 Hz corresponds to 10 lifetime measurements per second, or one lifetime measurement every 0.1 s.

A schematic diagram of the interfaced CE/MHF system is shown in Figure 4.1. The CE (Beckman P/ACE 5000) is equipped with a UV-visible detector. A phase-modulation fluorescence lifetime instrument (Model 4850 MHF, Spectronics Instruments) is interfaced to the CE for on-the-fly fluorescence lifetime

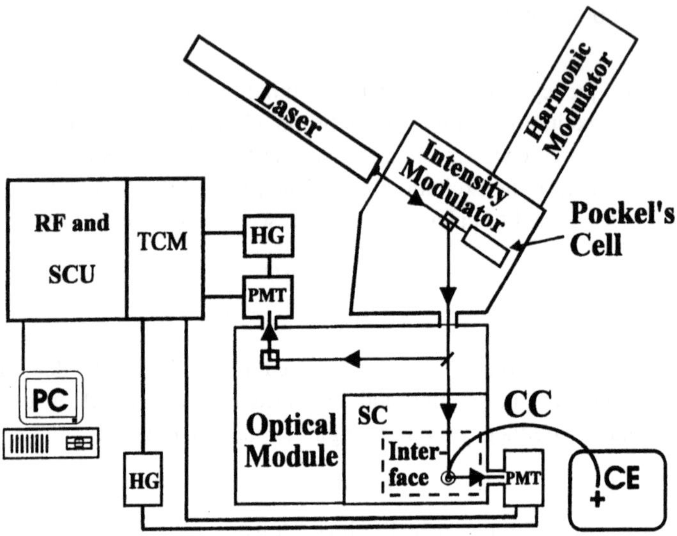

Figure 4.1. Schematic diagram of model 4850MHF fluorescence lifetime instrument interfaced to a model P/ACE 5000 CE system. SC: MHF sample chamber; HG: MHF harmonic generator; TCM: translator command module; PC: 386 personal computer; SCU: sampling control unit; CC: capillary column.[14]

measurements. An air-cooled argon ion laser provides excitation at 488 nm or 514 nm that is purified by passing it through a bandpass filter. The capillary is routed through the MHF sample chamber and terminates in the run buffer that is placed in the MHF sample compartment. To complete the circuit, an electrical wire is used to connect the ground of CE with the copper electrode in the run buffer in the MHF. In this configuration, the capillary can be as short as 40 cm and the detection window is free from loop strain. It is secured onto a specially designed capillary mount to keep it motionless.

A Pockels cell is used in the MHF instrument to electronically modulate the excitation beam intensity. The diameter of the excitation beam is reduced with a focusing lens into a spot smaller than the diameter of the CE capillary. The emission signal is collected by a microscope objective. The capillary is tilted by approximately 30° relative to the vertical axis of the microscope objective to reduce scattered light. The emission is selected through an appropriate long pass filter combined with a holographic filter corresponding to the excitation wavelength to further reduce the contribution from scattered laser light.

A scattering solution of kaolin generally is used as the lifetime reference for on-the-fly detection. In batch experiments, the sample turret in MHF alternates between the sample and the reference solutions. For on-the-fly detection in CE, the sample turret is removed, which disables the automatic reference lifetime acquisition. Instead, the reference is measured separately. This is done by placing neutral density filters in the emission beam instead of wavelength selection optical filters and measuring the capillary column filled with the run buffer. The neutral density filters are necessary to reduce the intensity of the scattered light. After the reference lifetime measurement is finished, the neutral density filters are replaced with appropriate filters for wavelength selection of the sample emission.

4.2. Analysis of On-the-Fly Lifetime Data

Each electrophoretic run results in a large amount of data. For example, a 20 min run may result in as many as 12,000 phase-modulation multifrequency profiles. Each profile is a Fourier-transformed, binary computer file containing phase and modulation data for each of 20 or more modulation frequencies. Software has been developed to extract the fluorescence intensity from the dynamic data to produce an intensity electropherogram that can be imported into programs for data analysis and display.

The lifetime data is generally analyzed using nonlinear least-squares analysis (NLLS) that fits the data using an *a priori* model in which the lifetimes are either allowed to float or are fixed to predetermined values. The recovered lifetime profile contains the lifetimes and the fractional contribution of each lifetime component to the total intensity at that point in the electropherogram. Contributions from scattered light and background luminescence can be resolved as needed by adding appropriate components to the model.

Resolution of overlapping electrophoretic peaks is achieved using NLLS in which the lifetime components in the fitting models may be allowed to float or may be fixed to values that were pre-determined for the individual components in the mixture under identical experimental conditions. The recovered fractional intensities at each point in the lifetime electropherogram are then multiplied by the total intensity at that point to construct the electropherogram of each individual component.

4.3. Lifetime Detection of Labeled DNA Primers

In the first studies of multiplex detection of DNA using fluorescence lifetime in CE,[18,19] overlapping electrophoretic peaks were artificially generated using sequential injections, in order to simulate actual sequencing results. On-the fly fluorescence lifetime detection is shown for each of four different dye-labeled M13 DNA primers in Figure 4.2. Recovered lifetimes for M13-forward primers

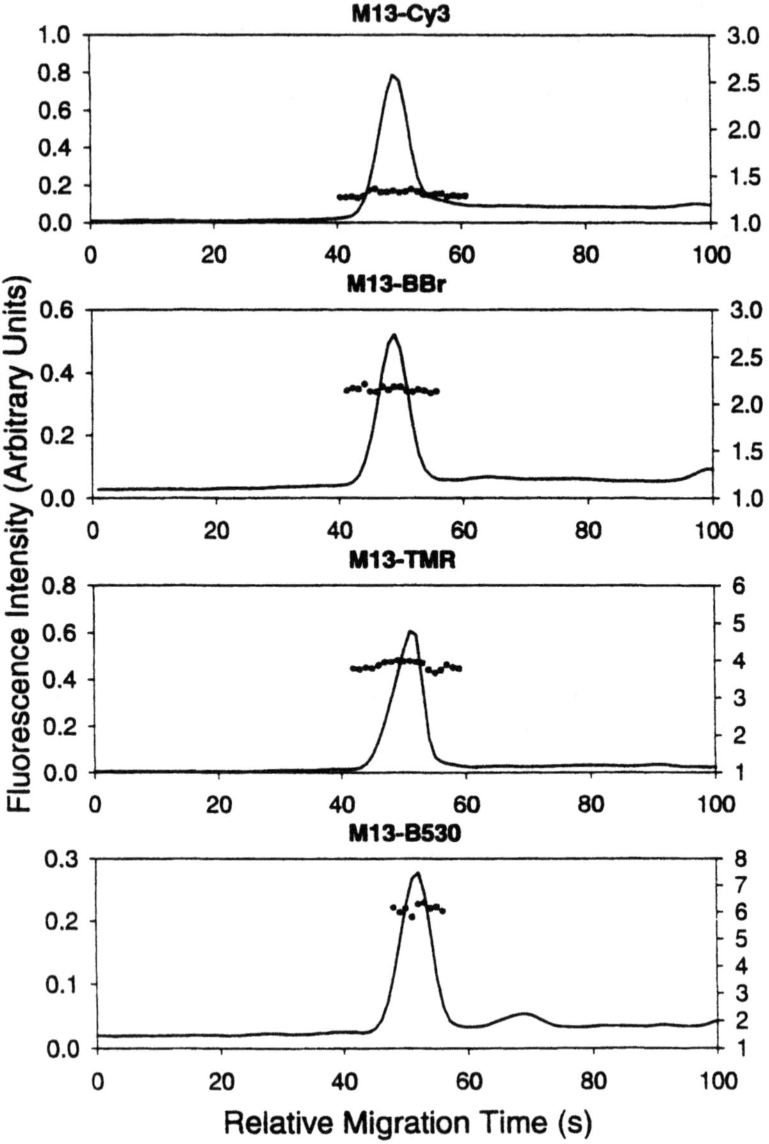

Figure 4.2. Fluorescence intensity (solid line) and lifetime (dotted line) electropherograms of dye-labeled M13 primers.[19]

(d(5′-TGTAAAACGACGGCCAGT-3′) that were labeled with the dyes Cy3, BODIPY-FL-Br_2 (BBR), tetramethylrhodamine (TMR), and BODIPY530/550 (B530) were 1.3 ns, 2.1 ns, 3.7 ns, and 5.0 ns, respectively. Intensity and lifetime electropherograms of the four dyes that were sequentially injected in a single run are shown in Figure 4.3. Nonlinear least squares data analysis was used to recover lifetime profiles at intervals of 1 s or 2 s across the electropherograms.

It has been reported that dyes may exhibit biexponential decay upon tethering to DNA, which is attributed to electron transfer between the dye and guanosine residues in the oligonucleotide. This results in two lifetimes, one longer and one shorter than the lifetime of the free dye in solution.[20,21] Interestingly, biexponential decay has not been observed in frequency-domain lifetime detected CE of dye-labeled primers that did show biexponential decay in batch lifetime measurements. The absence of the dye–guanosine interaction may be due to the application of the strong electric field in the CE experiment, which could overcome the attraction between a positively charged dye such as TMR and the negatively charged oligonucleotide and cause the dye to stretch away from the oligonucleotide and towards the opposite electrode.

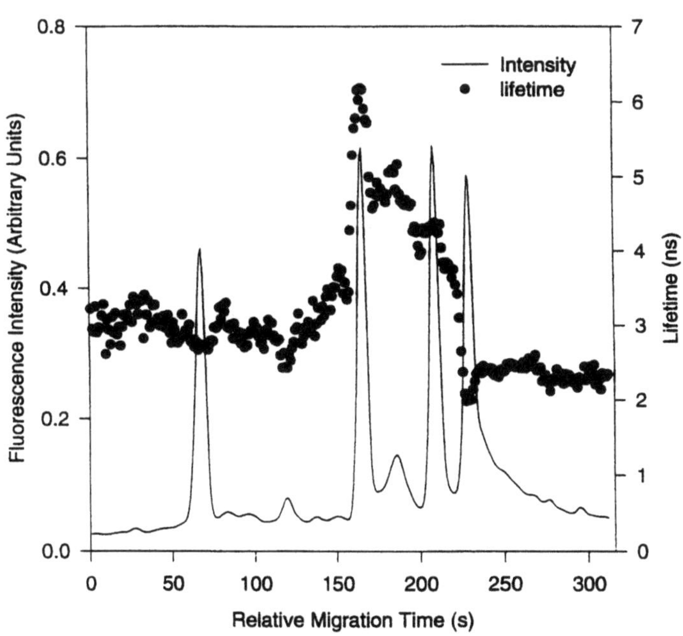

Figure 4.3. Fluorescence intensity (solid line) and lifetime (dotted line) electropherograms of sequential injections of dye-labeled M13 primers. Injection sequence: M13-BBr, M13-B530, M13-TMR, and M13-Cy3.[19]

Two- and three-component mixtures of dye-labeled primers were investigated. For a two-component mixture of M13-BODIPY-FL (M13-BOD) and M13-rhodamine green (M13-RG), five sequential injections were timed to artificially generate four overlapping peaks. The first and the last peaks were pure M13-BOD and M13-RG, respectively. The remaining peaks were resolved from the total intensity electropherogram using NLLS analysis in which the lifetimes were fixed to the lifetimes recovered for the peaks of the individual labeled primers. A third lifetime component was fixed to zero to account for scattered light. Lifetime resolution was able to recover the individual contributions of both labeled primers even when overlap was almost complete (Figure 4.4). The recovered peak-to-height ratio of the two labeled primers in each of the sequential injections was consistent with the ratio for a single injection of the same mixture, for which electrophoretic separation of the components was complete.

A three-component experiment was performed using the same approach as for the two-component system. In this case, five sequential injections were made of a mixture of M13-BOD, M13-NBD-aminohexanoic acid (M13-NBD), M13-TMR, and M13-RG. The order of migration is M13-BOD, M13-NBD, and co-

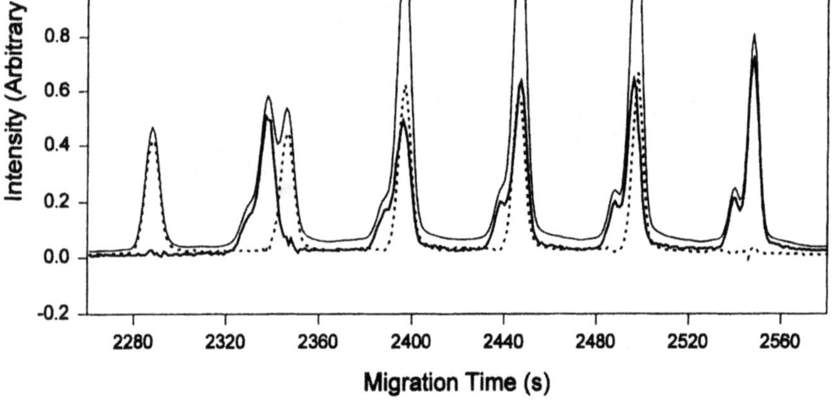

Figure 4.4. Lifetime-resolved intensity peaks of individual dye-labeled M13 primers from sequential injections of a mixture of M13-BOD and M13-RG. Fine solid line: total intensity; dotted line: intensity of M13-BOD; thick solid line: M13-RG.[18]

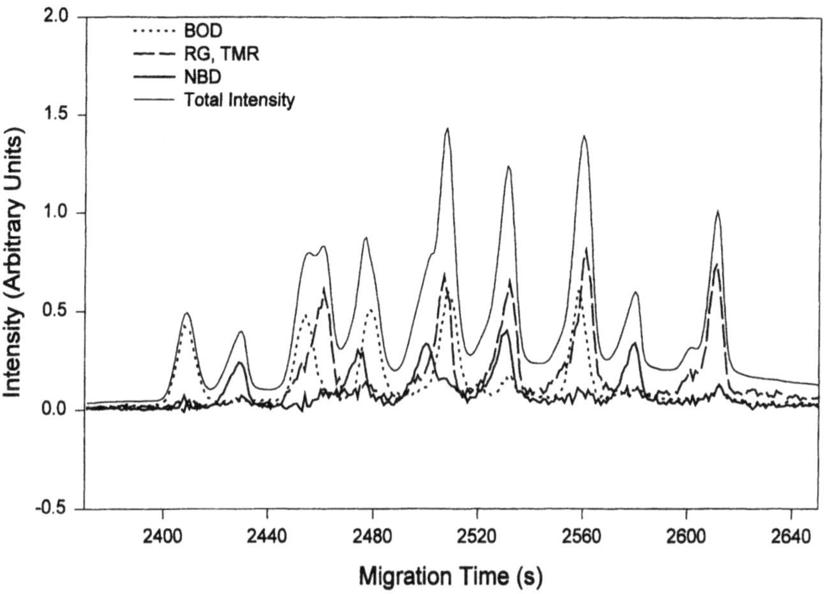

Figure 4.5. Lifetime-resolved intensity peaks of individual dye-labeled M13 primers from five sequential injections of a mixture of M13-BOD, M13-RG, M13-TMR, and M13-NBD. Fine solid line: total intensity; dotted line: intensity of M13-BOD; dashed line: overlapping intensities of M13-RG and M13-TMR; thick solid line: intensity of M13-NBD. M13-TMR and M13-RG labeled primers could not be resolved since their lifetimes were indistinguishable.[18]

migrating M13-RG and M13-TMR. M13-RG and M13-TMR are treated asa single lifetime component in the NLLS analysis because they have unresolvable lifetimes and migrate together. Results are shown in Figure 4.5 for identification and resolution of peaks using NLLS with three fixed lifetime components for the dye-labeled primers and a fourth component fixed to zero to account for scattered light. The fifth electrophoretic peak contains all three lifetime components. These components were successfully resolved by NLLS. Not surprisingly, the noise level is higher than in the two-component experiment.

Run buffer composition can be used to modify fluorescence lifetimes of dye-labeled primers.[22] In an attempt to resolve M13-RG and M13-TMR, 10% DMSO was added to the run buffer. The observed lifetimes of M13-TMR and M13-RG both decreased but to different extents (M13-TMR decreases to 3.5 (± 0.03) ns and M13-RG decreases to 3.3 (± 0.03) ns), allowing them to be distinguished by a separation of more than three standard deviations (Figure 4.6).

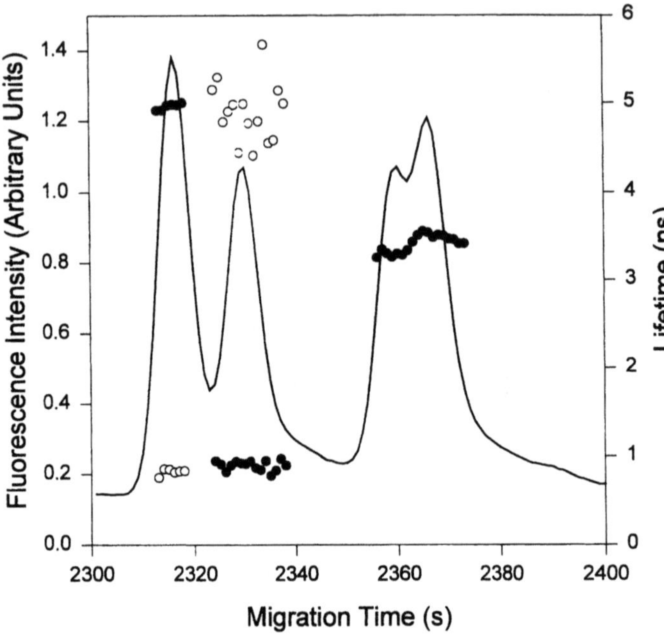

Figure 4.6. Electropherogram of mixture of (in order of migration) M13-BOD, M13-NBD, M13-RG, and M13-TMR using run buffer containing 10% DMSO. Solid circles are lifetimes of major component (fractional contribution >90%) and open circles are lifetimes of minor components, recovered from NLLS fits to a two-component model.[18]

4.4. DNA Sequencing

Fundamental to all DNA sequencing schemes is differentiation among the four bases—adenine (A), guanine (G), cytosine (C) and thymine (T)—that comprise the DNA sequence. Automated sequencers have commonly employed two approaches to this problem for real time detection. In the first, a single label is used in four base-specific reactions and the products of each of these reactions are separated along four different lanes. The sequence is then reconstructed by comparison of the positions of the bands among the four base-specific lanes. This approach offers the simplicity of a single label but is subject to inaccuracies and uncertainties in the ordering of the fragments due to distortions and physical irregularities along the individual lanes that make the necessary comparisons among lanes difficult. The second approach is the multiplex approach, such as in a "four-label, single-lane" scheme, in which each of the four bases is signaled by a unique label and the separation is performed along a single separation lane. Since all of the fragments migrate along a single lane, the ordering uncertainties of the four-lane approach are eliminated.

Radioactive labels were used in the earliest sequencing studies, but the difficulties associated with their use and disposable, as well as their short shelf lives, have encouraged the search for alternatives. Fluorescent labels are the most successful alternative to radioactive labels in sequencing and numerous other applications for several reasons. They are relatively innocuous, stable, and easy to detect. The excellent sensitivity and low detection limits of fluorescence measurements rival those of radioactive measurements. The multidimensional nature of fluorescence allows selective detection in multiplex methods based on absorption and emission characteristics, intensity changes, and fluorescence lifetime. Processes such as energy transfer can be used to further refine or optimize detection schemes.

Single-color, four-lane detection of FAM-labeled primer was used in high-speed separations of DNA sequencing reactions by capillary gel electrophoresis[23] and in capillary array gel electrophoresis using laser-excited confocal fluorescence detection.[24,25] FAM and JOE were similarly employed.[26] Four-color multiplex approaches initially used the dyes fluorescein, Texas Red, TMR and NBD-fluoride.[1,27] NBD-fluoride was later replaced with a more intense fluorescein dye that has a narrower emission band.[2] Other four-dye sets that have been used include four succinylfluorescein dyes,[28] the four Applied Biosystems dyes [FAM, JOE, TAMRA, and x-rhodamine (ROX)], and four rhodamine dyes (rhodamine 110, rhodamine 6G, TAMRA, and ROX) which were used in sequencing by ultrathin slab gel separations.[29]

As discussed above, fluorescence lifetime offers advantages over color for multiplex detection and analysis in CE. These advantages are particularly relevant to the case of DNA sequencing in which an electropherogram is generated from a series of DNA fragments that increase in length by one nucleotide at a time. In concept, the analysis is purely qualitative since only the identity of the terminal base of each fragment, and not the amount of the fragment, need be determined. However, since there are four possible bases and a virtual continuum of CE peaks, accurate base calling requires four different signals that can be readily distinguished from each other and resolved in regions of peak overlap. The discrete, monoexponential nature of fluorescence decay is ideally suited to the task.

Fluorescence lifetime detection in DNA sequencing using time-domain measurements and near-IR dyes has been described.[16] Dye-labeled fragments were identified by either pattern recognition of the decay profile or lifetime recovered from monoexponential maximum likelihood estimator. Since a decay pattern in on-the-fly, time-domain lifetime detection is constructed only from a small number of photon counts, it is difficult to extract multiple lifetimes from the decay pattern even when the actual decay is more complex.[16,17,30–33]

Frequency-domain fluorescence lifetime detection in DNA sequencing, referred to as "four-decay" detection, was recently demonstrated using a set of four dyes that are excited at 488 nm.[34] The dyes include Cy3, fluorescein-dTMR, RG, and BOD. Fluorescein-dTMR is an energy transfer BigDye[35,36] that contains the 5-carboxy isomer of 4′-aminomethylfluorescein as the donor dye and dTMR as the acceptor dye. The dTMR is 4,7-dichloro-substituted TMR that has a narrower emission spectrum than TMR. A rigid amino acid linker, 4-aminomethylbenzoic acid, separates the dyes in the BigDye molecule. Fluorescein-dTMR has high energy transfer efficiency when excited at 488 nm. There was essentially no interference from the donor in four-decay detection. However, mobility correction is required for the fluorescein-dTMR-labeled fragments since the dye is much larger than the three other dyes. This was achieved by using sequential injections of the four sets of dye-labeled sequencing reaction products in which the fluorescein-dTMR-labeled fragments were injected first and allowed to migrate for a short time before sequentially injecting the other dye-labeled fragments. This approach improves peak resolution since the injection time for each dye-labeled fragment is shortened to about one fourth of that for the mixture, and intensity matching among the four dye-labeled fragments could be done by controlling the injection time for each set of fragments. However, since the proportional effect of the dye on mobility decreases as fragment length increases, sequential injection can only compensate for the mobility difference within a certain range of the separation.

Two methods were used for base calling. In the first, bases were identified directly from the lifetimes recovered from one-component NLLS analysis of the lifetime data. The recovered lifetimes were compared to predetermined lifetimes obtained for sequencing products of each individual base under identical conditions. In the second method, lifetime-resolved electropherograms were obtained for each terminal base using NLLS analysis of the on-the-fly lifetime data in which the lifetime components in the fitting models were fixed to predetermined values obtained for sequencing products of each individual base under identical conditions. The recovered fractional intensity of each lifetime component at each point in the electropherogram of the mixture was then multiplied by the total intensity at that point to reconstruct the electropherogram for each of the individual bases in the mixture.

Before attempting the full four-dye system with sequential injection, sequencing of a mixture containing only Cy3-labeled T-terminated fragments, RG-labeled C-terminated fragments, and BOD-labeled G-terminated fragments was performed. The mixture was separated using linear polyacrylamide (LPA) gel. The lifetime results recovered using one-component NLLS analysis are shown in Figure 4.7. Figure 4.8 shows an expanded view from 2800 s to 3300 s. The base calling of T, C, and G was based on the lifetimes of the peaks recovered from the one-component fit. The Cy3-labeled T-terminated fragments have lifetimes of 0.9 ns, the RG-labeled C-terminated fragments have lifetimes of 2.3 ns, and BOD-

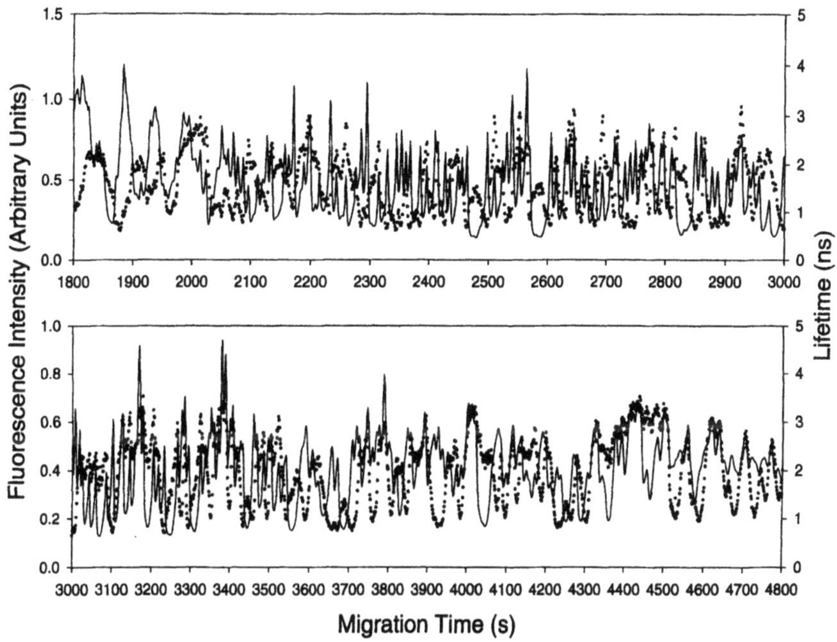

Figure 4.7. Fluorescence lifetime (dotted line) and intensity (solid line) electropherograms of a mixture of Cy3-labeled T-terminated fragments, RG-labeled C-terminated fragments, and BOD-labeled G-terminated fragments.[34]

labeled G-terminated fragments have lifetimes of 3.3 ns. From 51 to 300 bases, 19 base calling errors were found, equivalent to an accuracy of 90%. The gel gave poor separation for DNA fragments shorter than 50 bases and longer than 300 bases.

Base calling was also performed using the lifetime-resolved intensity electropherograms for each of the bases. Data smoothing was performed to remove high frequency noise from the lifetime analysis. The use of resolved peaks extended the base calling range to 320 bases. Base calling from the smoothed electropherograms (shown in Figure 4.9) gave only 3 errors, equivalent to an accuracy of 98.5%. The improved read length and accuracy are due to improved base calling for severely overlapping peaks.

A full, four-component sequencing run was then performed using sequential injections of A (fluorescein-dTMR), C (RG), T (Cy3), and G (BOD) fragments onto a 5% linear polyacrylamide gel-packed column. Results are shown in Figure 4.10. Figure 4.11 shows an expanded view of the results from 3500 s to 3900 s. Mobility correction was successful within this range of fragment length. Cy3-labeled T-terminated fragments have lifetimes of 1.7 ns, fluorescein-dTMR-

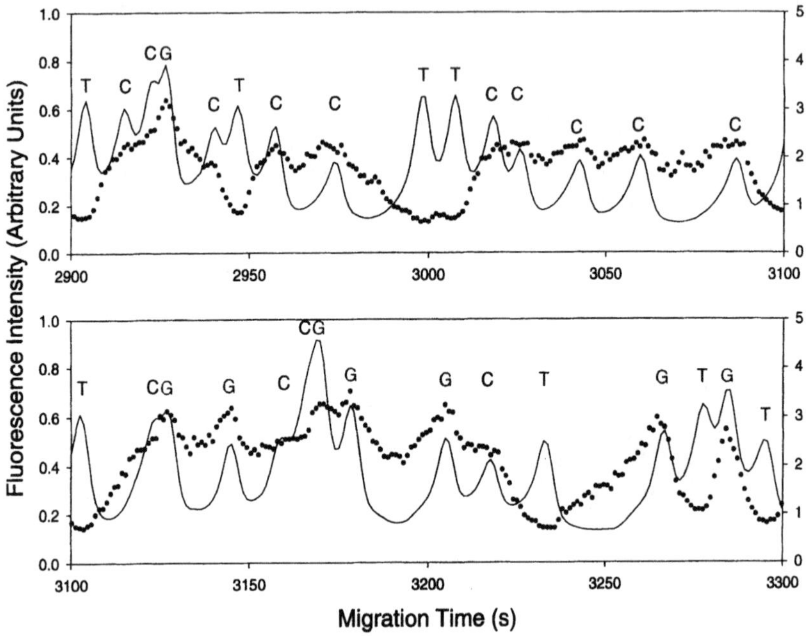

Figure 4.8. Expanded scale of a section of the electropherograms in Figure 4.7.[34]

labeled A-terminated fragments have lifetimes of 2.5 ns, RG-labeled C-terminated fragments have lifetimes of 2.9 ns, and BOD-labeled G-terminated fragments have lifetimes of 3.5 ns.

Base calling from 41 to 220 bases was based directly on lifetimes of the peaks. Eight errors occurred, equivalent to an accuracy of 96%. Base calling using lifetime-resolved peaks was unsuccessful due to difficulties in four-component NLLS analysis, even when the lifetimes were fixed to predetermined values. Improved approaches are needed for data analysis of four-component systems, which is generally problematic for any real system with noise and background.

Restriction Fragment Length Polymorphism Analysis

Another DNA application to which on-the-fly fluorescence lifetime detection has been applied is restriction fragment length polymorphism (RFLP).[37] In RFLP, DNA is cut by restriction enzymes that recognize certain base sequences in the DNA and clip it near the recognition site. When DNA from different sources is cut using the same restriction enzyme, DNA from each source will be "digested" into

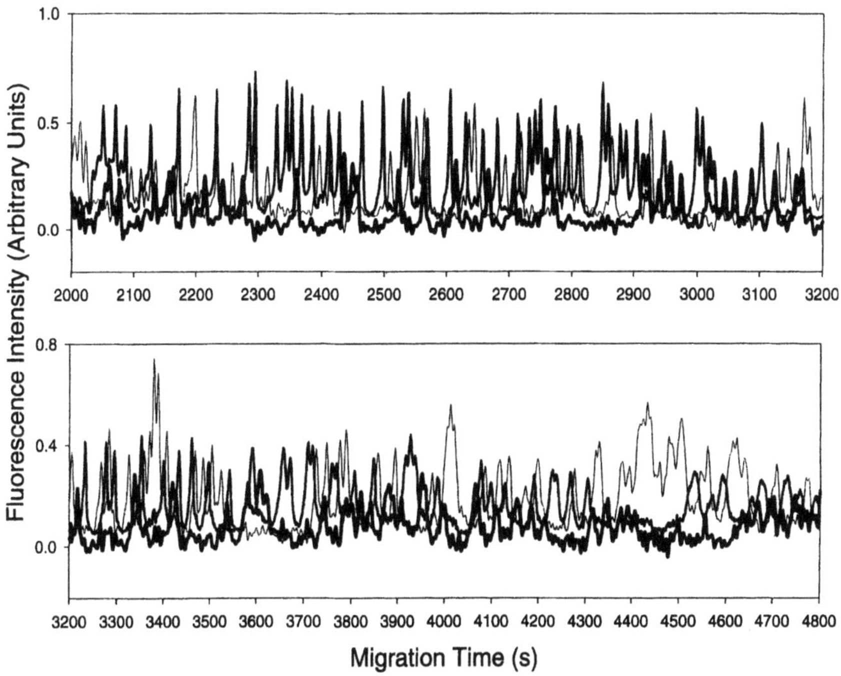

Figure 4.9. Lifetime-resolved intensity peaks of individual dye-labeled DNA sequencing fragments from a mixture of Cy3-labeled T-terminated fragments (dark gray), RG-labeled C-terminated fragments (black), and BOD-labeled G-terminated fragments (light gray).[34]

a series of fragments. The fragment pattern of the DNA depends upon the number of bases between recognition sites and is therefore a unique characteristic, or fingerprint, of DNA from that source.

Multiplex detection allows a dye-labeled digest and a dye-labeled ladder (size standard) to be simultaneously run in the same capillary. The peaks of different origin are distinguished using the fluorescence lifetimes of intercalating dyes that bind by inserting between the base pairs in the fragments. Intercalation is a simple alternative to the more labor intensive process of covalent attachment for labeling DNA fragments.

Fluorescent intercalating dyes have been used as fluorescent indicators of DNA fragments in CE.[38–40] For monomeric intercalating dyes such as TO and YO, it often is necessary to include the dye in the run buffer in order to obtain adequate signals due to the relative instability of the dye-DNA complex under electrophoretic conditions.[39] Unfortunately, free dye in solution may contribute significant background signal despite its low quantum yield. Furthermore, addition of dyes to the run buffer precludes multiplex analysis since it would not be possible to

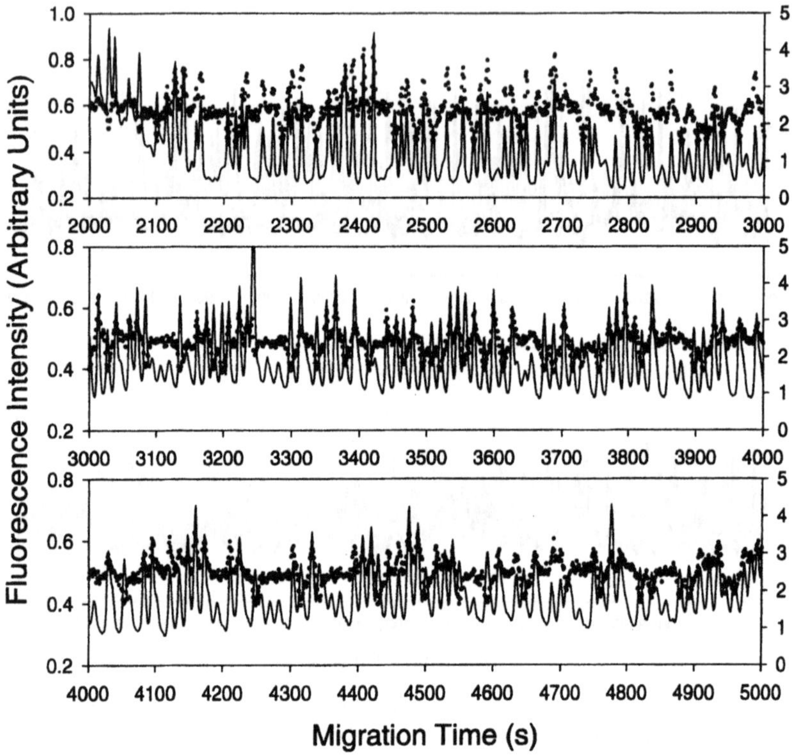

Figure 4.10. Fluorescence lifetime (dotted line) and intensity (solid line) electropherograms of sequential injections of A (fluorescein-dTMR), C (RG), T (Cy3), and G (BOD) fragments.[34]

selectively label DNA fragments from different sources. DNA complexes formed with some dimeric dyes such as YOYO and TOTO are stable enough to allow detection without addition of dye to the run buffer; however, multiplex analysis can still be difficult due to exchange of dyes between fragments.[38,41]

The dye-DNA complexes formed by the monomeric intercalating dyes that were used for fluorescence lifetime detection are sufficiently stable to provide measurable signals without adding dye to the run buffer. The dyes, including oxazole yellow (YO) and three novel intercalating dyes (referred to as Dye 1, Dye 2, and Dye 4),[42–44] are shown in Figure 4.12. It was also necessary that dye exchange among DNA fragments be slow enough to allow multiplex analysis during a CE run. As shown in Figure 4.13, good separation and lifetime detection were obtained for individual separations of Dye 1-labeled pBR322 DNA-BstNI digest fragments and of a YO-labeled 100 bp ladder using modified polyacrylamide gel. Attempts at multiplex detection of a mixture of the digest fragments

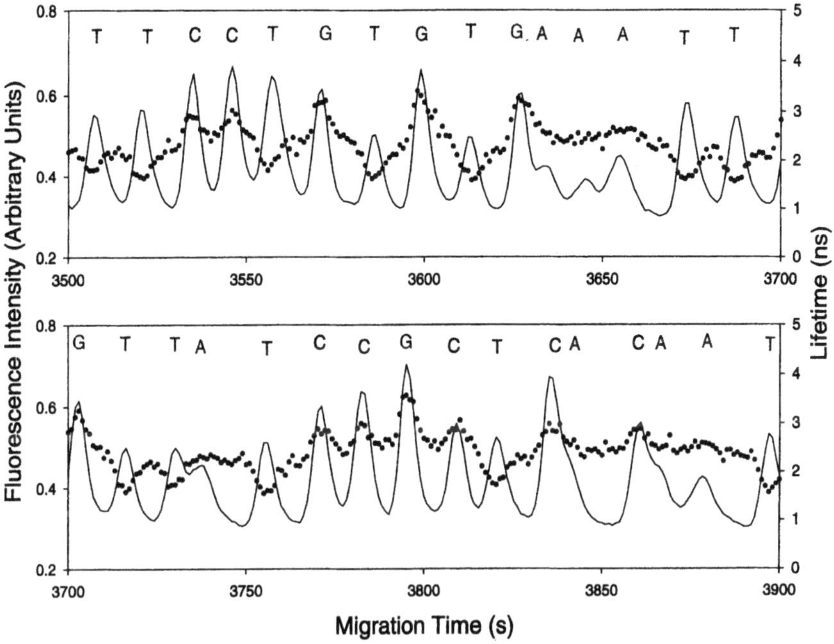

Figure 4.11. Expanded scale of one section of the electropherograms in Figure 4.10.[34]

and the ladder, however, indicated that one of the dyes was displacing the other from the DNA in the mixtures. Fluorescence emission spectra of digest mixtures in batch mode confirmed the occurrence of dye displacement.

In order to slow the rate of exchange of dyes between DNA fragments, hydroxyethylcellulose (HEC) was used instead of modified polyacrylamide gel as the sieving matrix. Both high molecular weight (HMW, 90,000–105,000) and low molecular weight (LMW, 24,000–27,000)) HEC gels were investigated. Results for separation of a dye-labeled 100 bp DNA ladder using four different gel buffers showed that the TBE/1% HMW HEC/0.3% LMW HEC provided the best peak resolution. This gel buffer was then used for multiplex detection of Dye 4-labeled pBR322DNA-BstNI digest and a reference fragment (Dye 2-labeled 200 bp). Results are shown in Figure 4.14. The broad peak near 800 s (peak 1) is due to unresolved fragments from the BstNI digest. Peaks 2, 4, 5, 6, and 7 are from the digest, and peak 3 is from the Dye 2-labeled 200 bp fragment.

In Figure 4.14, only five contiguous lifetime points centered at the peak apex are shown for each peak, in order to focus on the data that would be used for peak identification. Due to the relatively high background contribution in this run, a two-component model in which both lifetime components were allowed to vary in

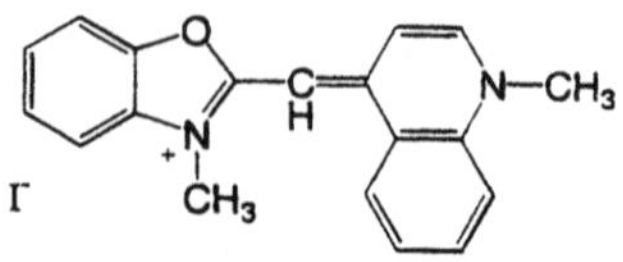

YO

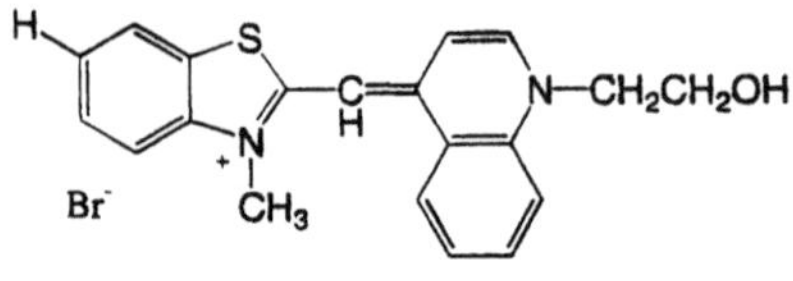

Dye 1

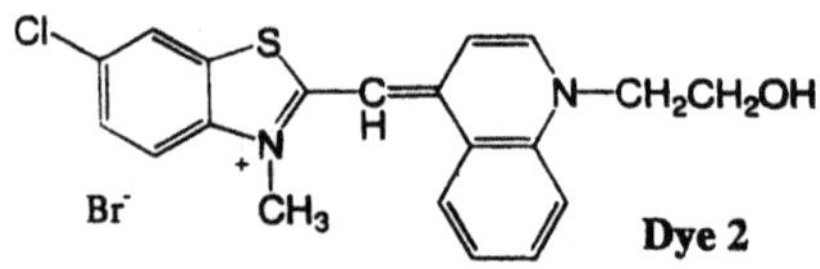

Dye 2

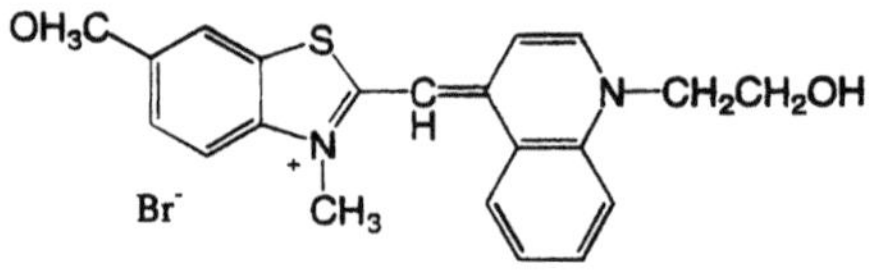

Dye 4

Figure 4.12. Structures of intercalating dyes.

order to account for the background signal gave better results than a one-component model. The recovered lifetimes of the peaks are the expected values (1.0 for the digest peaks and 0.5 ns for the 200 bp fragment peak) and show no indication of dye exchange between fragments.

4.5. Conclusions

The studies described in this chapter illustrate potential applications of on-the-fly fluorescence lifetime detection in CE for DNA analysis. The combined advantages of speed, high resolution, and small sample volume provided by CE and the accuracy and signal resolution provided by fluorescence lifetime in no more time than is needed for fluorescence intensity detection alone make frequency-domain lifetime detection a promising approach to problems in DNA analysis. Development of more dyes that offer a wider range of lifetimes and reduction in background contributions to the fluorescence signal will help to fully realize the potential of this approach.

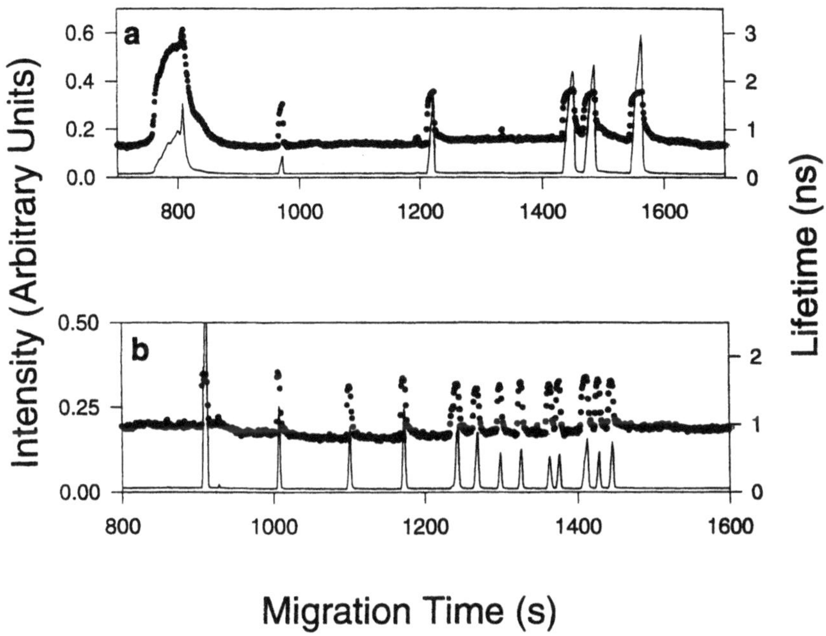

Figure 4.13. Fluorescence lifetime (dotted line) and intensity (solid line) electropherograms of DNA digests: (a) Dye 1-labeled pBR322DNA-BstNI (1 dye per 95 bp); (b) YO-labeled 100 bp DNA ladder (1 dye per 950 bp).[37]

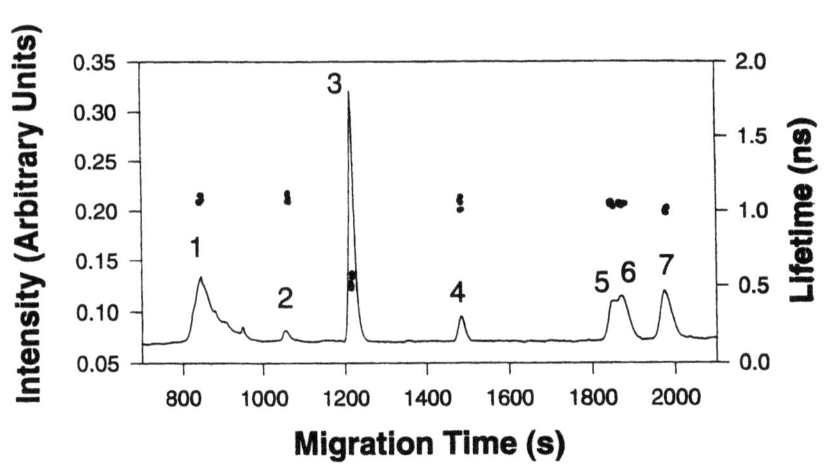

Figure 4.14. Electropherogram of mixture of Dye 4-labeled pBR322DNA-BstNI digest (1 dye per 8 bp) and Dye 2-labeled 200 bp DNA fragment (1 dye per 79 bp). Peaks 1, 2, 4, 5, 6, and 7 are from the pBR322DNA-BstNI digest and peak 3 is from the 200 bp fragment. Solid line is fluorescence intensity. Five contiguous lifetime points (solid circles) centered at the peak apex are shown for each peak.[37]

References

1. Smith, L. M., Fung, S., Hunkapiller, M. W., Hunkapiller, T. J., and Hood, L. E. *Nucleic Acids Res. 13*, 2399, 1986.
2. Connell, C., Fung, S., Heiner, C., Bridgham, J., Chakerian, V., Heron, E., Jones, B., Menchen, S., Mordan, W., Raff, M., Recknor, M., Smith, L. M., Springer, J., Woo, S., and Hunkapiller, M. W. *BioTechniques 5*, 342, 1987.
3. Smith, L. M. *Genome 31*, 929, 1989.
4. Lee, L. G., Spergeon, S. L., Heiner, C. R., Benson, S. C., Rosenblum, B. B., Menchen, S. M., Graham, R. J., Constantinescu, A., Upadhya, K. G., and Cassel, J. M. *Nucleic Acid Res. 25*, 2816, 1997.
5. Metzker, M. L., Lu, J., and Gibbs, R. A. *Science 271*, 1420, 1996.
6. Seeger, S., Bahteler, G., Drexhage, K. H., Arden-Jacob, J., Deltau, G., Galla, K., Han, K. T., Mueller, R., Keollner, R. A., Sauer, M., Schulz, A., and Wolfrum, J. *Ber. Bunsenges. Phys. Chem. 97*, 1542, 1993.
7. J. R. Lakowicz. *Principles of Fluorescence Spectroscopy, 2nd Ed.*, New York: Plenum, 1999.
8. Cobb, W. T., and McGown, L. B. *Anal. Chem. 62*, 186, 1990.
9. Cobb, W. T. Ph.D. dissertation, Oklahoma State University, Stillwater, OK, 1989.
10. Smalley, M. B., Shaver, J. M., and McGown, L. B., *Anal. Chem. 65*, 3466, 1993.
11. Smalley, M. B., and McGown, L. B., *Anal. Chem. 67*, 1371, 1995.
12. Mitchell, G., and Swift, K. In *Time-Resolved Laser Spectroscopy in Biochemistry II*, Lakowicz, J. R., ed. SPIE Vol. 1204; Bellingham, WA: SPIE—The International Society for Optical Engineering, 1990, Part 1, pp. 270–274.
13. Spencer, R. D., and Weber, G. *Ann. NY Acad. Sci. 158*, 361, 1969.
14. Li, L., and McGown, L. B. *Anal. Chem. 68*, 2737, 1996.
15. Li, L., and McGown, L. B., *Electrophoresis 21*, 1300, 2000.
16. Desilets, D. J., Kissinger, P. T., and Lytle, F. E. *Anal. Chem. 59*, 1830, 1987.
17. Soper, S. A., LeGendre, B. L., Jr., and Williams, D. C. *Anal. Chem. 67*, 4358, 1995.
18. He, H., Nunnally, B. K., Li, L., and McGown, L. B. *Anal. Chem. 70*, 3413, 1998.
19. He, H., Dissertation, Duke University, Durham, NC, 2000.
20. Edman, L., Mets, U., and Rigler, R. *Proc. Natl. Acad. Sci. USA 93*, 6710, 1996.
21. Vamosi, G., Gohlke, C., and Clegg, R. M. *Biophys. J. 71*, 972, 1996.
22. Li, L., and McGown, L. B. *J. Chromatogr. A 841*, 95, 1999.
23. Drossman, H., Luckey, J. A., Kostichka, A. J., D'Cunha, J., and Smith, L. M., *Anal. Chem. 62*, 900, 1990.
24. Huang, X. C., Quesada, M. A., and Mathies, R. A. *Anal. Chem. 64*, 967, 1992.
25. Huang, X. C., Quesada, M. A., and Mathies, R. A. *Anal. Chem. 64*, 2149, 1992.
26. Ruiz-Martinez, M. C., Berka, J., Belenkii, A., Foret, F., Miller, A. W., and Karger, B. L. *Anal. Chem. 65*, 2851, 1993.
27. Smith, L. M., Sander, J. Z., Kaiser, R. J., Hughes, P., Dodd, C., Connell, C. R., Heiner, C., Kent, S. B. H., and Hood, L. E. *Nature 321*, 674, 1986.
28. Prober, J. M., Trainor, G. L., Dam, R. L., Hobbsk F. W., Robertson, C. W., Zagursky, R. J., Cocuzza, A. J., Jensen, M. A., and Baumeister, K. *Science 238*, 336, 1987.
29. Kostichka, A. J., Marchbanks, M. L., Brumley Jr., R. L., Drossman, H., and Smith, L. M. *Bio/ Technology 10*, 78, 1992.
30. Bachteler, G., Drexhage, K.-H., Ardern-Jacob, J., Han, K.-T., Kollner, M., Muller, R., Sauer, M., Seeger, S., and Wolfrum, J. *J. Lumin. 60–61*, 511, 1994.
31. Bachteler, G., Drexhage, K.-H., Ardern-Jacob, J., Han, K.-T., Kollner, M., Muller, R., Sauer, M., Seeger, S., and Wolfrum, J. *J. Lumin. 62*, 101, 1994.

32. Flanagan, J. H., Jr., Owens, C. V., Romero, S. E., Waddell, E., Kajn, S. H., Hammer, R. P., and Soper, S. A. *Anal. Chem. 70*, 2676, 1998.
33. Lieberwirth, U., Arden-Jacob, J., Drexhage, K. H., Herten, D. P., Muller, R., Neumann, M., Schulz, A., Siebert, S., Sagner, G., Klingel, S., Sauer, M., and Wolfrum, J. *Anal. Chem. 70*, 4771, 1998.
34. He, H., and McGown, L. B. *Anal. Chem. 72*, 5865, 2000.
35. Lee, L. G., Spergeon, S. L., Heiner, C. R., Benson, S. C., Rosenblum, B. B., Menchen, S. M., Graham, R. J., Constantinescu, A., Upadhya, K. G., and Cassel, J. M. *Nucleic Acids Res. 25*, 2816, 1997.
36. Rosenblum, B. B., Lee, L. G., Spurgeon, S. L., Khan, S. H., Menchen, S. M., Heiner, C. R., and Chen, S. M. *Nucleic Acids Res. 25*, 4500, 1997.
37. McIntosh, S. L., Nunnally, B. K., Nesbit, A. R., Deligeorgiev, T. G., Gadjev, N. I., and McGown, L. B. *Anal. Chem. 72*, 5444, 2000.
38. Glazer, A. N., and Rye, H. S. *Nature 359*, 859, 1992.
39. Zhu, H., Clark, S. M., Benson, S. C., Rye, H. S., Glazer, A. N., and Mathies, R. A. *Anal. Chem. 66*, 1941, 1994.
40. Carlsson, C., Jonsson, M., and Akerman, B. *Nucleic Acids Res. 23*, 2413, 1995.
41. Clark, S. M., and Mathies, R. A. *Anal. Chem. 69*, 1355, 1997.
42. Deligeorgiev, T., Gadjev, N., Drexhage, K.-H., and Sabnis, R. *Dyes and Pigments 29*, 315, 1995.
43. Deligeorgiev, T. G., Zaneva, D. A., Katerinopoulos, H. E., and Kolev, V. N. *Dyes and Pigments 41*, 49, 1999.
44. Deligeorgiev, T. G., Zaneva, D. A., Kim, S. H., and Sabnis, R. W. *Dyes and Pigments 37*, 205, 1998.

5

Fluorescent Nucleoside Analogues as DNA Probes

Mary E. Hawkins

5.1. Introduction

The availability of a large selection of fluorescent probes has revolutionized our approaches to studying biochemical reactions and interactions. These probes have become so commonly used and important to many areas of research that it is hard to imagine working without them. This chapter will focus on a more recently developed subclass of DNA probes, nucleoside analogues, which are different from conventional probes in ways that provide opportunities for much more direct approaches in DNA research. In particular, the focus will be on these analogues used as probes incorporated into DNA. The source of their behavioral differences may not be apparent at first glance. An understanding of the differences between the carbon linker type fluorophores and the nucleoside analogues, and their physical relationships within the DNA will make this clear.

The value and usefulness of the rich variety of fluorescent probes currently available is indisputable. They are, however, poorly suited for the type of experiments defined in this chapter. Because the majority of fluorophores are larger than and structurally dissimilar to purines or pyrimidines, they must be placed on a linker arm at some distance from the site of interest on the DNA. While this placement allows them to be used without perturbing the system, in many cases it removes them from the subtle interactions which are often the target of such investigations.

The nucleoside analogues discussed within this chapter include only those probes that incorporate into an oligonucleotide through a deoxyribose linkage, are formulated as phosphoramidites to allow site-selective insertion, and participate in base-stacking (and sometimes base-pairing) interactions within the DNA. It is the base-stacking and base-pairing features that allow us to monitor subtle changes in DNA structure, binding, and composition. Probes that fit these require-

Mary E. Hawkins • Pharmacology and Experimental Therapeutics Section, National Cancer Institute, Bethesda, Maryland 20892.

Topics in Fluorescence Spectroscopy, Volume 7: DNA Technology, edited by Joseph R. Lakowicz, Kluwer Academic / Plenum Publishers, New York, 2003

ments exhibit dramatic fluorescence changes in part due to base-stacking or base-pairing interactions. It has been demonstrated experimentally that subtle changes that disrupt base stacking can clearly be monitored through changes in fluorescence intensity.[1–4] In many cases, the products of a reaction do not need to be separated prior to analysis, a definite advantage over tagging methods.

In monitoring binding through anisotropy measurements, these analogues display relatively little movement not associated with the motion of the DNA.[5] In the case of conventionally attached probes, flexibility of linker arms allows movement of the fluorophore in ways that are independent of the movement of the DNA being studied. This higher level of independent motion leads to more complex results.

The ideal is to have nucleoside analogues that are highly fluorescent and structurally as similar as possible to native nucleosides. It is preferable to have a probe with fluorescence properties, such as excitation and emission maxima, that differ from that of native DNA. Additionally, the probe must be stable, soluble, and fluorescent in an aqueous environment and at pH levels tolerated by enzyme activities and DNA interactions. The ability to place the probe in an oligonucleotide site-specifically using an automated DNA synthesizer is also very helpful.

A new group of nucleoside analogues based on pteridine structures is currently under development which will add to existing fluorescent nucleoside analogues including 2-aminopurine and ethenoadenosine phosphoramidites. The range of structure and fluorescence characteristics encompassed by these probes will help to expand the different types of research that can be done utilizing them. Examples are provided within this chapter detailing applications of fluorescent nucleoside analogues using the pteridine probes. A brief discussion of 2-AP and ethenoadenosine is also included. Some probes being offered commercially as fluorescent nucleoside analogues will not be included, since in many cases the fluorescent moiety is not positioned in the same relationship within the oligonucleotide.

This chapter will define and give examples of the various ways that the subtle relationship between the nucleoside analogue probes and the DNA can be used to observe minute changes within the DNA, often in a real time format. Although some information will be given about fluorescent analogues other than the pteridines, this is by no means a complete listing or a comprehensive review of the subject.

5.2. Pteridine Nucleoside Analogues

5.2.1. Background

Pteridines are a class of bicyclic planar compounds, some of which are highly fluorescent and structurally similar to purines. Initially 18 pteridine nucleosides

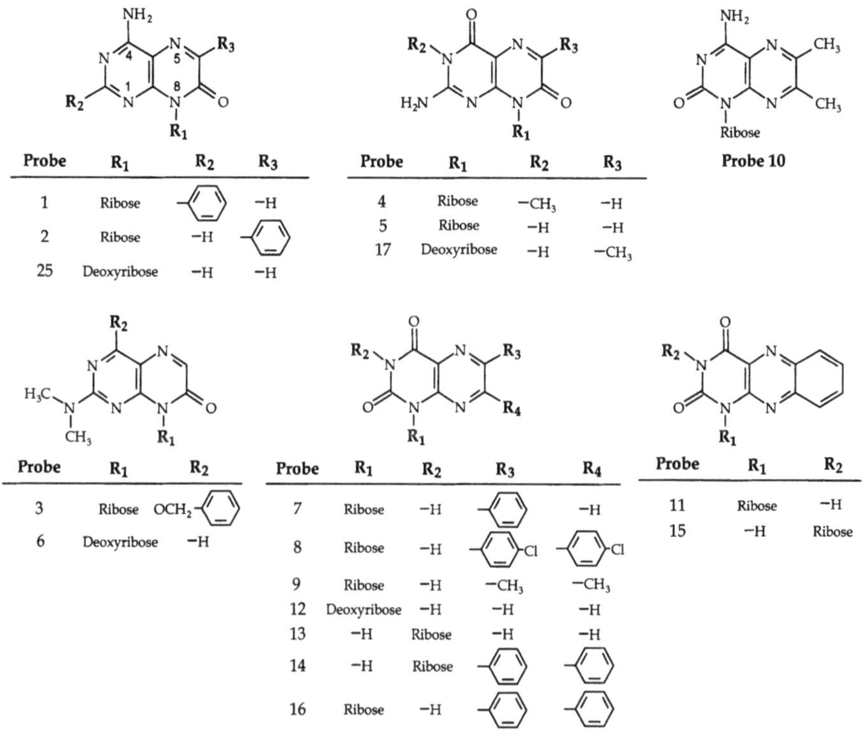

Figure 5.1. Chemical structures of the 18 pteridine nucleoside analogues. Compounds are grouped according to similarities in their chemical structures. The 7 pteridine compounds in the top row were selected for more detailed study based on their favorable Q_{rel}. (Reprinted with permission from reference 6.)

were screened in a search for potential nucleic acid analogues with the necessary characteristics[6] (Figure 5.1). The pteridines in the lower half of Figure 5.1 were ruled out for further studies since they were not fluorescent enough, not stable, not soluble in aqueous media, or bore no resemblance to native bases.

The pteridines in the top portion of Figure 5.1 are organized according to structural features. Probes 1, 2, and 25 are similar in structure to adenosine while probes 4, 5, and 17 are similar to guanosine. Of the seven that were chosen for further studies, probes 5 and 25 have proven to be quite unstable, while probes 1 and 2 are less favored because of structural disadvantages.

Table 5.1 lists fluorescence properties for the compounds shown in the top portion of Figure 5.1. 3-MI (probe 4) and 6-MI (probe 17) exhibit similar Q_{rel} (0.88–0.70) and mean lifetimes (τ_m ranges 5.63–6.54 ns).

At this time two of the four compounds actively being investigated are adenosine analogues, 6MAP and DMAP, which have been developed since the initial screening, and two are guanosine analogues, 3-MI and 6-MI, probes 4 and

Table 5.1. Fluorescence Properties of Pteridine Nucleoside Analogues[a]

Probe	Ex_{max}	Em_{max}	Q_{rel}	τ_i (ns)	α_i	$\%I_i$	τ_m	$\langle\tau\rangle$
1	354	444	0.41	$\tau_1 = 1.91$	$\alpha_1 = 0.19$	$I_1 = 9.9$	3.84	3.64
				$\tau_2 = 4.05$	$\alpha_2 = 0.81$	$I_2 = 90.9$		
2	358	440	0.16	$\tau_1 = 0.76$	$\alpha_1 = 0.67$	$I_1 = 59.6$	0.87	0.85
				$\tau_2 = 1.05$	$\alpha_2 = 0.33$	$I_2 = 40.4$		
25	334	443	0.27	$\tau_1 = 2.37$	$\alpha_1 = 0.19$	$I_1 = 11.6$	3.92	3.80
				$\tau_2 = 4.13$	$\alpha_2 = 0.81$	$I_2 = 88.4$		
4	348	430	0.88	$\tau_1 = 3.54$	$\alpha_1 = 0.02$	$I_1 = 1.2$	6.54	6.51
				$\tau_2 = 6.58$	$\alpha_2 = 0.98$	$I_2 = 98.8$		
5	348	430	0.87	$\tau_1 = 1.81$	$\alpha_1 = 0.37$	$I_1 = 14.6$	5.63	4.65
				$\tau_2 = 6.26$	$\alpha_2 = 0.63$	$I_2 = 85.4$		
17	340	431	0.70	$\tau_1 = 5.45$	$\alpha_1 = 0.20$	$I_1 = 17.5$	6.38	6.35
				$\tau_2 = 6.58$	$\alpha_2 = 0.80$	$I_2 = 82.5$		
10	336	400	0.54	$\tau_1 = 3.16$	$\alpha_1 = 0.97$	$I_1 = 92.8$	3.52	3.31
				$\tau_2 = 8.14$	$\alpha_2 = 0.03$	$I_2 = 7.2$		

[a] τ_i, lifetime for each component of a multiexponential model; α_i, preexponential for each component of a multiexponential model; $\%I_i$, percent fluorescence intensity for each component of a multiexponential model; $\langle\tau\rangle$, species-concentration-weighted lifetime; τ_m, intensity-weighted lifetime. (Reprinted with permission from reference 42.)

17 of the original set (Figure 5.2). 6MAP and DMAP are similar to probes 1 and 2 except that the phenyl groups are replaced by methyl groups.[7] The four compounds shown in Figure 5.2 are highly fluorescent, Q_{rel} of these ranging from 0.39 to 0.88. The adenosine analogues, 6MAP and DMAP are at the beginning of their development, but in initial studies they appear be quite similar to the guanosine analogues with respect to stability and quenching effects. Q_{rel} for 6MAP and DMAP are 0.39 and 0.48, respectively. 6MAP appears to participate in hydrogen bonding with thymidine as evidenced by melting temperatures that are very similar to those of controls. DMAP is currently under investigation.

The overlaid absorption spectra, corrected excitation spectra, and emission spectra of 3-MI and 6-MI are shown in Figure 5.3.

5.2.1.1. Effect of pH on Fluorescence Emission

Emission spectra of 3-MI and 6-MI over a pH range from 5.0 to 8.0 are shown in Figure 5.4. Varying the pH had little effect on the emission spectrum of 3-MI, whereas the emission spectrum of 6-MI shifted 10 nm to the red when the pH was increased from 7.0 to 8.0.

Similar pH titrations of DAS on 3-MI and 6-MI were performed. The lifetime components for 3-MI remained unchanged over the pH range from 7.0 to 9.0,

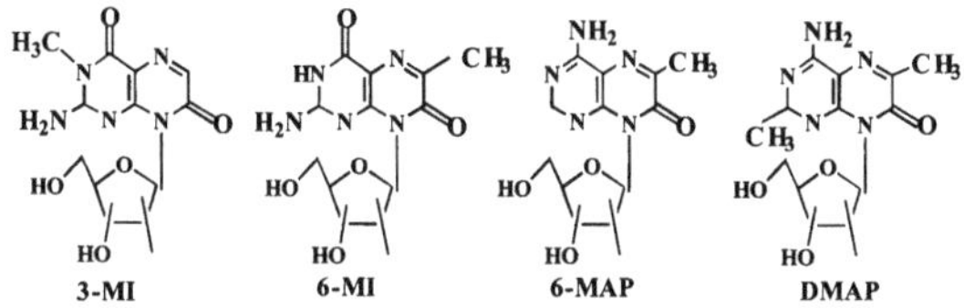

Figure 5.2. Guanosine and adenosine analogues currently under development. (Reprinted with permission from reference 42.)

while with 6-MI there is a blue shift in the longest lived DAS and a smaller increase in the shorter lived component. These DAS demonstrate a pH-dependent equilibrium between a minimum of two emitting species.

5.2.1.2. Fluorescence of Pteridine Analogue-Containing Oligonucleotides

3-MI and 6-MI were incorporated into the oligonucleotide strands shown in Table 5.2 and examined in single-and double-stranded forms (paired to cytidine in the complementary strand). Relative quantum yield (Q_{rel}) measurements of 3-MI-containing oligonucleotides (21-mers) in which the fluorophore is substituted for guanosine at several positions are also listed in Table 5.2.

In PTER8 the fluorophores are incorporated into a purine-rich segment of the oligonucleotide (Tables 5.2 and 5.3) and fluorescence of both probes is substantially quenched, whereas in PTER9 the fluorophores are surrounded by pyrimidines resulting in less quenching of the fluorescence signal. These were found to be the two extreme conditions for neighboring base quenching of the fluorophores.

With 3-MI, double strand formation led to less additional quench over that seen in the single-stranded oligonucleotide. In the PTER9 oligonucleotide, degree of quench relative to unincorporated 3-MI was 64% and 68% in the single- and double-stranded forms, respectively. For 6-MI, the degree of quench in PTER9 increased from 56% to 64% on going from the single- to the double-stranded form. The emission spectrum of 6-MI shifts 7 nm to the red when it is incorporated into a single- or double-stranded oligonucleotide compared to its unincorporated monomer form. The emission spectrum of 3-MI shifts only 2 nm when incorporated into an oligonucleotide.

The lifetimes, mean lifetimes, and amplitudes for 3-MI and 6-MI incorporated into single-stranded and double-stranded oligonucleotides are also listed in Table 5.3. The dominant component of the intensity decay curves becomes more complex and components of the decay curve with shorter lifetimes becomes more prominent. For example, 3-MI in PTER8 has a tri-exponential decay curve with the predominant component having a lifetime of 0.21 ns.

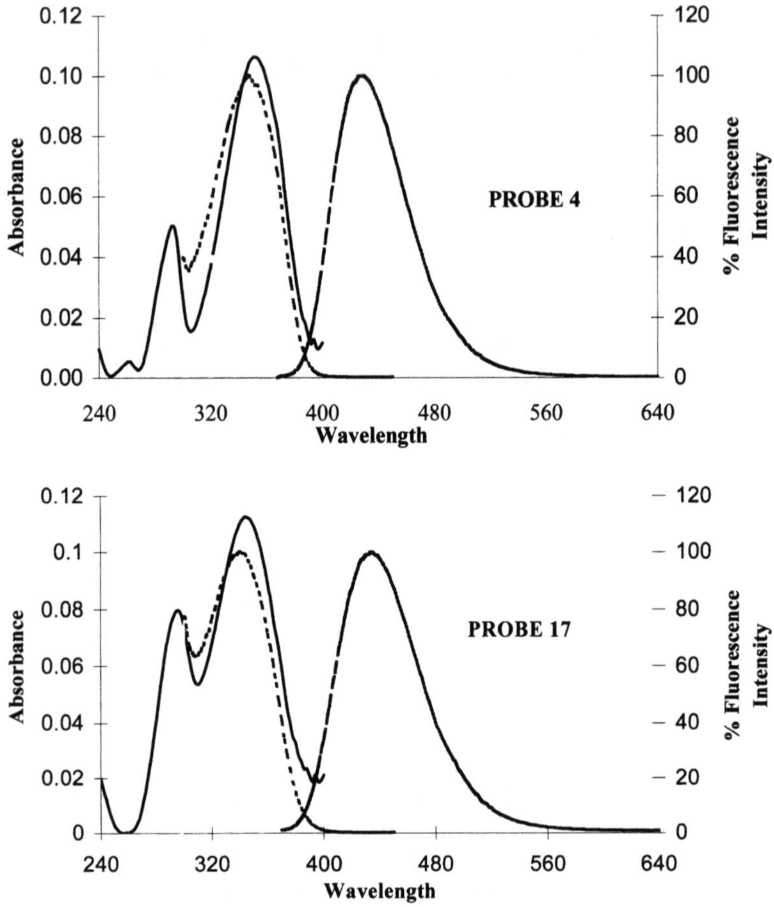

Figure 5.3. Overlay of absorption spectra, corrected excitation spectra, and emission spectra for probes 4 (3-MI) and 17 (6-MI). Excitation and emission are expressed as a percent of maximum. Solid line is absorption, dashed line is excitation, and dotted line is emission. (Adapted with permission from reference 6.)

5.2.1.3. Melting Temperatures

Melting temperatures (T_m) were measured on a series of probe-containing oligonucleotides as well as on the identical strands containing no probe and strands containing a single-base pair mismatch at the identical position to the fluorophore (Table 5.4). For 3-MI-containing oligonucleotides, Tm depression was approximately equivalent to a single base-pair mismatch in the same position.

These probes are quite stable and tolerate normal handling through synthesis

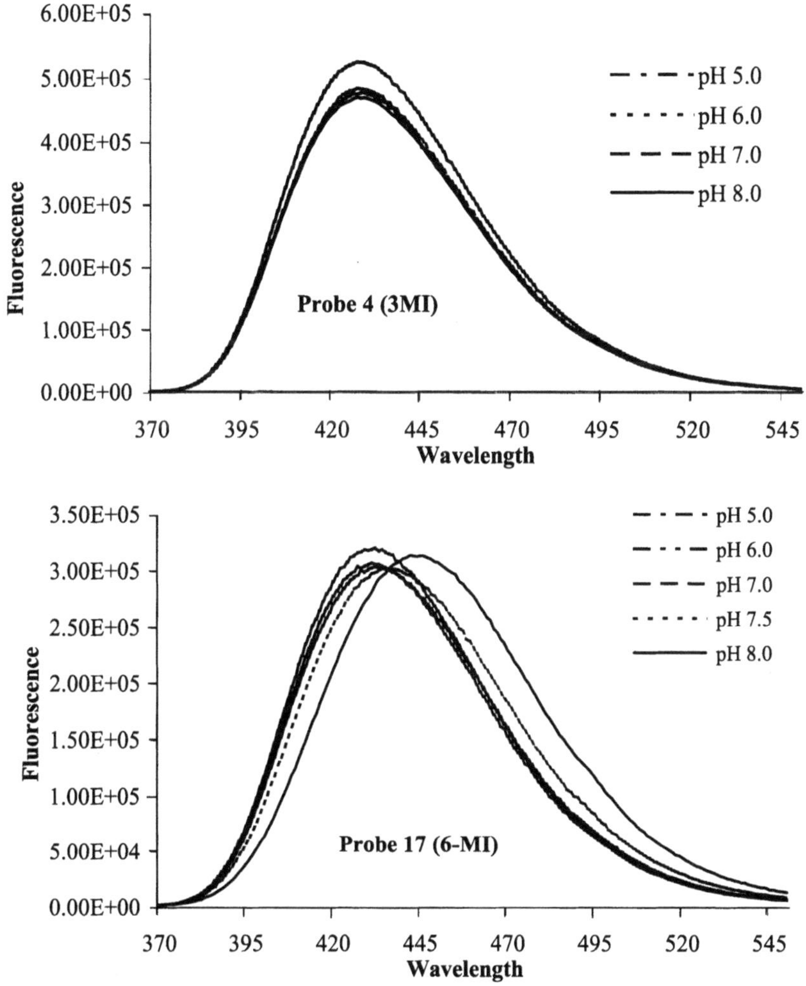

Figure 5.4. Fluorescence emission spectra of Probe 4 (3-MI) and Probe 17 (6-MI) at pH 5.0 to 8.0. (Adapted with permission from reference 6.)

as phosphoramidites and incorporation into oligonucleotides through the normal phosphodiester linkage using a standard DNA synthesizer. 3-MI requires no variations from standard de-blocking methods. 6-MI must be manually de-blocked using a 1M solution of 1,8-diazabicyclo[5.4.0]undec-7-ene (DBU) while still attached to the column. Details are available on the website for TriLink BioTechnologies, Inc. at www.trilinkbiotech.com or through the author. Typically,

Table 5.2. Oligonucleotides Containing 3-MI[a]

Name	Sequence	Quantum yield
PTER1	5′-GT**F** TGG AAA ATC TCT AGC AGT-3′	0.13
PTER2	5′-GTG T**F**G AAA ATC TCT AGC AGT-3′	0.10
PTER3	5′-GTG TG**F** AAA ATC TCT AGC AGT-3′	0.03
PTER4	5′-GTG TGG AAA ATC TCT A**F**C AGT-3′	0.06
PTER5	5′-GTG TGG AAA ATC TCT AGC A**F**T-3′	0.14
PTER7	5′-ACT GCT AGA **F**AT TTT CCA CAC-3′	0.04
PTER8	5′-ACT GCT A**F**A GAT TTT CCA CAC-3′	0.05
PTER9	5′-ACT **F**CT AGA GAT TTT CCA CAC-3′	0.29

[a]Position of the fluorophore is denoted by the **F**. The relative quantum yield for those containing 3-MI are listed at right. (Reprinted with permission from reference 42.)

Table 5.3. Fluorescence Properties of Probe-Containing Oligonucleotides in Single and Double Strands[a]

Probe	Pter		Q%	τ_i	α_i	$\%I_i$	τ_m
3-MI	8	ss	96	$\tau_1 = 2.35$	$\alpha_1 = 0.41$	$I_1 = 21.2$	5.27
				$\tau_2 = 6.06$	$\alpha_2 = 0.59$	$I_2 = 78.8$	
3-MI	8	ds	96	$\tau_1 = 0.21$	$\alpha_1 = 0.70$	$I_1 = 10.0$	4.60
				$\tau_2 = 2.88$	$\alpha_2 = 0.16$	$I_2 = 32.5$	
				$\tau_3 = 6.35$	$\alpha_3 = 0.13$	$I_3 = 57.5$	
3-MI	9	ss	64	$\tau_1 = 2.54$	$\alpha_1 = 0.31$	$I_1 = 17.8$	4.74
				$\tau_2 = 5.22$	$\alpha_2 = 0.69$	$I_2 = 82.2$	
3-MI	9	ds	68	$\tau_1 = 1.86$	$\alpha_1 = 0.33$	$I_1 = 14.5$	4.81
				$\tau_2 = 5.31$	$\alpha_2 = 0.67$	$I_2 = 85.4$	
6-MI	8	ss	96	$\tau_1 = 0.30$	$\alpha_1 = 0.81$	$I_1 = 25.6$	3.90
				$\tau_2 = 2.21$	$\alpha_2 = 0.14$	$I_2 = 31.3$	
				$\tau_3 = 7.28$	$\alpha_3 = 0.06$	$I_3 = 43.1$	
6-MI	8	ds	—	$\tau_1 = 0.21$	$\alpha_1 = 0.83$	$I_1 = 33.0$	2.59
				$\tau_2 = 1.20$	$\alpha_2 = 0.13$	$I_2 = 30.4$	
				$\tau_3 = 5.89$	$\alpha_3 = 0.03$	$I_3 = 36.6$	
6-MI	9	ss	56	$\tau_1 = 3.45$	$\alpha_1 = 0.25$	$I_1 = 14.1$	6.51
				$\tau_2 = 7.01$	$\alpha_2 = 0.75$	$I_2 = 85.9$	
6-MI	9	ds	64	$\tau_1 = 0.26$	$\alpha_1 = 0.88$	$I_1 = 43.4$	2.79
				$\tau_2 = 1.04$	$\alpha_2 = 0.09$	$I_2 = 17.0$	
				$\tau_3 = 6.30$	$\alpha_3 = 0.03$	$I_3 = 39.6$	

[a]Measurements were taken in 10 mM Tris buffer, pH 7.5 at room temperature. See Table 5.2 for oligonucleotide sequence and position of fluorophore within the strand. Studies were performed on single-stranded fluorophore-containing oligonucleotide (single) and on the identical oligonucleotide annealed to its complementary strand (double). The degree of quench of the fluorescence emission is relative to the deoxyribo-pteridine nucleoside analogue (monomer) Q%. Quantum yields of probes 4 (3-MI) and 17 (6-MI) are listed in Table 5.2. Abbreviations: τ_i, lifetime for each component of a multiexponential model; α_i, preexponential for each component of a multiexponential model; $\%I_i$, percent fluorescence intensity for each component of a multiexponential model; τ_m, intensity-weighted lifetime. (Reprinted with permission from reference 42.)

Table 5.4. Effects of the Position of the Fluorophore on the Melting Temperatures for 3-MI and 6-MI[a]

Oligonucleotide	3-MI	6-MI	Mismatch
PTER1	56.0°C		
PTER2	55.6°C		
PTER3	53.8°C	63.6°C	54.4°C
PTER4	52.0°C		54.4°C
PTER5	58.8°C	62.6°C	
PTER7	58.4°C		
PTER8	50.4°C	61.6°C	
PTER9	54.6°C		

[a]The site of fluorophore incorporation is shown in Table 5.2. T_ms were measured in 10 mM Tris, pH 7.5 with 10 mM NaCl. The T_m of the control double strand was 63.2°C. (Reprinted with permission from reference 6.)

probe-containing oligonucleotides are purified through polyacrylamide gel electrophoresis and ethanol precipitation. Although incorporation into DNA significantly quenches the fluorescence signal, the selected fluorophores start with quantum yields as high as 0.88 and can still be easily detected within oligonucleotides. Changes in the local structure of fluorophore-containing oligonucleotides can be monitored by measuring changes in fluorescence properties.

The significant quenching seen upon incorporation of 3-MI and 6-MI into single-stranded oligonucleotides is related to the proximity of purines to the fluorophore in the strand. The majority of quench is observed upon incorporation into the oligonucleotide with a less severe additional quench upon annealing the fluorophore-containing oligonucleotide to its complementary strand, suggesting that quenching originates mostly in base stacking interactions rather than base pairing. For PTER8, containing the fluorophore surrounded by purines, the fluorescence signal is almost completely quenched (96%) in the single-stranded form, and little additional quench is detected in the double-stranded form. For PTER9 containing 6-MI, quench increases from 56% to 64% with annealing, and for PTER9 containing 3-MI there is a smaller increase in quench from 64% to 68% when the labeled strand is annealed to its complement.

The T_m studies suggest that 3-MI does not participate fully in base pairing in double-stranded oligonucleotides because the T_m depression in 3-MI-containing oligonucleotides is approximately equivalent to that of a single base pair mismatch. In contrast, the T_ms of double-stranded 6-MI -containing oligonucleotides are very similar to the T_ms of the controls, suggesting that 6-MI may participate in base pairing. The shift in emission spectra of 6-MI going from single-stranded to double-stranded form is also consistent with base pairing. A shift of this type was not observed for double-stranded 3-MI (Figure 5.4).

The effects of quenchers or energy transfer interactions may also be monitored through lifetimes. For example, if a fluorophore within an oligonucleotide is shielded because of an interaction between the oligonucleotide and a protein, the lifetime will not be affected by the introduction of extrinsic dynamic quenchers (e.g., Cs^+ or I^-). However, if the probe is in a more exposed position within the oligonucleotide, then the lifetime will be affected by the addition of dynamic quenchers. Similarly, a probe may be dynamically quenched by collisions with intrinsic quenchers, e.g., neighboring groups such as carbonyl oxygen.

In its monomer form, 3-MI has one major lifetime component of 6.58 ns (99%) and a $\langle\tau\rangle$ of 6.51 ns. Upon incorporation into PTER8, the decay curve becomes more complex and two distinct lifetimes of 2.35 ns (41%) and 6.06 ns (59%) are required to describe the decay curve, resulting in a $\langle\tau\rangle$ of 4.54 ns. Annealing this fluorophore-containing oligonucleotide to its complementary strand forming the double-stranded PTER8 increases the complexity even further, requiring a third, dominant short-lived component (0.21 ns) to fit the decay curve (Table 5.3), and the $\langle\tau\rangle$ drops to 1.45 ns. This latter change may provide a convenient means of monitoring whether an oligonucleotide (or a segment of an oligonucleotide) containing one of the fluorophores is in the single- or double-stranded state. In PTER9 the degree of quench associated with incorporation of 3-MI into the oligonucleotide is less, and there is no increase in the complexity of the decay curve and no substantial change in $\langle\tau\rangle$ in the double strand.

For 6-MI, incorporation into PTER8 is also associated with substantial quenching, an increase in the complexity of the decay curve from two to three components, and a much shorter $\langle\tau\rangle$ (6.35 ns for the monomer and 0.95 ns in the single-stranded oligonucleotide). The double-stranded, 6-MI-containing PTER8 oligonucleotide has a slightly shorter $\langle\tau\rangle$ (0.53 ns) than the single-stranded oligonucleotide. The change in the shape of 6-MI's decay curve and the decrease in $\langle\tau\rangle$ resulting from annealing the fluorophore-containing single-stranded PTER9 to its complementary strand is more dramatic (Table 5.3). Based on the Tm measurements, 6-MI appears to participate in base-pairing interactions, while 3-MI does not. This may in part explain the differences in the observed changes in fluorescence properties of 3-MI- and 6-MI-containing oligonucleotides.

Comparing changes in $\langle\tau\rangle$ and relative quantum yield (expressed as percentage quench in Table 5.3) in the monomer and the fluorophore-containing single- and double-stranded oligonucleotide provides insight into the mechanism of the quench resulting from incorporation of the pteridine nucleoside into an oligonucleotide. Static quench should not be accompanied by a change in $\langle\tau\rangle$, while pure dynamic quench is associated with proportional changes in quantum yield and $\langle\tau\rangle$. Disproportionate changes in quantum yield and $\langle\tau\rangle$ suggest that quenching is due to a combination of static and dynamic events. The tertiary structure of the oligonucleotide could expose the fluorophore to collisional events from its surroundings (dynamic quenching) or could position the fluorophore in contact with other bases or backbone quenchers within the oligonucleotide (static quenching). In the 3-MI-containing PTER8 single-stranded oligonucleotide, the fluorophore is surrounded by purine bases and the Q_{rel} is 96% quenched compared with the monomer. The $\langle\tau\rangle$ of 3-MI is also shorter in the single-stranded oligonucleotide, but the change in $\langle\tau\rangle$ is only 30% (6.51 to 4.54 ns), suggesting that static quench arising from the surrounding purines is the primary mechanism involved in the quenching of the fluorescence signal. A more detailed analysis and discussion of the quenching mechanisms of 3-MI containing single and double strands has been done by Driscoll *et al.*[8]

The pH titration of the emission spectra and DAS of 3-MI and 6-MI yielded a shift in the emission spectrum of 6-MI between pHs of 7.0 to 8.0, and an increase in one of 6-MI's lifetime components over the pH range from 7.0 to 9.0. These DAS also provide a clear signature of an excited state reaction.[9] This property of 6-MI could be exploited to measure local pH and buffering. The shift in the emission spectrum of 6-MI was not seen with 3-MI and suggests that there is a protonation at the 3-position, which is protected by the methyl group in 3-MI. A minimum of two emitting species (e.g., tautomers) would be needed to explain these DAS.

In the following sections, we will demonstrate a number of applications that utilize these highly fluorescent analogues in ways that take advantage of their unique properties.

5.3. Applications Using Pteridines

5.3.1. Integrase Assay

One way to use these analogues is to simply monitor changes in fluorescence intensity when some action physically removes them from base-stacking interactions. The following example demonstrates this using the endonucleolytic cleavage activity of a protein from HIV-1.[1]

Integrase is a retrovirally-encoded protein responsible for the integration of viral DNA into the host cell's genome, a critical step in the life cycle of a retrovirus. This protein functions in a step-wise manner, the first step being cleavage of a dinucleotide from both of the 3′-ends (3′-processing reaction). Cleavage is followed by other steps, resulting in the integration of the HIV-1 genome into the host cell's DNA. The fluorescent pteridine nucleoside analogue based assay focuses on the cleavage step of this reaction.

The integrase catalyzed 3′-processing (endonucleolytic cleavage) and strand transfer reactions have previously been studied *in vitro* using recombinant protein and short double-stranded oligonucleotides (21-mers) with sequences identical to either the U5 or U3 terminus of HIV-1 DNA as substrates.[10–13] The substrate is ^{32}P-labeled and the reaction products are analyzed by polyacrylamide gel electrophoresis and autoradiography to detect cleavage (19-mer) or strand transfer (30-mer) products.

In the fluorescence-based assay utilizing the pteridine nucleoside analogue, 3-MI, the probe is site-specifically inserted into the cleavage site of a double-stranded oligonucleotide substrate. Cleavage of the 3′-terminal dinucleotide containing 3-MI results in a measurable real time increase in fluorescence intensity. Figure 5.5 shows a schematic of the integrase cleavage step with release of 3-MI (F) from the oligonucleotide.

Fluorescence intensity changes resulting from the integrase 3′-processing reaction are depicted in Figure 5.6. This assay is based on an *in vitro* cell-free and virus-free system in which purified recombinant HIV-1 integrase protein is incubated with a short duplex oligonucleotide (21-mer), with a sequence identical to the U5 terminus of viral DNA.[10–12,14] 3-MI is substituted for guanosine at the cleavage site of the oligonuceotide substrate. As part of the 3′processing reaction, the terminal 3′-dinucleotide, containing 3-MI, is cleaved from the oligonucleotide and the quenching effect of the neighboring adenosine is abrogated resulting in a marked increase in the fluorescence intensity of the reaction mixture.

The increase in fluorescence intensity was linear for up to 20 min. The rate of the increase in fluorescence intensity is proportional to the rate of the integrase 3′-processing reaction and, because the change in fluorescence intensity is dependent on release of the 3-MI-containing dinucleotide, this assay is specific for the integrase 3′-processing reaction. The potential increase in fluorescence intensity from the release of the 3-MI-containing dinucleotide is also illustrated by the marked difference in the fluorescence emission profiles of the 3-MI-containing oligonucleotide and the 3-MI-containing dinucleotide shown in Figure 5.7.

GTGTGGAAAATCTCTAGCA**F**T - 3′
CACACCTTTTAGAGATCGTCA - 5′

↓ Integrase

GTGTGGAAAATCTCTAGCA – 3′ + **F**T – 3′
CACACCTTTTAGAGATCGTCA - 5′

Figure 5.5. Integrase 3′-processing reaction with 3-MI-containing substrate. The cleavage reaction of integrase releases the 3-MI containing 3′-terminal dinucleotide from the strand. 3-MI is denoted as **F**. (Reprinted with permission from reference 42.)

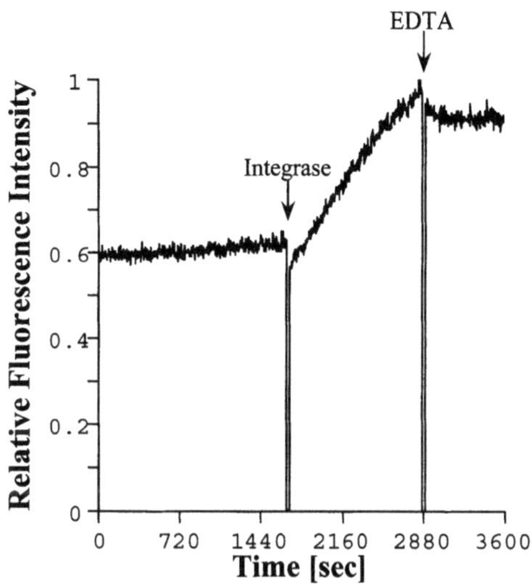

Figure 5.6. Change in fluorescence intensity over time during incubation of HIV-1 integrase with the 3-MI-containing oligonucleotide substrate. The real time kinetic trace of fluorescence intensity depicts the blank rate prior to addition of integrase to the reaction mixture (0 to 1700 s), the increase in fluorescence intensity following addition of integrase to the reaction mixture (1700 to 2880 s), and cessation of the reaction following the addition of EDTA (>2880 s) which chelates manganese, a required cofactor. The fluorescence emission was measured at 400 nm following the excitation at 360 nm. (Reprinted with permission from reference 1).

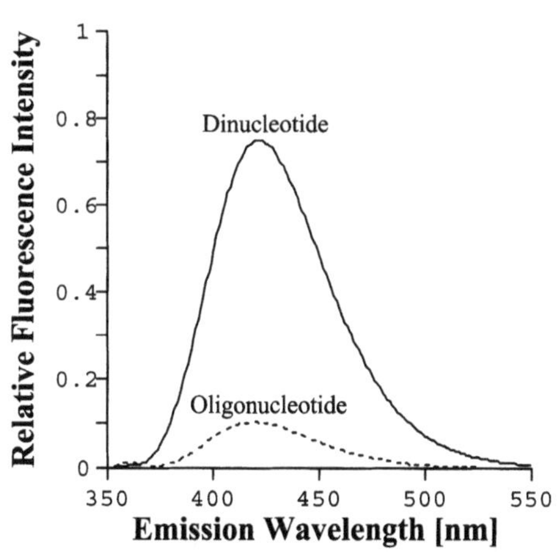

Figure 5.7. Comparison of the fluorescence emission spectra for the 3-MI-containing substrate oligonucleotide and the 3-MI-containing dinucleotide cleavage product. Excitation was at 360 nm. The oligonucleotide and the dinucleotide solutions were analyzed at equimolar concentrations as measured by absorbance of the solution at 360 nm. (Reprinted with permission from reference 1.)

The removal of the fluorophore from a single- or double-stranded oligonucleotide through the action of any endo- or exonuclease may be presumed to be easily monitored through changes in fluorescence intensity as well. This increase has also been demonstrated using other enzymes, including Pl nuclease and Exonuclease III.

5.3.2. Bulge Hybridization Probes

The unique properties of the pteridine nucleoside analogues provide us with a very convenient method for detection of hybridization of oligonucleotides. Since the quench in fluorescence intensity is largely attributable to base stacking interactions, it was assumed that any disruption in the base stacking arrangement would have a direct influence on the fluorescence intensity. Disruption in the helical arrangement of bases in an oligonucleotide can be caused by various means. This application explores the effects of forcing a one, two, or three base bulge within a double strand.

The application of the fluorophores as bulge hybridization probes utilizes the probe incorporated into an oligonucleotide that is perfectly complementary to a known target sequence, except for the addition of the fluorophore. When annealing occurs, the probe is squeezed out of the strand since it does not have a base-pairing partner. Since this partially relieves fluorescence quenching induced by interactions with neighboring bases, a direct increase in fluorescence intensity is observed (Figure 5.8). This technique is used to determine the presence or absence of specific DNA sequences in a mixture without the need for separation of annealed or labeled products. Fluorescence increases upon bulge formation range from −1.5-fold to 27-fold depending on the sequence. Although some increase in fluorescence intensity is realized with 2 or 3 base bulges, the highest signal increase is seen with the single base bulge.

The most productive bulges occur within a group of adenosines. Since this is also the most quenched environment for 3-MI, it provides a very low background.

The degree of increase revealed by annealing the two strands is dependent on the identity of the bases nearest to the fluorophore. The combination of selectivity and the fact that the products do not need to be separated for analysis make this a very rapid and attractive technique.

This bulge hybridization technique has been used to measure the products of polymerase chain reactions (PCR) without the need for separation techniques (Hawkins *et al.*, unpublished). In this approach, we synthesized a probe-containing strand that was complementary to a section of the region of product between (but not overlapping) the primers. An investigation of the advantages and disadvantages of using multiple probes per strand revealed that each 3-MI bulge depresses the Tm of the double strand by 1–2°C. We designed a PCR probe strand

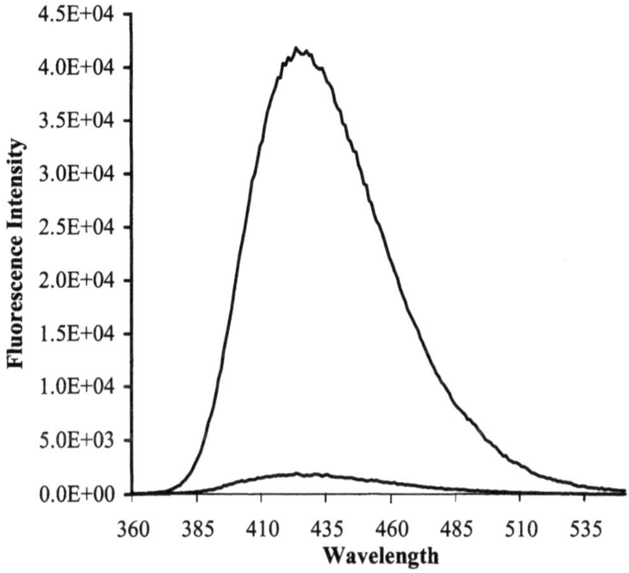

Figure 5.8. Difference in fluorescence intensity resulting from bulge formation as measured by integration of the area under each peak. The two 3-MI-containing strands are equimolar. Hybridization was done by heating the strands to 95°C for two minutes and allowing them to cool to room temperature. Excitation was at 350 nm. The larger peak is the scan of the double-stranded material of the sequence 5′-agcagttgctaaagaaFaattgaacacgctcggacttgc-3′. This is measured to be a 21-fold increase.

containing two 3-MI molecules designed to form bulges when annealing occurs. The probe strand is added to the PCR mixture prior to amplification at which time no complementary strands are present. As new product is created in the process of the amplification, complementary strands (targets) become available for the probe to bind resulting in increasing fluorescence intensity. Products containing no template (blank) and a series of known template concentrations were scanned at the conclusion of 30 cycle. Positive PCR reactions exhibit up to threefold increase in fluorescence intensity over the blank and showed a linear correlation between the amount of template and the amount of fluorescence increase. Results were confirmed using standard agarose gel electrophoresis.

5.3.3. Anisotropy

Studies done at Wesleyan University by Mukerji *et al.*[15] have utilized steady state fluorescence anisotropy using 3-MI fluorescence to monitor the binding of the HU protein to duplex DNA. HU, a multifunctional histone-like protein found only in prokaryotes, does not recognize or bind to a particular sequence of DNA, but does bind preferentially to supercoiled and bent DNA. This protein contains no Trp or Tyr residues; therefore, to probe the binding interaction by fluorescence spectroscopy it was necessary to label the DNA duplexes with 3-MI. This probe was specifically incorporated at one site into a 13-bp and a 34-bp DNA duplex. The binding affinity of the protein as well as the conformation of DNA were addressed in these experiments.

Anisotropy measurements were performed at a constant concentration of DNA with increasing concentrations of HU protein. Under these conditions the binding curves (Figure 5.9) yielded a Kd value of 1.51×10^{-6} M for the 34-bp duplex (3MI-34) and a value of 3.16×10^{-6} M for the 13-bp duplex (3MI-13). As shown by the fits, the interaction is well described by a binding equation that assumes a 1:1 interaction in both cases. The observed difference in binding constants for the two duplexes is larger than the error associated with the fit and is suggestive that the flanking regions are relevant for the interaction, since binding to 3MI-34 is slightly stronger than binding to 3MI-13. Previous studies[16] had suggested that HU binds to 9 bp of DNA; however, fluorescence anisotropy measurements are more consistent with a 1:1 interaction for both the 13- and 34-bp duplexes. Interestingly, in gel mobility shift assays (GMSA) performed with

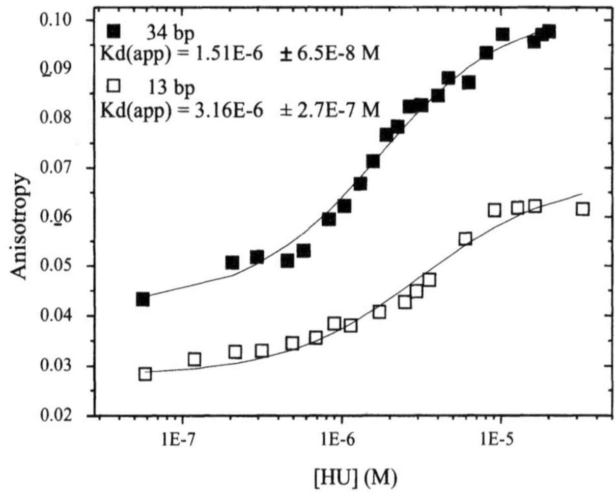

Figure 5.9. The binding of HU to 13 bp and 34 bp duplex oligomers as monitored by fluorescence anisotropy is shown. Excitation wavelength was 330 nm and anisotropy was monitored from 409 to 417 nm. Plotted values are an average with an error of ± 0.0025. Samples were in a 15 mM KCl, 2 mM spermidine, 0.1 mM EDTA, 10 mM Tris pH = 7.6 buffer. Temperature was maintained at 4°C. Kd(app) values were determined by assuming a 1:1 binding interaction with the fits as shown.

3MI-34, formation of 1:1 and higher order HU:DNA complexes (2:1, 3:1, 4:1, etc.) was observed. In this case the Kd(app) determined by GMSA is 5.0×10^{-8} M, which is significantly smaller (~30-fold) than the value determined by anisotropy. The observation of multiple HU:DNA complexes by GMSA has been previously reported with other DNA duplexes of comparable length.[17] For 3MI-13, however, only one complex is observed in the GMSA, and the determined Kd(app), 1.30×10^{-6} M, is similar to that determined by fluorescence anisotropy (see above). Despite the differences in the absolute determination of the Kd(app), both the GMSA and fluorescence measurements are suggestive that HU has a higher binding affinity for 3MI-34 relative to 3MI-13.

Fluorescence stoichiometry experiments performed with 3MI-34 are consistent with the anisotropy experiments and are indicative of a binding stoichiometry of at most 2:1 HU:DNA. Thus, we suggest that the numerous complexes observed in the GMSA result not from higher order protein:DNA complexes but altered migration patterns as a consequence of DNA bending and the geometry of the bend. HU has striking primary and secondary structural similarities with a homologous protein, integration host factor (IHF), and the mode of interaction with DNA may be similar to that of IHF. IHF forms a 1:1 complex with DNA and protein binding induces the DNA to form a bend of ~160.[18] Similarly, HU exhibits a marked preference for bent DNA and facilitates circularization of DNA.[19] Upon binding to HU, 3MI-34 exhibits an increase in fluorescence intensity, which is consistent with increased solvent exposure of the nucleoside analogue. This increase in fluorescence intensity is strongly suggestive of a structural deformation of the DNA that leads to an unwinding of the helix and possible unstacking of the DNA bases. This result is also supported by the observed change in anisotropy upon protein binding (see below). No fluorescence intensity change is observed for 3MI-13 upon binding to HU, which is indicative of 3-MI remaining in a stacked configuration. We suggest that the relatively short length of this DNA is prohibitive for bending. The absence of a fluorescence intensity change is consistent with the observation of only one band in the DNA binding gel and is suggestive of little or no change in the conformation of 3MI-13 upon binding to HU.

The anisotropy of 3MI-13 alone is 0.02 compared to a value of 0.04 for 3MI-34; thus, the $r(0)$ values reflect the length difference of the two duplexes. The overall anisotropy change upon protein binding is similar for the two duplexes, but is more pronounced in the case of 3MI-34 (0.054 vs. 0.039). This increment, while larger, may in fact be reduced by an increase in lifetime upon binding. The relationship between the correlation time (ϕ) and lifetime (τ) is shown in a form of the Perrin equation shown below.

$$\bar{r} = \frac{\bar{r}_0}{1 + \bar{\tau}/\bar{\phi}} \tag{5.1}$$

Consequently, this contrast probably would be greater if changes in had been measured instead of changes in anisotropy. If both measurements monitor the binding of 1 HU dimer to DNA, as suggested by the stoichiometry experiments and the analysis of the binding curve, the bound correlation time change should be the same. The larger anisotropy change for 3MI-34 potentially results from a change in conformation of the DNA upon binding. This is supported by the concomitant increases in fluorescence intensity and anisotropy. The intensity increase is suggestive of local unstacking of the DNA bases or unwinding of the helix, since the quantum yield of 3-MI is larger in single-stranded DNA relative to double-stranded. Thus, in contrast to the GMSA, fluorescence experiments (intensity, anisotropy, and stoichiometry) are more consistent with a 1:1 HU:DNA inteaction and are also strongly indicative of a change in DNA conformation upon binding, such that in the 34 bp duplex 3-MI becomes solvent exposed. The absence of a fluorescence intensity increase upon HU binding to 3MI-13 is suggestive that it is too short to experience a similar change in conformation. These fluorescence binding experiments provide relevant information regarding the protein-DNA interaction especially with respect to DNA conformation, and are suggestive of a new interpretation of the GMSA data.

5.3.4. Intracellular Transport of Oligonucleotides

Studies are being carried out at Mt. Sinai in New York by Basil Hanss *et al.* (unpublished) using 3-MI as a probe for oligonucleotide transport. Cellular uptake of oligonucleotides has important therapeutic implications. Genes can be transferred to cells using plasmid DNA, protein synthesis can be interrupted with ribozymes or antisense oligodeoxynucleotides, and, in emerging technology, genetic defects caused by point mutations can be corrected with RNA/DNA hybrid oligonucleotides.[20] All of these promising approaches require internalization of oligonucleotides (ODNs) at appropriate intracellular locations. Overall oligonucleotide uptake by most tissues *in vivo* is relatively low, the highest being in kidney, liver, and brain.[21–25] The aim of this study is to define the molecular mechanism of oligonucleotide internalization.

The molecular mechanism of ODN uptake is not completely understood, although it is known that uptake is saturable and energy requiring in most tissues.[25–29] A considerable body of literature exists describing the kinetics of ODN transport and, based on those studies, three transport mechanisms have been proposed: receptor mediated endocytosis, fluid phase endocytosis (i.e., pinocytosis), and transport through an oligonucleotide conducting channel.

Kinetics studies, for the most part, have used ODNs labeled with either a radioisotope or a fluorescent tag, most commonly end labeling with FITC or a similar fluorophore. Evidence in the literature reports different rates of uptake for radio-labeled and FITC-labeled ODNs.[30] In order to further explore the source of this difference, 3-MI will be incorporated into an oligonucleotide.

Three general areas of investigation will be optimized for use with 3-MI.[1] Oligonucleotides containing 3-MI (3MI-ODN) to measure ODN uptake/efflux kinetics. An extremely bright oligonucleotide containing six 3-MI molecules in a 22-mer is being used to increase sensitivity (5′-Ftc cFc tcF tct Fct cFt tcF c-3′). Currently, radiolabeled ODNs are being used to study uptake kinetics, trans-epithelial flux, and the molecular mechanism of ODN transport across both the apical and basolateral membranes of renal epithelia. The model systems consist of primary cultures of human renal tubule cells and isolated perfused renal tubules. Addition of 3MI-ODN into these studies will allow real time measurement of ODN flux using confocal fluorescence microscopy.[2] The intracellular fate of ODNs in renal tubule cells is unclear; yet their therapeutic efficacy is absolutely dependent upon trafficking to the correct intracellular compartment. Use of 3MI to follow trafficking will provide a more native-like model than that previously used.[3] Finally, to be effective for antisense purposes, ODNs must hybridize to target mRNAs; however, little is known about the *in vivo* binding of ODNs to target sequences and it is difficult to predict which sequences will bind efficiently to the target mRNA. Pteridine nucleoside analogue-containing ODNs are being used to develop an *in vivo* hybridization assay to address these important issues.

5.3.5. DNA Conformation

Laws and others at Mt. Sinai in New York,[7] have studied nucleoside analogues using time-resolved fluorescence resonance energy transfer. Studies done by Parkhurst *et al.*[31–33] have greatly enhanced our understanding of DNA conformation and dynamics. Typically, resonance energy transfer data provides information about the distribution of distances between donor and acceptor fluorophores. If the donor and acceptor are attached to specific sites on a macromolecule, this distance distribution can provide invaluable insights into structure and conformational dynamics. Since nucleoside analogues exist in a very restricted conformation within the DNA, the donor–acceptor distance distributions experienced by the nucleoside analogue will arise predominantly from the conformation and dynamics of the DNA itself. In the studies by Parkhurst, measurements reflect the distances between probes positioned in a region external to the helix. Using the nucleoside analogues should give considerably different results since the measurement is being taken between points within the helix. An ongoing study is using 6-MI to function as an acceptor. The use of this pteridine nucleoside analogue will greatly improve the quality of the information that is available from this class of experiments.*

5.4. 2-Amino Purine

5.4.1. Background

The potential for using fluorescent nucleoside analogues in DNA research was suggested in an article in 1969 by Ward *et al.*[34] In this paper the fluorescence characteristics of 2-amino purine (2-AP) were defined, demonstrating that it would be practical for study because of a number of qualities. A quantum yield of 0.68 in an aqueous solvent at pH 7.0 and an emission maximum of 370 nm identify this compound as a powerful tool for studies of DNA. These authors also successfully incorporated 2-AP into oligonucleotides using *Escherichia coli* RNA polymerase and substituting 2-AP-triphosphate for ATP. The fluorescence changes seen upon incorporation into an oligonucleotide made it apparent that this probe could be used to monitor changes in an oligonucleotide structure or environment. The synthesis of 2-AP as a phosphoramidite by McLaughlin *et al.* in 1988[35] has allowed the site-specific incorporation of this fluorophore into oligonucleotides through standard automated synthesis. Since that time 2-AP has been used in various studies. Its behavior in oligonucleotides is similar to that of the pteridine nucleoside analogues in that it is quenched upon incorporation and fluorescence properties change dramatically when events or other molecules impact the structure of the oligonucleotide. Several examples of research using this probe are briefly described.

5.4.2. Applications

In a structural study of DNA containing 2-AP Nordlund et al have analyzed the movement and flexibility of DNA.[4] This study utilized the temperature-dependent time-resolved fluorescence of 2-AP. By comparison of the single exponential decay of the monomeric probe and the multiexponential decay of the

*Deoxyribose, ribose, and phosphoramidite forms of the pteridines, 3-MI, 6-MI, and 6-MAP are available through Toronto Research Chemicals, Inc. (TRC), Toronto, Canada. Custom oligonucleotides containing 3-MI, 6-MI, and 6-MAP are available through TriLink BioTechnologies, Inc., San Diego, CA.

probe within the oligonucleotide, conclusions could be drawn about conformational states of the probe. Data from this study leads to the conclusion that 2-AP within the oligonucleotide exists in four or more conformational states. These can be described as various levels of base stacking or unstacking within the helix.

Another study of structural dynamics of DNA was conducted by Guest *et al.*[5] in which the strength of base-pairing interactions in various mismatches was examined. This study analyzed internal rotation of 2-AP within a double strand using fluorescence amplitude and correlation times. The degree of annealing of the two oligonucleotides was studied using tumbling correlation times and the emission lifetimes were used to determine the excited-state quenching of 2-AP. The findings demonstrate the usefulness of 2-AP to measure differences in the strength of base-pairing interactions as a function of base motion.

In a similar study by Hochstrasser *et al.*[2] 2-AP was used to label substrates for monitoring base-pairing interactions and local melting of DNA by measuring fluorescence decay characteristics. Once again, the complexity of the decay curve between a 2-AP-containing single strand and a 2-AP-containing double strand was used to draw conclusions about the single-strandedness of the 2-AP position. Variations in the apparent fraction of paired 2-AP bases led the authors to conclude that the Klenow fragment can melt a DNA duplex terminus resulting in a single-stranded conformation.

In a very simple and direct application of 2-AP for monitoring helicase activity Raney *et al.*[3] used an oligonucleotide substrate containing seven 2-AP's spaced at regular intervals. Fluorescence intensity which is quenched 2-fold within the oligonucleotide is restored when the substrate is unwound by helicase. By measuring fluorescence intensity changes over real time, the rate of helicase driven unwinding of DNA was measured.

Frey *et al.*[36] used 2-AP in a stopped flow study to examine the kinetics of the polymerization reaction of Klenow fragment. This experiment also utilizes the quenching affect of the double-strand conformation on 2-AP.

The measurement of very subtle movements between DNA and protein is presented in a paper by Beechem *et al.*[37] In this study 2-AP was site-specifically incorporated into a primer terminus and then changes in fluorescence were observed as the T4 polymerase was allowed to bind. This protein has distinctly different binding sites for polymerase and exonuclease activities. A variety of factors can shift the predominant binding from one site to the other. After determining that fluorescence quenching or hyperfluorescence were dependent on the site to which the primer was bound, the authors were able to investigate transitions between polymerase site and exonuclease site bound primer. This is a good demonstration of the way the fluorophore can reflect subtle changes in its environment.

It is apparent from the number and diversity of applications using 2-AP, only a small sampling of which are noted here, that there are many ways to observe the changes experienced by the probe when changes occur in the oligonucleotide environment.

5.5. 1,N^6-Ethenoadenosine

Another potentially useful fluorescent nucleoside analogue is 1,N^6-ethenoadenosine. This compound along with ethenocytidine have been used in research since their development in 1972. The fluorescence quantum yield of 0.5 and the maximum emission wavelength of 410 nm make ethenoadenosine a particularly attractive probe.[38,39] A great deal was learned about the fluorescence characteristics of chloroacetaldehyde-modified dinucleosides. By exposing dinucleosides to chloroacetaldehyde under specific conditions a series of dinucleotides were modified to contain ethenoadenosine or ethenocytidine. Fluorescence analysis of the resulting compounds revealed that the fluorescence was quenched from the interactions with neighboring bases. It was apparent from these experiments that neighboring purines resulted in a greater quench than neighboring pyrimidines.[40] Now that these probes have been synthesized as phosphoramidite forms allowing incorporation into DNA on an automated synthesizer,[41] more can be learned about their behavior within the oligonucleotide strand. Since the etheno-compounds are unstable under standard deprotecting methods, it is necessary to use fast base deprotecting groups and a short treatment with ammonia.

5.6. Summary and Outlook

The focus of this chapter was to point out the unique properties of fluorescent nucleoside analogue probes and to examine how one can take advantage of those properties to observe quite subtle interactions within DNA. There are a growing number of nucleoside analogues that are highly fluorescent, stable, and available as phosphoramidites to facilitate site-specific incorporation into oligonucleotides. Some of these probes apparently participate in base pairing (2-AP, 6-MI, 6-MAP) and some are comparable to a single-base-pair mismatch as measured by melting temperatures. All four of the pteridine analogues, when incorporated into an oligonucleotide migrate through a polyacrylamide gel at a rate indistinguishable from the equivalent sequence as a control. This would suggest that there are no gross configurational changes in the oligonucleotide due to the presence of the pteridine. Table 5.5 lists some of the fluorescence properties of several probes for comparison.

Table 5.5. Fluorescence Properties of Some Nucleoside Analogues

Probe	Ex_{max}	Em_{max}	Q_{rel}	Probe	Ex_{max}	Em_{max}	Q_{rel}
3-MI	348 nm	431 nm	0.88	6-MI	340 nm	430 nm	0.70
6-MAP	330 nm	435 nm	0.39	2-AP	303 nm	370 nm	0.68
DMAP	330 nm	437 nm	0.48	Etheno-A	300 nm	410 nm	0.50

Studies using nucleoside analogues in oligonucleotides have demonstrated that subtle changes in DNA often can be observed using a variety of fluorescence properties. They provide us with a new window on DNA structure and its relationship with the environment. In many cases the probes can be placed so close to the site of interest that changes at that site can be witnessed in a real time format using various fluorescence techniques.

The ideal probe, one that closely resembles the natural bases and is highly fluorescent, however, is quite elusive. As with most things in science, the problem is much more complex than one might expect. The characteristics that make guanosine recognizable from the point of view of a protein, for example, extend beyond the hydrogen bonding moieties. The size, shape, and electronic distribution within guanosine all contribute to its being recognized. Nothing can mimic this perfectly without being guanosine and consequently being only faintly fluorescent. The objective is to create the most native-like, stable and highly fluorescent probes possible. In the search for an ideal analogue, something must always be sacrificed in order to achieve acceptable levels of stability and fluorescence. There is substantial structural difference between fluorescent nucleoside analogues and native nucleosides, and yet similarities in overall shape, size, and linkage allow us to place a fluorophore within the DNA without totally disrupting the tertiary structure. With a variety of different analogues available, it will be possible to select the one that is most appropriate for a given application.

The difference between the conventional fluorophores and the analogues is only a matter of subtlety. There is no dispute about the potential and usefulness of the hundreds of conventional fluorophores now available for research. Having a variety of analogues capable of being placed within in a specific sequence and at different sites will give researchers one more way to approach the study of DNA.

Acknowledgments

I would like to thank Dr. Ishita Mukerji and Kristi Wojtuszewski at Wesleyan University for sharing their research on the HU protein; Basil Hanss at Mt. Sinai for his contribution on oligonucleotide transport; and Dr. William Laws and Ed Radhofsky for their contribution on DNA conformation. I also wish to thank Drs. Jay Knutson and Frank Balis for helpful comments and stimulating discussions.

References

1. Hawkins, M. E., Pfleiderer, W., Mazumder, A., Pommier, Y. G., and Balis, F. M. *Nucleic Acids Res. 23*, 2872–2880, 1995.
2. Hochstrasser, R. A., Carver, T. E., Sowers, L. C. and Millar, D. P. *Biochemistry, 33*, 11971–11979, 1994.
3. Raney, K. D., Sowers, L. C., Millar, D. P., and Benkovic, S. J. *Proc. Natl. Acad. Sci. USA 91*, 6644–6648, 1994.
4. Nordlund, T. M., Wu, P., Andersson, S., Nilsson, L., Rigler, R., Graslund, A., McLaughlin, L. W., and Gildea, B. *SPIE Time-Resolved Laser Spectrosc. Biochem. II 1204*, 344–353, 1990.
5. Guest, C. R., Hochstrasser, R. A., Sowers, L. C., and Millar, D. P. *Biochemistry 30*, 3271–3279, 1991.
6. Hawkins, M. E., Pfleiderer, W., Balis, F. M., Porter, D., and Knutson, J. R. *Anal. Biochem. 244*, 86–95, 1997.
7. Hawkins, M. E., Pfleiderer, W., Jungmann, O., and Balis, F. *Anal. Biochem. 298*, 231–240, 2001.
8. Driscoll, S. L., Hawkins, M. E., Balis, F. M., Pfleiderer, W., and Laws, W. R. *Biophys. J. 73*, 3277–3286, 1997.
9. Davenport, L., Knutson, J. R., and Brand, L. *Faraday Discuss. Chem. Soc. 81*, 81–94, 1986.
10. Brown, P. O., Bowerman, B., Varmus, H. E., and Bishop, J. M. *Cell 49*, 347–356, 1987.
11. Brown, P. O., Bowerman, B., Varmus, H. E. and Bishop, J. M. *Proc. Natl. Acad. Sci. USA 86*, 2525–2529, 1989.
12. Fujiwara, T., and Craigie, R. *Proc. Natl. Acad. Sci. USA 86*, 3065–3069, 1989.
13. Vink, C., Banks, M., Bethell, R., and Plasterk, R. H. A. *Nucleic Acids Res. 22*, 2176–2177, 1994.
14. Katzman, M., Katz, R. A., Skalka, A. M., and Leis, J. *J. Virol. 63*, 5319–5327, 1989.
15. Wojtuszewski, K., Hawkins, M. E., Cole, J. L., and Mukerji, I. *Biochemistry 40*, 2588–2598, 2001.
16. Broyles, S. S., and Pettijohn, D. E. *J. Mol. Biol. 187*, 47–60, 1986.
17. Bonnefoy, E., and Rouviere-Yaniv, J. *EMBO J. 10*, 687–696, 1991.
18. Rice, P. A., Yang, S., Mizuuchi, K., and Nash, H. A. *Cell 87*, 1295, 1996.
19. Hodges-Garcia, Y., Hagerman, P. J., and Pettijohn, D. E. *J. Biol. Chem. 264*, 14621–14623, 1989.
20. Cole-Strauss, A., Yoon, K., Xiang, Y., Byrne, B. C., Rice, M. C., Gryn, J., Holloman, W. K. and Kmiec, E. B. *Science 273*, 1386–1389, 1996.
21. Cossum, P. A., Sasmor, H., Dellinger, D., Truong, L., Cummins, L., Owens, S. R., Markham, P. M., Shea, J. P., and Crooke, S. *J. Pharmacol. Exp. Ther. 267*, 1181–1190, 1993.
22. Meeker, R., Legrand, G., Ramirez, J., Smith, T., and Shih, Y. H. *J. Neuroendocrinol. 7*, 419–428, 1995.
23. Oberbauer, R., Schreiner, G. F., and Meyer, T. W. *Kidney Int. 48*, 1226–1232, 1995.
24. Ogawa, S., Brown, H. E., Okano, H. J., and Pfaff, D. W. *Regul. Pept. 59*, 143–149, 1995.
25. Rappaport, J., Hanss, B., Kopp, J. B., Copeland, T. D., Bruggeman, L. A., Coffman, T. M., and Klotman, P. E. *Kidney Int. 47*, 1462–1469, 1995.
26. Yakubov, L. A., Deeva, E. A., Zarytova, V. F., Ivanova, E. I., Ryte, A. S., Yurchenko, L. V., and Vlassov, V. V. *Proc. Natl. Acad. Sci. USA 86*, 6454–6458, 1989.
27. Wu-Pong, S., Weiss, T. L., and Hunt, C. A. *Cell. Mol. Biol. 40*, 843–850, 1994.
28. Iverson, P. L., Zhu, S., Meyer, A., and Zon, G. *Antisense Res. Dev. 2*, 221–222, 1992.

29. Crooke, R. M., Graham, M. F., Cooke, M. E. and Crooke, S. T. *J. Pharmacol. Exp. Ther 275*, 462–473, 1995.
30. Zhao, Q., Waldschmidt, T., Fisher, E., Herrera, C. J., and Krieg, A. M. *Blood 84*, 3660–3666, 1994.
31. Parkhurst, K. M., Brenowitz, M., and Parkhurst, L. J. *Biochemistry* 23, 7459–7465, 1996.
32. Parkhurst, K. M., and Parkhurst, L. J. *Biochemistry 34*, 293–300, 1995.
33. Parkhurst, K. M., and Parkhurst, L. J. *Biochemistry 34*, 285–292, 1995.
34. Ward, D. C., and Reich, E. *J. Biol. Chem. 244*, 1228–1237, 1969.
35. McLaughlin, L. W., Leong, T., Benseler, F., and Piel, N. *Nucleic Acids Res. 16*, 5631–5644, 1988.
36. Frey, M. W., Sowers, L. C., Millar, D. P., and Benkovic, S. J. *Biochemistry 34*, 9185–9192, 1995.
37. Beecham, J. M., Otto, M. R., Bloom, L. B., Eritja, R., Reha-Krantz, L. J., and Goodman, M. F. *Biochemistry 37*, 10144–10155, 1998.
38. Leonard, N. J. *CRC Crit. Rev. Biochem. 15*, 125–199, 1993.
39. Secrist, J. A., Barrio, J. R., and Leonard, N. J. *Science 175*, 646–647, 1972.
40. Tolman, G. L., Barrio, J. R., and Leonard, N. J. *Biochemistry 13*, 4869, 1974.
41. Srivastava, S. C., Raza, S. K., and Misra, R. *Nucleic Acids Res. 22*, 1296–1304, 1994.
42. Hawkins, M. E. *Cell Biochem. Biophys. 34*, 257–281, 2001.

6

Lanthanide-Labeled DNA

Paul R. Selvin

6.1. Overview

Lanthanide ions can have highly unusual emission characteristics in aqueous solution, including a long (millisecond) excited-state lifetime, sharply spiked emission spectra (<10 nm width), and a large Stokes shift (>150 nm). These characteristics, when using pulsed excitation in combination with time-delayed and wavelength-filtered detection, are advantageous for discriminating against background fluorescence, which tends to be short lived (primarily nanosecond) and broadly spread in wavelength. For this reason, lanthanide ions are of significant interest as alternatives to conventional fluorophores, particularly when autofluorescence is a problem. This is particularly true in high-throughput screening assays for drug development where autofluorescence commonly limits sensitivity with conventional probes and radioactivity has undesirable environmental, health, and cost considerations. Detection sensitivity of 10^{-12}–10^{-15} M can be achieved with lanthanides, exceeding sensitivity achievable with conventional fluorophores and approaching or equalling radioactivity. A number of companies have commercially available lanthanide-based assays although availability of the chelates remains an issue for many university researchers.

A second area of practical and fundamental interest is the use of lanthanide ions as donors in resonance energy transfer studies for the detection of binding between biomolecules or the measurement of nanometer-scale distances within and between biomolecules. It has recently been realized that lanthanide ions make excellent donors in energy transfer experiments, enabling distances up to 100 Å feasible with greatly improved accuracy compared to conventional fluorescent probes. Lanthanide-based resonance energy transfer (LRET) has been applied in both basic and applied studies, including DNA and DNA-protein complexes. Very recently, it has been realized that LRET can lead to new classes of DNA-dyes with tuneable excited-state lifetimes in the 10–500 msec time regime and tunable color

Paul R. Selvin • Physics Department and Biophysics Center, University of Illinois, Urbana, Illinois 61801 (selvin@uiuc.edu).

Topics in Fluorescence Spectroscopy, Volume 7: DNA Technology, edited by Joseph R. Lakowicz, Kluwer Academic / Plenum Publishers, New York, 2003

emission, from 500 nm to 700nm, which may lead to many advantages, including greatly multiplexed detection of simultaneous signals.

A number of reviews have been written on the use of lanthanide probes, typically in time-resolved measurements, including their use in DNA assays.[1–10] This chapter focusses on lanthanides applied to DNA, but examples of lanthanides used in protein studies are included because the logical extension to DNA studies is clear in many cases and indirect detection of DNA through the use of proteins (such as streptavidin) is common. Lanthanide chelates as detection reagents are first discussed, followed by their use in resonance energy transfer studies.

6.2. Lanthanide as Luminescent Labels

Many of the lanthanide atoms, also known as rare-earth atoms, are widely used as luminescent materials. For example, neodymium is used in lasers; erbium is used in telecommunication devices; and terbium, europium, and occasionally gadolinium are used in fluorescent bulbs. However, in aqueous systems only europium and terbium have significant luminescence, with dysprosium and samarium being used rarely and the others not at all. Europium emits primarily in the red and terbium primarily in the green. Even with europium and terbium, to get significant luminescence in aqueous solution two steps must be taken. First, water must be prevented from binding the lanthanide since water highly quenches its excited state. This is generally accomplished by encapsulating the lanthanide in a chelate which binds the lanthanide tightly and expels water. In addition, the chelate often serves as a means for attachment of the lanthanide to a biomolecule via some thiol- or amine-reactive functionality. Second, because the lanthanide absorption cross-section is extremely weak (typically $<1\ M^{-1}\ cm^{-1}$, compared to $\approx 100{,}000\ M^{-1}\ cm^{-1}$ for many conventional fluorophores), an antenna (or sensitizer) is placed near the lanthanide, usually via covalent attachment of the antenna to the chelate. This antenna, usually an organic chromophore, absorbs the excitation light and transfers the energy to the lanthanide, which then re-emits characteristic lanthanide luminescence.[11]

6.2.1. Representative Chelates

Figure 6.1 shows a number of luminescent lanthanide complexes, most of which have been attached to amino-modified nucleotides and others which have been made as phosphoramidites.[12,13] Luminescent cryptates were originally developed by Lehn *et al.* where the bipyridine antenna molecules are an integral part of the chelation complex.[14] Both Eu and Tb are luminescent, although they are pri-

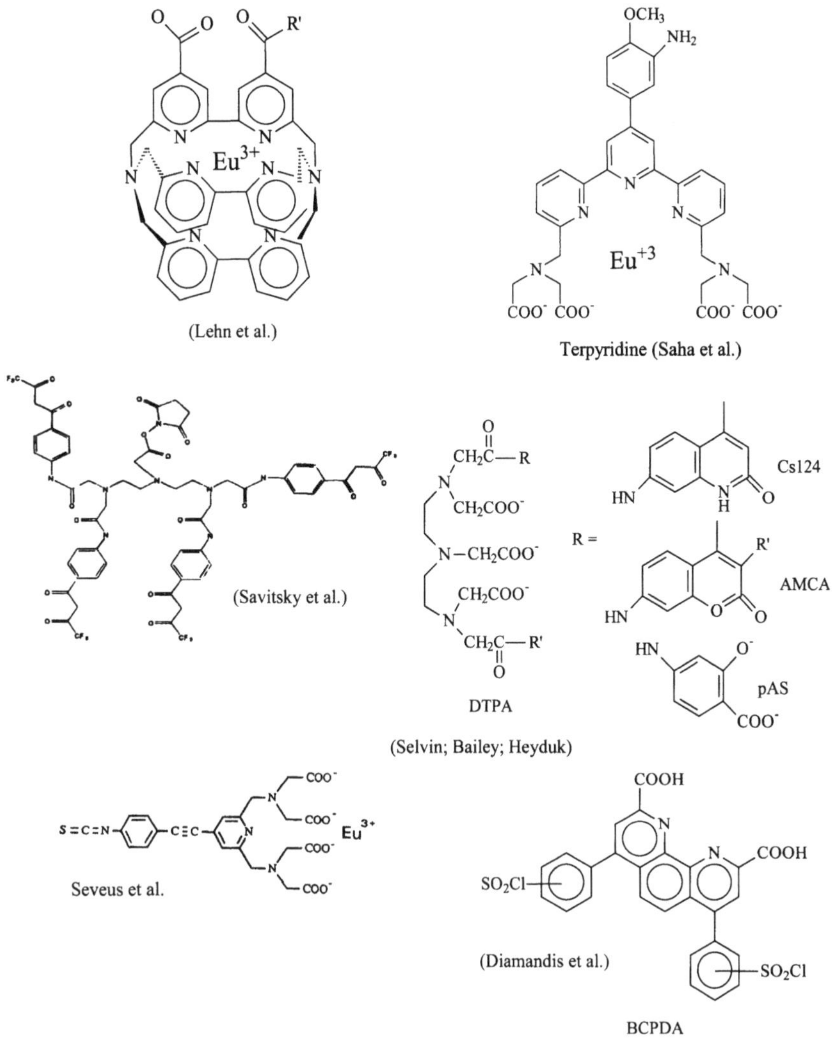

Figure 6.1. Some luminescent lanthanide chelate complexes. (See text for details.)

marily used with Eu because Tb has a back transfer of energy to the antenna molecule which leads to lanthanide quenching.[15] They are used in commercial assays by Cis-Biointernational (Cedex, France).[16–19] The chelate and sensitizers are also an integral unit in the terpyridine complex developed by Saha *et al.*, which is luminescent with Eu (now available from Amersham Biosciences, Piscataway, NJ),[20] and in the BCPDA complex (4,7-bis(chlorosulfophenyl)-1,10-phenanthroline-2,9-

dicarboxylic acid) developed by Diamandis *et al.* and originally marketed by CyberFluor (Toronto, Canada).[3]

Polyaminocarboxylate chelates, such as diethylenetriaminepentaacetic acid (DTPA), covalently attached to a number of possible antenna molecules, have been widely used primarily because of their ease of synthesis, good chelation properties, and the ability to independently tailor the chelate and sensitizer for the desired properties.[21–23] Paraaminosalicylate was the original antenna used[21] with DTPA and is luminescent with terbium but not europium. It is hydrophilic, which is advantageous when attaching many chelate-complexes to a biomolecule (see Section 6.2.3 below) but requires fairly short wavelength excitation ($\leqslant$308 nm) with limited brightness. The antenna molecule carbostyril 124[23] or its derivatives,[24] used with polyaminocarboxylate chelates, are brighter and can be excited at 340 nm or longer but are more hydrophobic (available from Panvera, Madison, WI).[23] They have also been shown to be excellent donors in resonance energy transfer applications (discussed below). Savitsky attached up to four β-diketone antenna molecules to a DTPA.[25] This system required the use of excess Eu and the presence of detergent and trioctylphosphinoxide but claimed sensitivity equal to the DELFIA system (dissociation-enhanced lanthanide fluorescence immunoassay), which is generally the most sensitive lanthanide-based system. In general, it is difficult to compare the brightness of various chelates because a side-by-side comparison has not been published, although some listing of sensitivities have been summarized[5] and are presented below.

Figure 6.2 shows representative emission spectra and lifetime for Tb^{3+} and Eu^{3+}. The relative intensities in each peak, particularly with Eu, are a function of chelate structure (symmetry), and the excited-state lifetimes also vary by chelate. All complexes require UV excitation, ranging from 280 nm to approximately 365 nm (see reference 26 for a discussion of the photophysics). The transitions from excited to ground state are spin and parity forbidden and hence long-lived. Formally these transitions are not called phosphorescence because they do not arise from a triplet to singlet state.[1] Importantly, lanthanide lifetimes in many chelates are not sensitive to oxygen concentration, in contrast to phosphorescent probes. The quantum yield for emission at room temperature is reported to be as high as 0.7 orders of magnitude higher than phosphorescent probes.[26,27] Emission is also highly spiked in wavelength, enabling color-discrimination against background signals. Their spiked spectra is also important when using lanthanides as donors in resonance energy applications because it enables the donor emission to be separated from acceptor emission (see below).

An alternative methodology is to use a nonluminescent lanthanide chelate as a label, instead of the luminescent labels discussed above (Figure 6.3). After specifically binding the label to the biomolecule of interest, the lanthanide is extracted from the chelate, usually by dropping the pH, and placed in an "enhance-

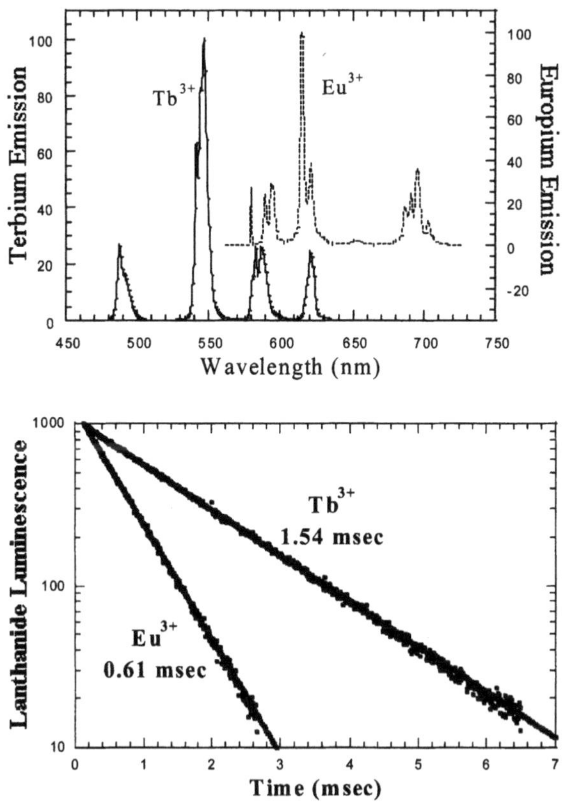

Figure 6.2. Representative emission spectra and excited-state lifetime. The particular chelate is Tb^{3+}- or Eu^{3+}-DTPA-cs124 (see Figure 6.1).

ment solution" where the lanthanide is luminescent. The enhancement solution typically contains a chelate in a micellar environment (to eliminate quenching effect of water) with organic chromophores (to act as antenna molecules). The commercially available DELFIA from Wallac Inc. utilizes this method and is based on $Eu(NTA)_3(TOPO)_2$, europium-naphthoyltrifluoroacetone-triocylphosphine oxide, in the presence of the nonionic detergent Triton-X. Tb can also be used at reduced sensitivity compared to Eu, and Dy and Sm can be used with yet a further reduction in sensitivity. One clear disadvantage of the system is that all spatial information is lost, precluding applications such as fluorescence *in situ* hybridization (FISH). A second disadvantage is that extra steps are required. An advantage of the system, however, is that the detection sensitivity is typically an order of magnitude or more better than directly luminescent chelates.

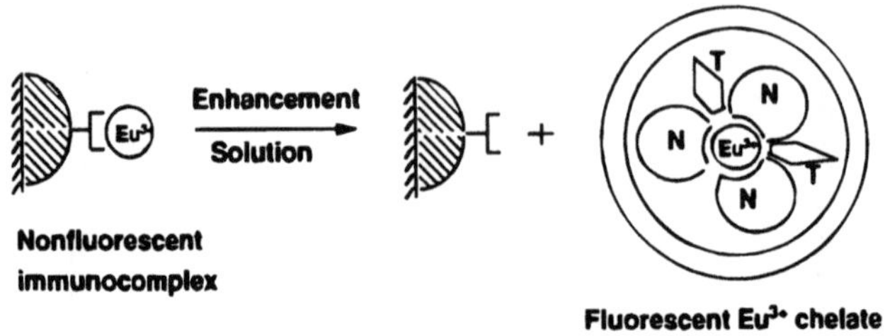

Figure 6.3. DELFIA enhancement methodology. A nonfluorescent lanthanide chelate is bound to analyte, and captured on a surface. An enhancement solution releases the Eu which is then bound in a luminescent complex containing a chelate-sensitizer, often napthyltrifluoracetate (NTA), in the presence of triocylphosphine oxide and Triton X-100. The resulting micelle is highly luminescent. (From reference 25.)

6.2.2. Sensitivity

An important figure of merit when discussing sensitivity is a comparison to radioactive probes. Another benchmark are ELISAs (enzyme-linked immunosorbent assays), the standard clinical assay for antibodies, which have sensitivities in the 10^{-9} to 10^{-11} M range. Detection of a single *conventional* (i.e., organic) fluorophore is feasible if background fluorescence is not a problem, whereas this is not possible for a single radioactive label or antibody in ELISA. (Detection of a single lanthanide probe is highly unlikely because the long lifetime limits photon flux to below detectable levels.) Hence fluorescence can be orders of magnitude more sensitive than radioactive probes or ELISA assays. However, under many conditions, including those commonly found in high-throughput drug screening applications, radioactivity is more sensitive than conventional fluorophores because contaminating background fluorescence often limits the sensitivity of conventional fluorophores. With lanthanides, however, temporal and spectral discrimination against background fluorescence, which tends to be widely spread in wavelength and nanoseconds in duration, is very effective and sensitivities approaching radioactivity are possible.

The lowest detection limit reported for a single chelate is 10^{-15} M but this was in ethanol.[28] Among practical systems, the DELFIA system can be quite sensitive, in part because luminescence is achieved in an optimized "enhancement solution." Detection levels of 10^{-12}–10^{-14} M and 10^{-16}–10^{-18} moles were reported in immunoassays,[29,30] which compares to roughly $0.05–2 \times 10^{-18}$ moles for radioactivity.[31] In DNA studies, Hurskainen *et al.* detected 5 pg (0.15×10^{-18} moles or 0.15 attomoles) of lambda phage DNA, labeled with nonfluorescent Eu chelates, by hybridizing to a lambda phage DNA-coated microtiter plate, followed by DELFIA enhancement solution. 7% of total nucleotides were labeled with nonfluorescent Eu chelate introduced by transamination of the DNA, followed by reaction with an amine-reactive chelate. For comparison, ^{32}P-labeled lambda phage DNA was hybridized against lambdaphage DNA spotted onto nitrocellulose filter. Hybridization sensitivity was 1 pg after overnight autoradiography at −70°C and 10 pg after Cerenkov counting (counting in liquid scintillation counter without scintillation cocktail). For oligonucleotides (approximately 20-mer) end-labeled with about 20 Eu-chelates, DELFIA detection in a hybridization-capture assay led to a sensitivity of 10^7 target molecules, or 17 attomoles.[32]

Among directly luminescent chelates, Savitsky *et al.*, in a brief report, claimed equal sensitivity to the DELFIA system.[25] The detection limit for the free terpyrdine complex of Saha *et al.* was reported to be 3×10^{-17} moles (presumably in 100 μL, or 0.3 pM) and 1.5×10^{-16} moles in 100 mL (1.5 pM) when bound to DNA.[20] Selvin *et al.* reported 2pM detection sensitivity of free Tb-DTPA-cs124 in an optical system which was not configured for maximum sensitivity.[23] Mathis and coworkers detected 2 attomoles of DNA on dot blots with Eu-cryptate, which they stated was similar to other nonisotopic methods.[19] and 1 pM of prostate-serum antigen in another Eu-cryptate assay.[18] They also compared the sensitivity of Eu-cryptates versus ^{32}P in a DNA-hybridization assay and found them equivalent, although the ultimate detection limit in each case was not reported and conditions of the assay necessarily differed in the two cases.[33] Specifically, they hybridized a cryptate- or ^{32}P-labeled 21-mer to a complimentary biotin-labeled DNA. They then added the ^{32}P labeled DNA to a steptavidin-coated microtiter plate, washed it, and read the remaining radioactivity. In parallel, they added the cryptate-labeled DNA to a multimeric streptavidin complex bound to a modified allophycocyanine. The cryptate transferred most of its energy to the APC and the time-delayed fluorescence from the APC was measured. (See Section 6.4. below for further discussion.) When using 0.2 pmoles of DNA in 50 μL, in both cases they found the same sensitivity. The cryptate has the advantage of being homogenous (requiring no washes) and nonradioactive, but cryptates require 0.4M NaF in the buffer for optimal fluorescence. Other chelates do not require this.

6.2.3. Multiple Labeling and Selected Applications

To increase detection sensitivity of biomolecules it is routine to label biomolecules with multiple fluorescent probes. However, organic dyes are self-quenching, i.e., the fluorescence intensity increase with the number of labels is less than linear. This is due to the small Stokes shift and possible direct interactions between dyes. Multiple labeling can also reduce the solubility of the labeled biomolecule and adversely affect its function. These factors limit the number of conventional fluorophores attached to an antibody, for example, to approximately six.

The large Stokes shift and excellent solubility of lanthanide chelates makes multiple labeling of biomolecules possible, although the effect on the biomolecules function is likely application specific. Takalo *et al.* showed that an IgG antibody could be labeled with 25 chelates with a linear increase in luminescence (Figure 6.4, triangles; circles are intensity per chelate), although antibody activity was not reported.[34] Kwiatkowski *et al.* introduced multiple chelates into DNA via a phosphoramidite technology.[12] Dahlén *et al.* synthesized a series of 14–28 base oligomers with 10, 25, or 40 amino-deoxycytidines on the 5′-end, which were then labeled with a europium chelate.[52] The 10- and 25-amine DNA was synthesized in high yield and could be completely labeled, while the yield on the 40-amine-labeled DNA was relatively low and could be labeled with a maximum of 26 chelates. In a solid phase assay the melting temperature and hybridization efficiency was unaffected by the chelate tails, although the kinetics were slowed—not due to the chelates but due to the extra sequence complexity from the addition Cs. The linearity of europium fluorescence vs. chelate number was not reported, although their data clearly show the oligos with larger numbers of chelates are brighter. A detection limit of 10^7 oligos was reported, which the authors say is at least as high as other oligonucleotide-based detection systems. Interestingly, on the same 25-chelate oligo the authors compared the sensitivity when using the directly luminescent europium chelates to the DELFIA method, where the europium was released from the luminescent chelate and then reconstituted into the

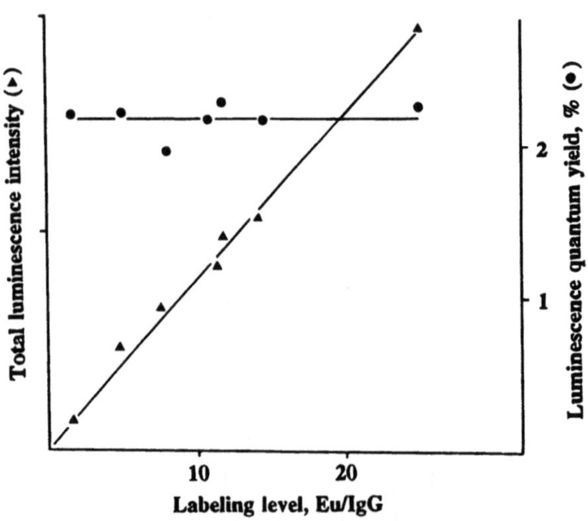

Figure 6.4. Multiple chelates can be attached to a biomolecule (antibody) with a linear increase in brightness, in contrast to conventional fluorophores which self-quench.

enhancement solution. Two attomoles (10^6 molecules) in the DELFIA method compared to 10^7 using the directly luminescent chelates were reported.

A common way to minimize adverse effects due to multiple labels is to produce a long tail which contains the chelates and then attach the tail to the biomolecule at only one point. (This method is also used in magnetic resonance imaging with nonluminescent gadolinium chelates to increase contrast and sensitivity.[35]) Bailey was the first to report such a method where they used a polylysine tail labeled with over 100 DTPA-paraaminosalicylate chelates, attached at a single point to an antibody[36] (see also Figure 6.1). A critical element appears to be complete acetylation of the polylysine to reduce excess positive charge which can lead to loss of function and nonspecific staining. Morrone has given a detailed procedure and has well characterized polylyine-DTPA-pAS complexes.[37] He showed that 80 Tb chelates could be attached to a polylysine tail which was then attached to a protein (IgG or streptavidin). The protein-chelate complexes were then used to specifically stain cellular organelles, showing that the complexes were functionally active (see Figure 6.8 and Section 6.3 below). Lamture and Wensel have used polylysine with the chelate-sensitizer dipicolinic acid.[38]

An alternative method for various levels of multiple labeling was reported by Diamandis *et al.* and commercialized by CyberFluor Inc. (Toronto, Canada) under the name FIAgen (reviewed in reference 2; see Figure 6.6). In one version,

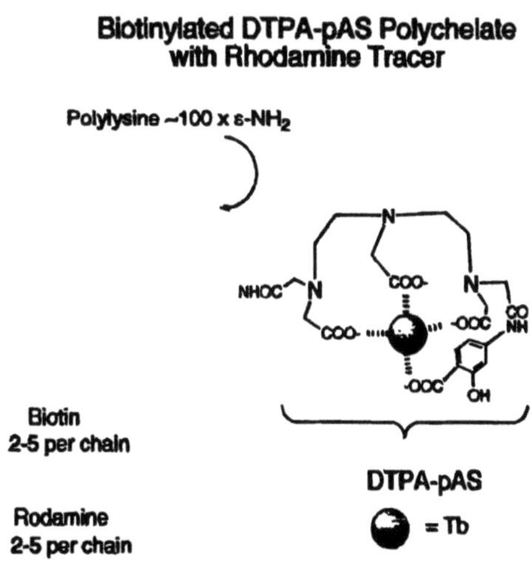

Figure 6.5. Multiple chelate-labeling can be achieved by attaching many (10–150) chelates on a polylysine tail, which is then conjugated to a biomolecule at a single point. If conventional fluorophores are also sparsely attached to the polylysine, direct comparison between conventional (including confocal) imaging and time-resolved imaging can be made. (Figure from reference 64.)

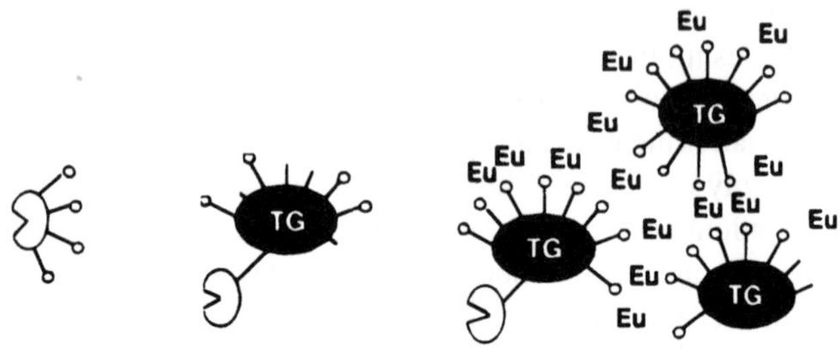

Figure 6.6. Multiple BCPDA europium chelates can be incorporated onto streptavidin-thyroglobulin complex (see text for details; figure from reference 2).

streptavidin was labeled with approximately 14 BCPDA molecules (see Figure 6.1). The sensitivity was 10^{-10}–10^{-11} M. In a second version, approximately 160 BCPDA molecules were bound to a thyroglobulin crosslinked to streptavidin yielding a detection limit for biotinylated targets of 10^{-11}–10^{-12} M. In a third version, these thyroglobulins were noncovalently bound together, yielding a detection limit of 10^{-12}–10^{-13} M. The complexes were also used in time-delayed imagining studies.[39]

Yet another method for signal amplification is to use an enzyme-amplification scheme to deposit many labels locally. Tyramide signal amplification (TSA), is a peroxidase-driven amplification system by which a large accumulation of biotin-tyramide is deposited locally and detected by fluorescently-labeled streptavidin. Signal amplification can be 100-fold. TSA has been used with both conventional fluorophores and with lanthanide in time-resolved detection, the latter being beneficial when autofluorescence is still a problem despite the signal amplification. For example, de Haas *et al.* used TSA in DNA and RNA *in situ* hybridization in strongly autofluorescent cells where the biotin-tyramide was detected with europium-labeled streptavidin.[40] Signals from the europium labels were sufficiently bright that they could be observed with the eye.

The Diamandis group has pioneered an alternative enzymatic amplification scheme to achieve the low detection limit of 1.5×10^5 analyte molecules in 100 μL, or 2.5×10^{-15} M.[41] The general idea is that the presence of an analyte leads to the capture of an enzyme which converts an organic molecule into a sensitizer, leading to lanthanide luminescence. Specifically, an analyte (anti-α-fetoprotein antibody) was immobilized to a surface via a coating antibody. A biotinylated antibody was then added, excess removed, and a streptavidin alkaline phosphatase (ALP) complex added and excess removed. Fluorosalicylic acid phosphate ester (FSAP) was added, which is not a terbium sensitizer but becomes one when the phosphate is removed by the ALP, forming FSA. The pH is then increased to 13 and Tb-EDTA is added. The Tb-EDTA (without sensitizer or with FSAP) is essentially nonfluorescent, but at pH 13 the FSA can ligate to Tb-EDTA, forming a luminescent complex. The result is that the presence of an analyte leads to many luminescent terbium complexes via the ALP-catalyzed creation of FSA sensitizers. The combination of enzyme amplification and time-resolved detection led to the excellent sensitivity, albeit at a cost of several wash steps. This methodology has been applied to DNA hybridization in both dot-blot and solid-phase microtiter-based assays, achieving detection limits for target DNA of 0.2 fmol,[42] 125 pg (pBR322 DNA; 6×10^{-17} moles),[43] 4 pg.[44]

Ioannou and Christopoulos have taken enzyme amplification one further step, combining TSA and ALP amplification to increase the signal by another 30-fold (and the signal to noise by 10-fold) compared to ALP amplification only.[45] DNA was immobilized, hybridized to biotinylated DNA, and bound to streptavidin horseradish peroxidase which catalyzed the deposition of biotinylated tyramide to the surface. Alkaline phosphatase-labeled streptavidin was then bound to the immobilized biotin and enzymatically converted the FSAP into FSA as described above.

PCR is of course a very powerful method of DNA amplification and has been used in conjunction with lanthanide luminescence. Diamandis *et al.*,[3] using the time-resolved enyzme amplification scheme described in combination with reverse transcription polymerase chain reaction (RT-PCR), measured prostate serum antigen (PSA) mRNA with high sensitivity: 160 PSA cDNA molecules in the preamplification sample were detected with a signal to background of 10, and mRNA corresponding to one PSA-producing cell in the presence of a million PSA-negative cells was detected with a signal to background of three.[46,47] Sensitive detection of PSA-expressing cells, particularly in the blood and lymphatic systems, is important for evaluating the stage of prostate cancer. (PSA is a 33-kD serine protease glycoprotein used as a marker for prostate cancer.) Diamandis's method involved using RT-PCR to amplify any PSA mRNA present, with digoxigenin (a hapten)-dUTP used in the amplification step. The denatured amplified DNA was hybridized to a 24-mer DNA previously immobilized on a streptavidin-coated microtiter plate via a biotin incorporated at the end of the 24-mer. An ALP-labeled antibody to digoxigenin was added and excess removed. Upon addition of FSAP, as described above, the ALP then forms fluorescent terbium complexes which are detected in time-resolved mode.

PCR and the DELFIA method have also been combined. For example, PCR in combination with DELFIA detection system was capable of detecting a single purified coxsackievirus A9 RNA from clinical samples[48]; five copies of HIV-1 DNA[49]; 50 PSA mRNAs and a single PSA mRNA-expressing cell.[50]. In the latter case, to enhance sensitivity, oligonucleotide probes for PSA mRNA contained 20 chelates at the end. PCR and the DELFIA methodology have also been combined to detect point mutations.[49,51]

One of the more interesting applications of PCR and lanthanide luminescence is a seven pseudo-color assay utilizing the DELFIA system to detect human papilloma virus (HPV) types. This application takes advantage of the sharply spiked spectra of lanthanides, which leads to very little spectral overlap when using more than one lanthanide and the ability to excite multiple lanthanides with single excitation wavelength.[13] HPV infection is implicated in cervical and other cancers. The cancer risks are correlated with HPV type, and hence detection of HPV type is important. The general scheme is outlined in Figure 6.7. PCR was used to amplify a portion of the HPV containing type-specific sequences and to introduce a biotin which was then used to immobilize the amplified DNA on a streptavidin-coated solid support. DNA probes specific to HPV type and containing 10–20 (nonfluorescent) chelates labeled with a unique combination of Eu, Tb, or Sm were then hybridized to the target DNA. (This combinatorial approach has

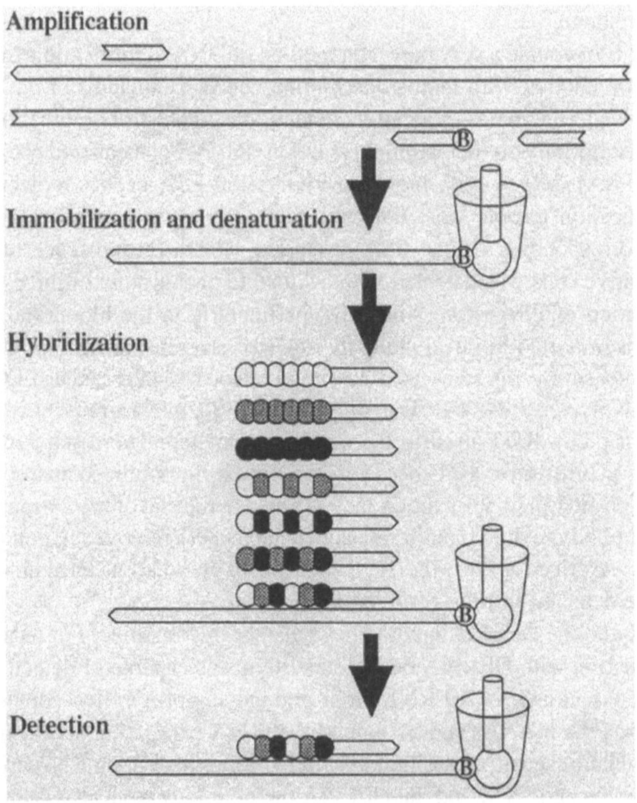

Figure 6.7. Methodology for 7-pseudo color detection of multiple human papilloma virus types using PCR amplification and multiple lanthanide chelates. See text for details. (Figure from reference 13.)

been used with great success with regular probes as well.[52,53]) After free probes were washed away, an enhancement solution was added and Eu, Tb, and Sm luminescence was measured at 615 nm, 545 nm, and 642 nm with various time delays. For the Eu-labeled probes, 100 amol of target DNA could be detected, with poorer sensitivity with Tb and Sm. All seven types of HPV strain were correctly identified. The group has also applied this technology to the detection of diabetes-related HLA alleles.[54]

6.3. Imaging

One of the most demanding applications of lanthanides is in imaging. In scanning microscopy, the slow emission rate of lanthanides greatly slows data acquisition: with a millisecond lifetime, it is necessary to wait at least this long on each pixel, leading, for example, to an acquisition time of over 5 minutes for a 500 × 500 image. Nevertheless, Morrone used scanning x-ray microscopy to take very high resolution (75 nm) images of lanthanides specifically labeled in cells.[37] To increase signal levels, Morrone used a terbium-loaded polylysine tail (see above), which also contained trace rhodamine (or fluorescein). Importantly, this dual-label approach enabled him to correlate the images obtained with x-ray excited terbium luminescence to confocal images of rhodamine emission and showed that they were essentially identical, except for the higher resolution afforded by the x-ray excitation. Although the published figures where for the proteins actin and tubulin (Figure 6.8A), similar results were achieved using FISH (Figure 6.8B, C), splicing factor (a protein involved in removal of introns and splicing of mRNA form exons), and NuMA (nuclear mitotic spindle apparatus, a protein needed in cell division) (M. M. Morrone, personal communication). These latter examples are particularly important because labeling is within the nucleus where nonspecific labeling can be the most troublesome because trivalent lanthanide can bind directly to DNA if they are pulled out of their chelate.

In wide-field imaging, lanthanides and time-resolved detection can be used to great advantage in samples where autofluorescence reduces contrast with conventional probes.[55] Pulsed excitation and time-delayed detection is used to discriminate against the autofluorescence, which tends to be nanosecond in duration. The time-delay can be achieved with a chopper,[39,56] a ferroelectric liquid crystal shutter,[57,58] an image-intensified CCD,[59] or possibly in the near future with CCDs containing an electronic shutter. The slow emission rate of the lanthanides still precludes rapid acquisition times, but in wide-field imaging acquisition times of tens of seconds may be acceptable.

Periasamy *et al.* detected a Eu-labeled FISH probe against human papillomavirus (HPV) 16 DNA.[60] The use of long-lived probes in such application is

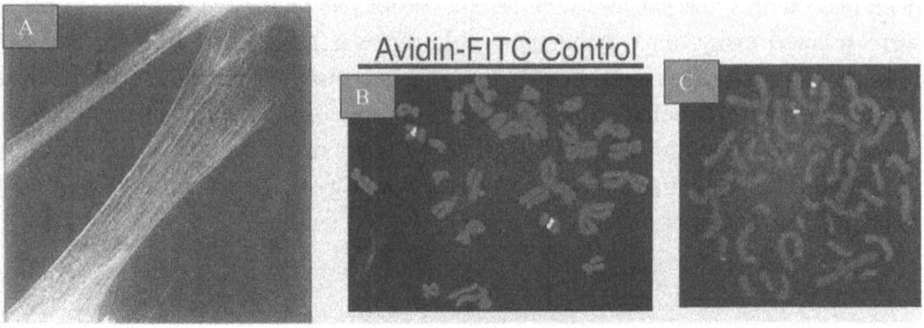

Figure 6.8. Specificity in labeling can be achieved with multi-chelate complexes. (A) Actin fibers in a Wi38 fibroblast cell labeled with biotinylated phalloidin, followed by streptavidin and a biotinylated polylysine-chelate containing approximately 80 chelates. Terbium luminescence was stimulated by soft x-ray excitation. (B) FISH using conventional fluorescein-labeled probes, a control for (C). (C) FISH using polychelate Tb probes containing trace rhodamine and visualized by exciting and detecting rhodamine fluorescence, similar to (B). A comparison of (B) and (C) shows that the polychelate probes are capable of specific hybridization to their target.

particularly advantageous for FISH on standard pap smears which have already been stained with conventional absorption dyes used by pathologists because these dyes tend to be highly autofluorescent and obscure FISH signals when using conventional fluorescent dyes. Figure 6.9 shows the images without and with time-delayed detection, where the latter dramatically increases contrast. In both cases excitation was in the UV. A more realistic comparison is to compare lanthanide probes (necessarily excited in the UV) using time-resolved detection with conventional fluorescent dyes excited in the visible without time-resolved detection. Seveus *et al.* made such a comparison for europium- and fluorescein-labeled streptravidin bound to biotinylated antibodies to a human colon cancer antigen.[61] The cells were highly autofluorescent due to glutaraldehyde fixation in a permanent mounting medium. Integration times of 10 seconds for the europium in time-resolved mode and 0.5 seconds for the fluorescein in prompt (conventional) mode led to similar signal strengths—16,000 for europium and 12,000 for fluorescein—but the background with europium was near the detector noise (5 counts per pixel), whereas with fluorescein the autofluorescent background was approximately 2000 counts. Hence the signal:background with europium was over 3000:1, whereas with fluorescein it was 6:1, or a 500-fold improvement using time-resolved detection of europium. An earlier paper by the same authors reported more modest gains when using time-resolved imaging of europium chelates in FISH for the detection of collagen mRNA and human papillomavirus sequences.[62] The limitations of this earlier work appeared to be due to a chelate which was not very photostable. More recently other workers have used time-

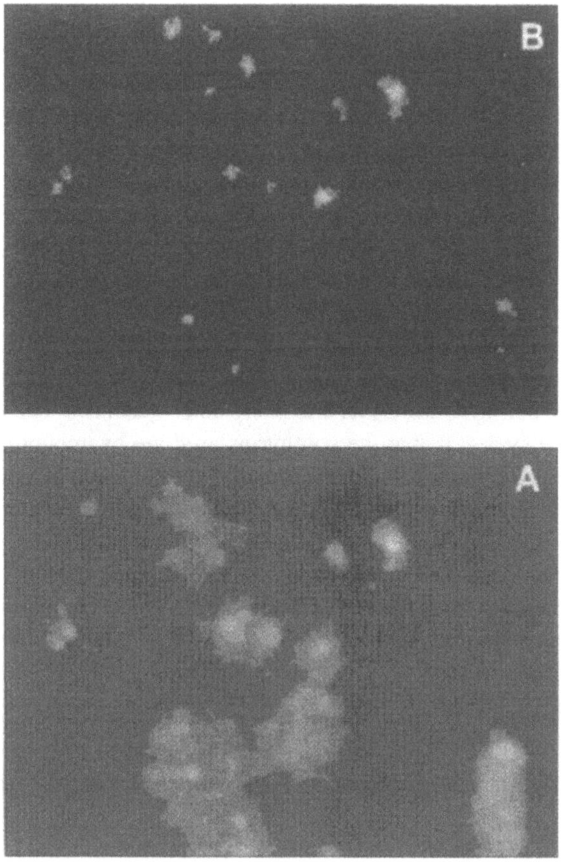

Figure 6.9. Comparison of time-resolved (B) and conventional, i.e., prompt or steady-state, fluorescence image of HPV DNA in fixed CaSki cells. The contrast is greatly enhanced by time-resolved detection (30 μsec delay after pulsed excitation with a window of 250 μsec). CaSki cells contained approximately 500 copies of HPV. 8kB DNA probes were labeled with biotin-11-dUTP via random-priming method and after hybridization to target, europium-conjugated avidin (Perkin Elmer Inc.) was added.

resolved imaging of a europium chelate to image PSAs where they state that photostability was not a problem.[63] Future applications of time-resolved imaging in FISH and other areas using lanthanides in samples which are highly autofluorescent look promising, although very low target number will likely require amplification schemes due to the inherently low intensity (long lifetimes) of the lanthanides.

6.3.1. Alternative Time-Resolved Probes

As mentioned, the millisecond lifetime and sharply spiked spectrum of lanthanides are highly advantageous for discriminating against background fluorescence. However, the long lifetime limits emission rate. To date, lanthanides also require UV excitation, which is particularly disadvantageous in microscopy. When this is an issue, microsecond dyes are attractive alternatives. Tanke *et al.* and Savitsky *et al.* have developed an impressive set of Pd and Pt porphoryn-based probes which appear to be unusually bright with many favorable spectral characteristics: extinction coefficients up to 200,000 M^{-1} cm^{-1} with an excitation peak at 380 nm and a smaller peak in the visible (530 nm), emission in the red (>650 nm), and quantum yields ranging from approximately 0.1 to 0.3.[57,64] They have used the porphyrins in FISH and many other systems.[57] The primary disadvantage of the system is that the porphyrins used are hydrophobic, they require deoxygenation for maximal emission, and the chemistry of coupling to biomolecules is not straightforward. (Tanke *et al.* have developed an amine-reactive NHS ester,[65] although a thiol-reactive porphoryn has not been reported.) Lakowicz *et al.* has published many papers reporting microsecond dyes based on ruthenium, rhenium, and osmium. These probes can be excited in the visible to red, be highly polarized, and have pM sensitivity. A review has been written summarizing this work.[66]

6.4. Lanthanides as Donors in Resonance Energy Transfer

Fluorescence resonance energy transfer (FRET) is a widely used technique to measure the distance between two points which are separated by approximately 10–100 Å. A number of excellent reviews on FRET have been written.[67–75] Lanthanide-based RET (LRET) is a recent modification of the technique with a number of technical advantages, yet relies on the same fundamental mechanism—subject to careful interpretation of various terms. A recent review of LRET has appeared,[8] as well as a summary of lanthanide luminescence[1] and its application in biology.[7,9] LRET has primarily been applied to proteins,[16,17,76–81] although there have been a number of applications to DNA.[82–88] Here we provide a brief summary of the relevant theory before highlighting some LRET applications to DNA.

In FRET, an excited fluorescent donor molecule attached at one point in the biomolecule transfers energy to an acceptor molecule attached at a second site through a nonradiative dipole-dipole coupling which is inversely proportional to the sixth power of the distance between the two dyes:

$$E = 1/[1 + (R/R_0)^6] \qquad (6.1)$$

E is the efficiency of energy transfer from donor to acceptor, i.e., E is the probability that an excited donor will transfer its energy to the acceptor instead of decaying via other pathways such as fluorescence or heat. R_0 is the distance at which half of the energy is transferred and is generally 20–50 Å, although it can be ⩾70 Å when using lanthanides and ⩾80 Å with far-red dyes.[89] R_0 depends on the spectral properties of the donor and acceptor and also on their relative orientation. (The orientation dependence of R_0, is often referred to as κ^2. This can be a significant source of uncertainty in determining distances via FRET—see Section 6.5 below.) By knowing R_0, which can be calculated or experimentally determined, and measuring E, the distance between the probes can be found. E can be measured because it reduces the donor's intensity and excited-state lifetime:

$$E = 1 - I_d/I_{da} = 1 - \tau_d/\tau_{da} \tag{6.2}$$

where the subscripts refer to donor's intensity or lifetime in the absence (I_d, τ_d) and presence of acceptor (I_{da}, τ_{da}). E can also be measured by comparing the amount of fluorescence from donor and from sensitized emission of the acceptor, i.e., the acceptor emission due only to energy transfer from the donor.

$$E = (I_{ad}/Q_a)/(I_{da}/Q_d + I_{ad}/Q_a) \tag{6.3}$$

where I_{da} is the integrated area under the donor emission curve in the presence of acceptor, I_{ad} is the integrated area of the sensitized emission of the acceptor (i.e., not including the fluorescence due to direct excitation of the acceptor, which may occur when exciting the donor) and Q_i is the quantum yield for donor or acceptor.

6.5. Advantages of LRET

We and others have shown that using luminescent terbium or europium chelates as donors and organic-based acceptors, yields many technical advantages over conventional FRET.[8,16,17,23,74,87,88] Terbium can readily transfer energy to dyes such as fluorescein, tetramethylrhodamine, Cy3, and Alexa dyes, while europium can transfer energy to Cy5, Alexa dyes, or allophycocyanine.

An example which highlights these advantages is shown in Figure 6.10, where energy is transferred from a europium chelate attached at the 5′-end of a DNA oligomer to an organic dye (Cy5) at the 5′-end of a hybridized strand. Figure 6.10A shows energy transferred measured spectrally. Pulsed excitation, followed by detection after a delay of approximately 100 μsec, is used to eliminate any prompt fluorescence, including direct excitation of the acceptor. Hence, any emission of the acceptor is due only to energy transfer from the (long-lived) donor. The donor–acceptor spectrum displays a large peak around 670 nm, which is the Cy5-sensitized emission. The signal to background is excellent because the donor

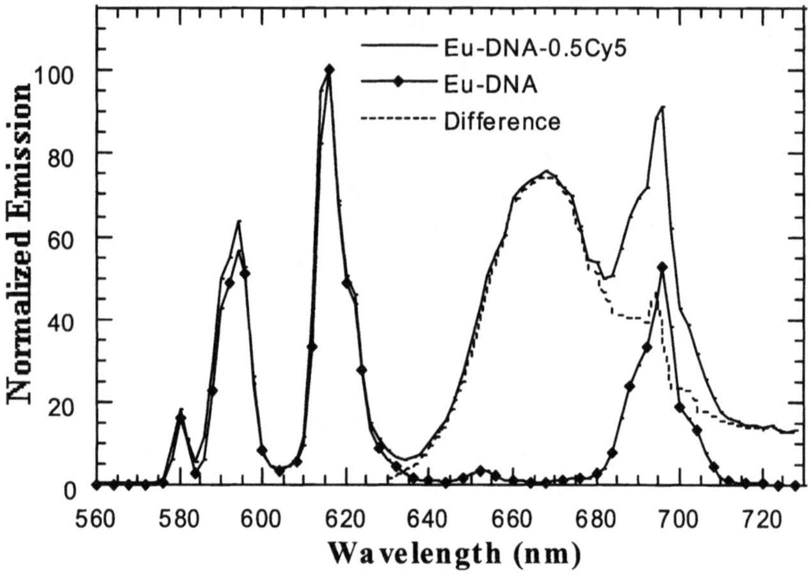

A

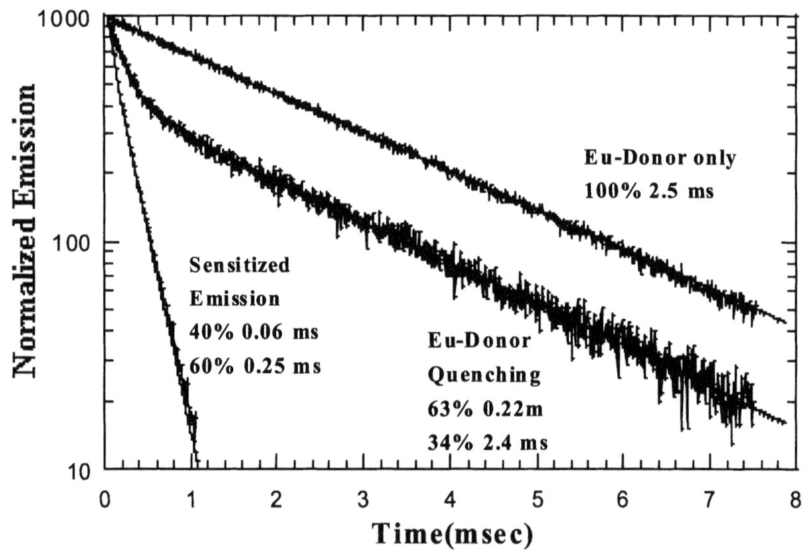

B

Figure 6.10. Time-delayed spectra (A) and time-resolved lifetime (B) measurements of LRET from a Europium donor to a Cy5 acceptor on a 10-mer DNA. (See text for details. Adapted from reference 88.)

is silent in this region and direct acceptor emission is eliminated. Indeed, the background is largely due to detector noise, which can be made extremely small with modern detectors.

Figure 6.10B shows the same system where energy transfer is measured temporally, following an excitation pulse. The europium donor-only is single exponential with a 2.5 msec lifetime. Sub-stoichiometry amounts of a Cy5-labeled complementary DNA were then added. Upon hybridization, the Eu^{3+} decay becomes biexponential, indicating two populations of donors. One is essentially unquenched (lifetime 2.4 msec) which is the europium not hybridized to a Cy5 strand. The other is highly quenched (lifetime 0.22 msec; $E = 1 = 0.22/2.5 = 0.91$), corresponding to a Eu-DNA hybridized to Cy5-DNA. (A control of Eu-DNA hybridized to an unlabeled strand shows no quenching.) The Eu-DNA-Cy5 complex leads to sensitized emission of Cy5, whose lifetime can be measured at 668 nm. This lifetime is determined by the lifetime of the Eu-donor "feeding" the Cy5. The 0.25 msec of the sensitized emission is in good agreement with the 0.22 msec Eu-lifetime component measured by donor-quenching. Note that even though a mixture of donor-only and donor-acceptor is present, the sensitized emission lifetime only arises from the donor–acceptor complex. This has important implications for systems where labeling is incomplete or unknown.

The advantages of LRET, compared to conventional FRET which relies on two organic fluorophores, are the following:

- The efficiency of energy transfer using lanthanides as donors is considerably higher compared to most organic-based donors, leading to larger signals and farther measurable distances. The R_0 is typically 50–70 Å with current chelates and as high as 90 Å with allophycocyanin as the acceptor. This is larger than most conventional FRET pairs. In addition, the R_0 can be optimized to the particular system by varying the amount of H_2O/D_2O because D_2O increases the lanthanide quantum yield,[90] thereby changing R_0. Lastly, the lanthanide quantum yield is insensitive to pH over a wide range (pH 5–9) and hence, these R_0s are maintained over a wide pH range if the appropriate acceptor is used.
- In LRET, energy transfer is primarily dependent on the distance between donor and acceptor, and not on their relative orientation, leading to more precise distance measurements. This is because donor Tb^{3+} emission is inherently unpolarized and hence donor orientation does not affect energy transfer efficiency.[91] Eu^{3+} can be polarized[92] but is often unpolarized when attached to biomolecules via conventional and relatively flexible linkers. With the donor emission unpolarized, κ^2 is constrained to be between 1/3 and 4/3, the extreme cases corresponding to a completely rigid acceptor or perpendicular or parallel to the radius vector, respectively. (In FRET, $0 < \kappa^2 < 4$, where $\kappa^2 = 0$ leads to $R_0 = 0$ and hence no energy transfer at any distance.) If one then assumes $\kappa^2 = 2/3$, as is often done

(donor and acceptor both unpolarized), then at most the distance inferred is in error by ±12%.[8,75]

• Because of the donor's unusually long lifetime (msec), LRET lifetime measurements can be made with high precision: lifetime measurements avoid problems of concentration determination and allow analysis of heterogeneous mixtures. The long lifetime also means it is likely that the donor's and acceptor's orientations are unpolarized because they have a millisecond to reorient, rather than the nanoseconds typical of organic lifetimes.

• In LRET, a 50- to 100-fold improvement in signal to background of the sensitized emission compared to conventional FRET has been achieved, largely through the reduction of background. In particular, the acceptor's sensitized emission can be measured with no background, either from donor luminescence or from direct excitation of the acceptor. Specifically, the emission of the acceptor due to excitation by the excitation light, which has nanosecond lifetime, is discriminated against using pulsed excitation and time-delayed detection. The donor emission is discriminated against by wavelength filtering—donor emission is highly spiked, with regions of darkness (see Figure 6.2). With the appropriate choice of acceptor, acceptor fluorescence is in this dark region. The improvement in signal to background enables long-distances to be measured (where the signal is weak) and easier handling of incompletely-labeled samples (because excess acceptor or donor do not contribute background).

• In LRET, the lifetime of the acceptor decay due to energy transfer, i.e., sensitized emission, can be measured (see "sensitized emission" lifetime curve in Figure 6.10B). This decay follows the donor's lifetime because it is being "fed" by the long-lived donor, which is the rate-limiting step. As soon as the acceptor receives a quanta of energy from the donor, it will, within its nanosecond lifetime, emit that energy as an emission photon. The ensemble decay of many such acceptor molecules is a signal at the acceptor's emission wavelength which decays with the donor lifetime in the donor–acceptor complex. The sensitized emission lifetime is insensitive to concentration, *and arises only from the fully labeled donor–acceptor complex.* Hence if there is incomplete labeling, with some biomolecules labeled only with donor or only with acceptor, these do not contribute to the sensitized emission signal. This significantly increases the number and type of samples which can be looked at with resonance energy transfer since complete labeling is often not possible and leads to difficulties in conventional FRET.

A description of an instrument optimized to perform LRET has been described recently.[93] A schematic of the instrument is shown in Figure 6.11. There are several commercially available instruments (Perkin-Elmer Corp., Norwalk CT; Packard Instruments, Meriden, CT; Spex-Instruments SA, Edison, NJ; Cis Bio International, Cedex France) which are capable of making micro- to millisecond time-resolved fluorescence measurements used in LRET.

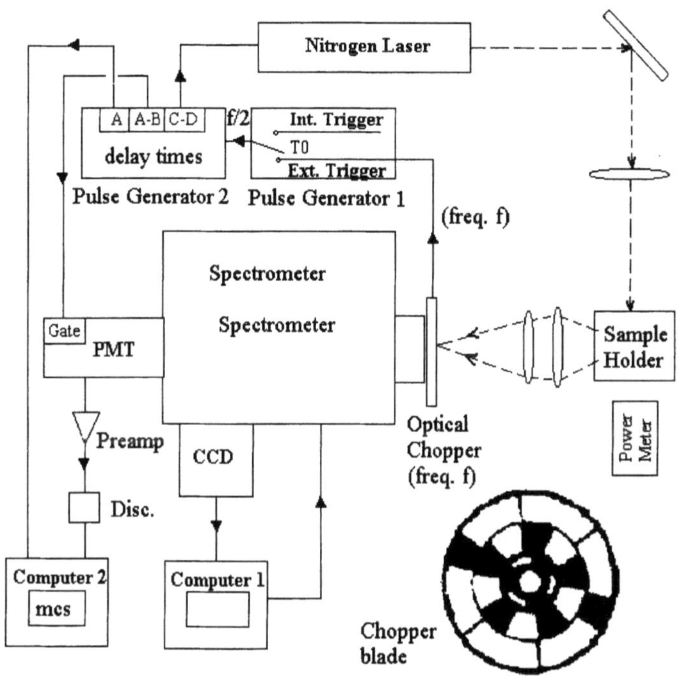

Figure 6.11. In LRET, a pulsed nitrogen laser (λ = 337 nm, 5 nsec full-width-half-max, 40–50 Hz repetition rate) excites the lanthanide chelate which then emits and/or transfers energy to an acceptor which also emits. An optical chopper directly in the detection pathway is timed such that all prompt fluorescence is eliminated. The spectrometer spreads the emitted light by wavelength and the light is detected either by a CCD to obtain an emission spectra with all wavelengths collected simultaneously, or by a photomultiplier tube which is connected to a multi-channel scalar to acquire time-decay of donor or acceptor (sensitized emission) luminescence at a particular wavelength, with 2 microsecond resolution. (See reference 93 for more details.)

6.6. LRET Applied to Protein–DNA Interactions

6.6.1. Protein-Induced DNA Bends

The Heyduk lab has been very active in using LRET to study DNA–protein interactions. In one of their earlier works they used LRET to measure protein-induced DNA bending (Figure 6.12).[85] The bending of DNA is essential to the packing of DNA into chromosomes, and is often involved in transcriptional regulation as well. Electrophoretic gel-shift assays are widely used to determine if a particular protein bends DNA—bent DNA runs differently in the gel than

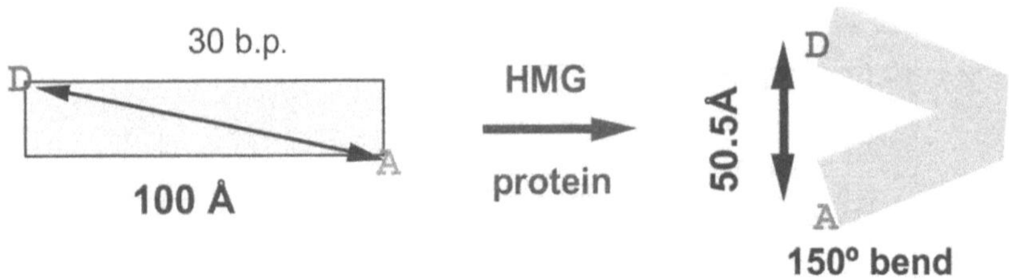

Figure 6.12. Some proteins, such as high mobility group proteins, can bend DNA. This can then be detected via LRET because the end-to-end distance changes. (Adapted from reference 85.)

straight DNA. However, LRET has a number of advantages, albeit at a cost of greater experimental complexity. In particular, LRET is a relatively direct measure of DNA bending and can be accomplished under a wide variety of solution conditions, including moderate to high salt conditions not attainable in gels. LRET, unlike FRET, is also capable of measuring the requisite distances, which can be up to 100 Å.

The Heyduk lab studied DNA bending induced by a class of proteins known as high-mobility-group (HMG) proteins. Except for the histone proteins, which bend DNA into sharply curved structures called nucleosomes, HMG proteins are the most abundant proteins associated with chromatin. One subset of HMG proteins binds to DNA with high affinity at specific sequences, and a second subset binds with relatively little specificity (aside from binding to A-T rich regions). A number of the sequence-specific HMG proteins have been shown to sharply bend DNA when bound to their high-affinity site, but before the Heyduk and coworkers' paper only indirect evidence existed that sequence nonspecific HMG proteins bent DNA.

The authors end-labeled complementary 30-base pair strands of DNA, one with a europium donor chelate, and the other with the acceptor dye Cy5. The idea is that in the absence of bends, the DNA adopts its normal B-form configuration and is therefore straight. The donor-acceptor distance is then approximately 100 Å (3.4 Å per base along the helix axis, and secondarily, a helix diameter of approximately 15 Å). In this case the donor-only lifetime was 603 ± 2 msec and the donor-acceptor lifetime was 586 ± 3 msec, corresponding to 2.7% ± 1% energy transfer or a distance of 100 Å ± 10 Å. Note that because the donor-only lifetime was rigorously single exponential, the lifetimes could be measured very accurately, and hence even this small amount of energy transfer could be measured. Upon addition of a protein that bends the DNA the ends come considerably closer, measured to be 50.7 Å ± 1.5 Å from a Eu lifetime of 226 ± 22 msec, or 62% ± 4% energy transfer. Figure 6.13 is representative data showing that as more protein is added and binds, a greater fraction of the DNA is bent, resulting in a larger fraction

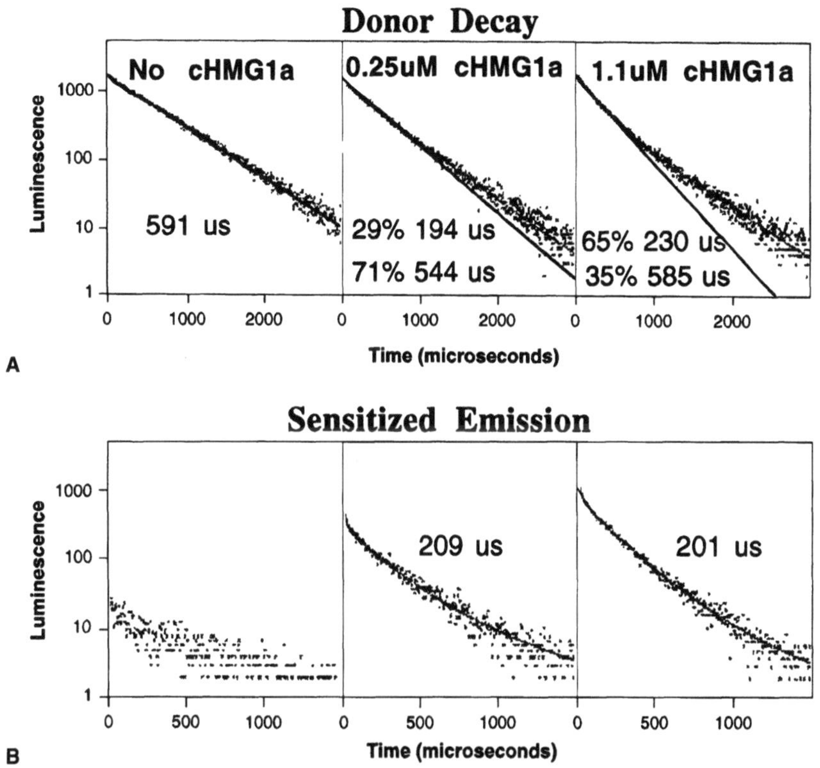

Figure 6.13. The europium donor, hybridized to a Cy5-containing complementary DNA is bent by HMG protein. The donor decay displays two components, a long-component corresponding to unbent DNA (from protein-free DNA) and a short component from bent DNA (from protein-bound DNA). This can be measured either by donor lifetime quenching (A) or sensitized emission (B). (Adapted from Figures 3 and 4 of reference 85.)

of the europium that is quenched. The longer lifetime (544–585 μs) is from Eu on unbent DNA (with Cy5 on the complementary strand); the shorter component (200 μs) is from DNA which is bent by the cHMG protein. There is good agreement between the short component of the donor lifetime and the sensitized emission lifetime.

More recently, Heyduk and coworkers have used LRET to study conformational changes and the architecture of macromolecular complexes involved in transcription initiation in prokaryotes.[84,94] The first step in transcription initiation in prokaryotes involves recognition of promoter DNA sequence by a multisubunit enzyme RNA polymerase (reviewed in reference 95). One of the subunits, σ^{70}, is involved in the initial recognition of the promoter DNA via direct protein-DNA contacts separated by ~17 base pairs. σ^{70} exits in the cell in two major forms: free and in complex with the remaining RNA polymerase subunits (core polymerase). However, only σ^{70} in complex with the core RNA polymerase is able to specifically recognize promoter DNA while the free σ^{70} does not bind to promoter DNA. Thus, the promoter recognition capabilities of σ^{70} are allosterically regulated by an interaction of σ^{70} with the core polymerase.

LRET experiments were used to investigate the nature of the regulation of σ^{70} promoter DNA binding activity.[94] The idea was to look for conformational differences in σ^{70} in the bound and free form which might affect its DNA recognition ability. The specific incorporation of the donor (DTPA-Eu-DTPA-AMCA-maleimide)[83] and the acceptor (Cy5 maleimide) into selected domains of σ^{70} was achieved by preparing a set of σ^{70} mutants with pairs of unique reactive cysteine residues engineered into the desired locations. Since both donor and acceptor were thiol-reactive and both labeling sites were cysteine residues, a mixture of donor-donor, donor-acceptor, and acceptor-acceptor labeling resulted. However, by using LRET the sensitized emission arising from only the donor–acceptor complex could be measured. Good quality determinations of τ_{ad} were possible in this case even though the donor-acceptor species constituted only approximately 25% of the mixture. Representative LRET data is shown in Figure 6.14. It would be difficult to perform these measurements with FRET utilizing classical fluorescence probes.

Comparison of distances measured for the free σ^{70} and the core-bound σ^{70} revealed that most distances were very significantly increased upon binding of σ^{70} to the core polymerase. (DNA was not present in these experiments since σ^{70} binding to the core polymerase is not DNA-dependent.) (See Figure 6.14.) One of the distances measured, between residues 442 and 366, allowed a direct comparison between a distance measured by LRET in solution and a distance between the same residues measured in the crystal, since both of these residues were present in a σ^{70} fragment for which a crystal structure is now available.[96] An excellent agreement between these distances was found—35 Å in the crystal structure versus 38 Å measured via LRET—providing a further validation of LRET results. In total, six distances between four sites in the σ^{70} protein were measured, making it possible to build three-dimensional models of the architecture of σ^{70} protein domains in free and core-bound protein.

These models revealed that an interaction of σ^{70} with the core polymerase induced repositioning of DNA binding domains of σ^{70} and the movement of N-terminal domain away from the DNA binding domains. These results provided a structural rationale for understanding the core-induced changes in DNA binding properties of σ^{70} protein, and were consistent with previous mutagensis experiments which suggested that σ^{70} promoter DNA binding activity is inhibited in free σ^{70} by the N-terminal domain of the protein. The regulation of DNA binding activity of σ^{70} by the core polymerase involves both "unmasking" the DNA-

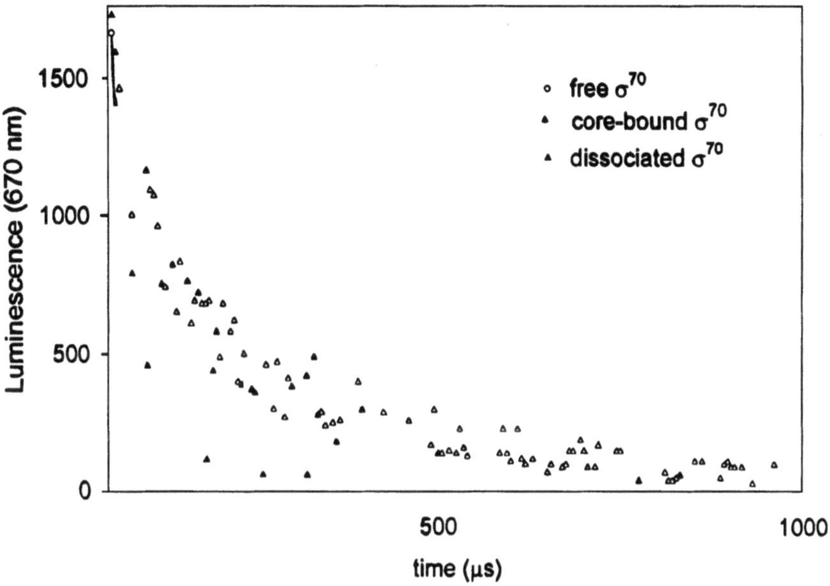

Figure 6.14. The effect of core RNA polymerase on σ^{70} measured by sensitized emission lifetime. Thiol-reactive Eu donor and Cy5 acceptor were labeled at positions A59C to R596C in σ^{70} and sensitized emission of free and core-bound σ^{70} was measured. The increase in sensitized emission lifetime upon core-binding indicates less energy transfer and an increase in distance between these sites. (From Figure 3D of reference 94.)

binding domains by the N-terminal domain of the protein and the repositioning of DNA binding domain such that in core-bound σ^{70} they have a spacing compatible with simultaneous interaction with promoter elements separated by ~17 bp DNA.

The Heyduk lab later went on to use LRET to measure distances between five sites in σ^{70} and both the 3′- and 5′-ends of bound DNA.[84] The LRET experiments were performed to obtain information regarding the architecture of the nontemplate ss DNA-σ^{70} complex.[84] Energy transfer, as measured by donor decay and sensitized acceptor signal, were easily detected in the presence of acceptor-labeled oligonucleotide in the case of core-bound σ^{70}, while no detectable energy transfer was observed in the case of free σ^{70} (as expected, since free σ^{70} does not bind DNA). LRET experiments were performed with all combinations of donor-labeled core-bound σ^{70} subunit and acceptor-labeled oligonucleotides. Thus, 10 distances were measured which allowed building a model of a complex between ss DNA and the σ^{70} subunit. This model revealed that ss oligonucleotide was bound with a defined polarity and was located across the α-helix containing regions of σ^{70} indicated by mutagenesis as important for σ^{70} function.

6.7. New DNA Dyes Based on LRET with Tuneable Emission Color and Excited-State Lifetime

One of the most important applications of fluorescence is simply the detection of fluorescently labeled targets. Multicolor fluorescence is widely used in biosciences for the simultaneous or sequential detection of multiple targets. Using conventional techniques, approximately five different colors can be detected on a single sample, and by using combinatorial methods, detection of more than two-dozen pseudocolors on a single sample has been achieved.[56] Multilifetime fluorescence is another possible method for discrimination of signals.[97] However, the excited-state range of organic-based fluorophores is limited, generally in the 1–10 nsec range, and not systematically tunable. Metal–chelate complexes can extend the lifetime regime to microseconds[98] and lanthanides, as we have seen, can have millisecond lifetimes.[1] As previously mentioned, these long-lived probes have the advantage that background fluorescence, which is typically nanosecond in lifetime, can be readily discriminated against. Another advantage is that instrumentation is considerably simpler in the micro- and millisecond time regimes than in the nanosecond regime. However the number of long-lived luminescent probes is limited, and millisecond probes have inherently limited emission rates (≈1000 photons/sec/molecule).

Recently it has been shown that using compound dyes in which a fluorescent donor molecule transfers energy to a nearby acceptor dye can yield enhanced emission characteristics including a single excitation wavelength and multiple emission wavelengths.[99,100] These dyes have been particularly useful for improving sensitivity in DNA sequencing applications.[101] In principle these dyes can be used with any biomolecules, although in practice they have been limited to DNA-binding dyes because the DNA provides a very convenient and robust scaffold for placing the donor-acceptor at well defined positions. These dyes relied on organic fluorophores having excited-state lifetimes in the nanosecond range, and hence enhanced lifetime characteristics have not been shown.

More recently we have reported on the development of compound dyes using long-lived lanthanides as donors and conventional organic fluoropohores as acceptors. These compound dyes have both tunable emission wavelengths, from 520 to 680 nm, and simultaneously tunable excited-state lifetimes, ranging from 50 to 500 μsec.[82] Future work is expected to extend these ranges further. We call these lifetime and color-tailored fluorophores (LCTF). By utilizing both the time-domain and wavelength-domain, these probes have the potential for quadratically increasing the number of detectable probes on a single sample.

To construct an LCTF, a lanthanide donor molecule (lifetime $\tau_0 \approx$ msec in the absence of acceptor) is placed a defined but controllable distance from an organic acceptor fluorophore. The lanthanide transfers energy to the acceptor, which then

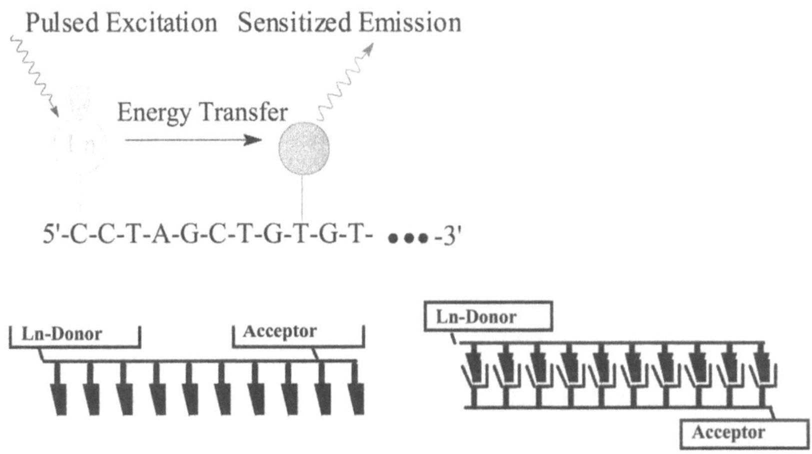

Figure 6.15. General outline for construction of lifetime- and color-tailored fluorophores based on a long-lived lanthanide chelate transferring energy to a conventional fluorophore. The probes can be attached to DNA on the same strand or on complementary strands, at the end or in the middle.

re-emits (Figure 6.15). The color and lifetime of this re-emission, i.e., the sensitized emission, are tuneable. The color is tuneable because when energy transfer (E) is chosen to be large emission color is largely determined by the acceptor fluorophore, which can be chosen for the desired emission color. The lifetime of the complex is tuneable because it is determined by the amount of energy transfer between the donor and acceptor. Specifically, the donor's lifetime in presence of acceptor (τ_{DA}) is reduced by energy transfer:

$$\tau_D = \tau_0(1 - E) \qquad \tau = \tau_0/[1 + (R_0/R)^6] \tag{6.4}$$

As discussed previously, the acceptor has an intrinsic fluorescence lifetime of a few nanoseconds, but because it is continually being excited via energy transfer from the donor, its emission intensity decays with a lifetime that follows the donor's lifetime, τ. The lifetime is readily tuned by altering the distance (R) and hence energy transfer efficiency between donor and acceptor.

As proof of principle of these lifetime and color-tailored compounds, we have used oligonucleotide duplexes as a rigid yet distance-adjustable scaffold for donor-acceptor pair attachment.[82] The DNA also acts as a means of attaching (hybridizing) the LCTF to target DNA. In principle, the potential exists for making LCTFs for attachment to protein and other targets. Note that this scheme is identical to when one uses LRET to measure distances, but here the donor and acceptor combination is considered as a "black-box" where the interest is in the input parameters (wavelength of excitation) and the output parameters (wavelength, intensity, and temporal decay of emission). The distance between donor and acceptor is simply a means to achieve a desired level of energy transfer and is not of interest (or not known) by the user.

We have attached donor-acceptor pairs to DNA in two ways. One method links a lanthanide chelate donor on the 5′-end of a DNA strand, and a fluorescein acceptor in the middle of the *same* strand via C-5 deoxyuridine tethering (Figure 6.15). In this method, by choosing an appropriate DNA sequence, the LCTF can bind and label a unique complementary DNA sequence (in a target DNA, for example) with a probe of defined color *and* lifetime—e.g., in FISH. The other method (Figure 6.15) tethers a lanthanide chelate donor on the 5′-end of one DNA strand, and places the acceptor on the 5′-end of a complementary strand, as described previously.[86–88] This attachment mechanism may be useful in chip-based DNA diagnostics and sequencing, because an oligonucleotide can be conveniently synthesized from a solid support with an incorporated acceptor fluorophore or the support itself may contain an acceptor.

Figure 6.16 shows the delayed emission spectra of terbium and fluorescein placed on the same strand, separated by either 8 or 10 base pairs (Tb-8-FL, Tb-10-FL, respectively) and hybridized to a complementary unlabeled DNA target; also shown is the terbium-only spectra. The data is collected beginning 30 μsec after

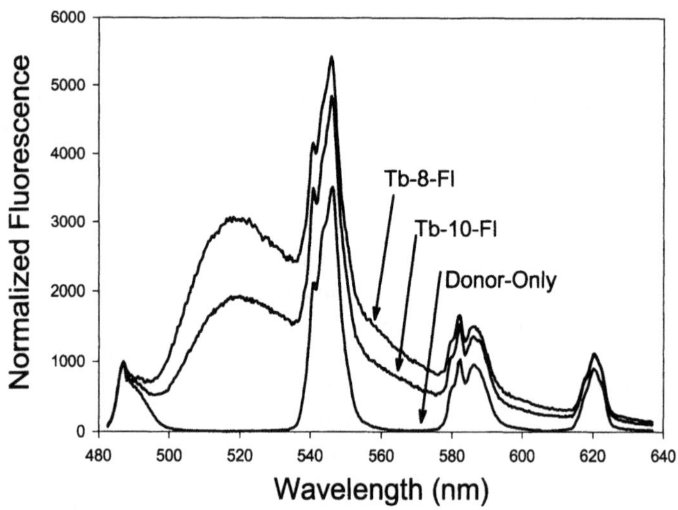

Figure 6.16. Time-delayed emission spectra of terbium-fluorescein LCTF placed on the same DNA strand, separated by either 8 or 10 base pairs and hybridized to unlabeled complementary strand. Emission is collected 30 msec after the excitation pulse; all fluorescein emission (peak at 520 nm) is due to energy transfer from terbium. Energy transfer is larger for the Tb-8-Fl, yielding a larger amount of fluorescein sensitized emission.

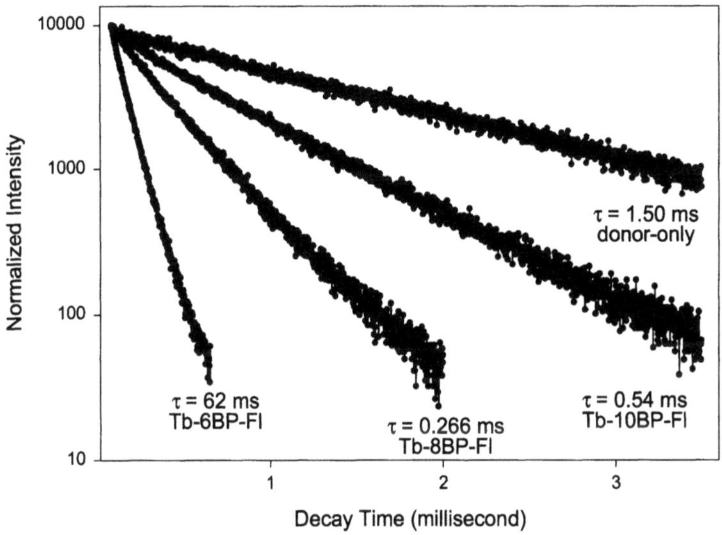

Figure 6.17. Lifetime decay of fluorescein sensitized emission (520 nm) after receiving energy from terbium donor. The lifetime can be tailored depending on distance between donor and acceptor. Donor and acceptor placed on same strand and hybridized to unlabeled complementary strand.

pulsed excitation of the sample and consequently any prompt (nsec) fluorescence of the acceptor has decayed away. The large broad peak around 520 nm is therefore due to fluorescein emission arising only from energy transfer. With greater energy transfer, the Tb-8-FL displays more sensitized emission than the Tb-10-FL. (Tb-6-FL, data not shown, has sufficiently short lifetime that a significant fraction of the signal was lost in the 30 μsec delay time.)

Figure 6.17 shows the temporal decay of the sensitized fluorescein emission at 518 nm for the Tb-6-FL, Tb-8-FL, Tb-10-FL, and Tb-DNA-donor-only reference. These decays were curve fit to give coefficient-weighted averages of 62, 266, 540, and 1500 μs lifetimes for Tb-6-FL, Tb-8-FL, Tb-10-FL, and donor-only, respectively. This result clearly shows that the sensitized emission lifetime can be experimentally tuned in the microsecond time regime. Each LCTF is best fit to a 3-exponential decay, likely caused by the fluorescein acceptor experiencing three different local conformations. Despite this complexity, the average lifetimes are extremely reproducible, with repeat measurements on the same sample within 1% and between different samples within 10%. Hence a target DNA labeled with one of these LCTFs can readily be distinguished from a target DNA labeled with a different LCTFs, simply based on lifetimes.

Figure 6.18 shows that different emission colors with similar lifetimes can also be generated by appropriate choice of acceptors, donors, and donor-acceptor

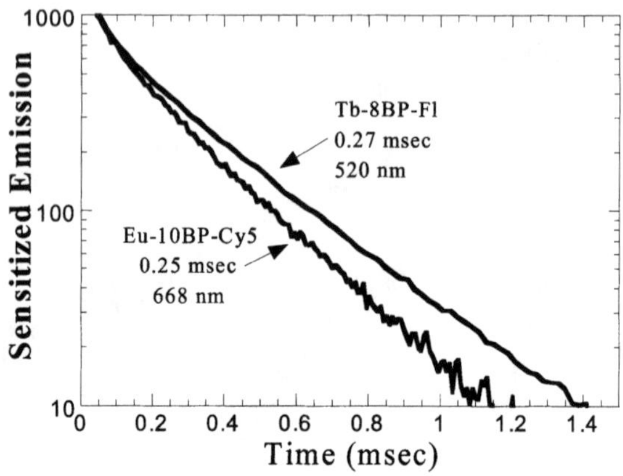

Figure 6.18. Similar lifetimes with different emission colors can be generated. The donor was placed on the 5′-end of DNA, the acceptor on the 5′-end of complementary DNA, and the DNAs hybridized. The data is from the sensitized emission lifetime of the acceptor.

distances. Two sets of LCTFs were synthesized. One has Eu^{3+}-chelate as donor on the 5′-end of one strand and Cy5 as the acceptor on the 5′-end of the complementary strand, with 10 base pairs separating the donor and acceptor. The other has a Tb^{3+}-chelate as donor on the 5′-end of one strand and fluorescein as the acceptor on the 5′-end of the complementary strand, with 8 base pairs separating the donor and acceptor. (The fluorescein was also placed internally on the complementary strand using an abasic fluorescein phosphoramidite with similar results (data not shown).) Figure 6.18 shows that the two sensitized emissions have very similar lifetimes, 250 μs for Eu-10-Cy5 pair and 270 σ for Tb-8-Fl pair. However, Cy5 fluoresces around 668 nm (red color) whereas fluorescein emits around 520 nm (green color). These results demonstrate that probes with similar lifetimes, but different emission colors, can be generated.

Figure 6.19 shows the ability to discriminate between a mixture of two probes with the same emission color (detection at 520 nm) but different lifetimes. A Tb-6-Fl DNA ($\langle\tau\rangle = 62$ μsec) was mixed with various amounts of Tb-10-Fl ($\langle\tau\rangle = 540$ μsec) and the resulting multi-exponential decay was curve-fit to a sum of 6-mer-only and 10-mer-only decays. The population of each species was determined by the relative amplitudes of each component of the fit and compared to the known amount determined by a simple titration based on absorption measurements at 260 nm. The linearity is quite good ($r = 0.993$). The slope differs from unity by 24%, but this is within the expected variability due to uncertainty in DNA concentration (approximately $\pm$ 10% for each strand) based on the absorbance titration.

The lifetime- and color-tailored dyes reported here open the possibility of quadratically increasing the number of resolvable probes on a single sample,

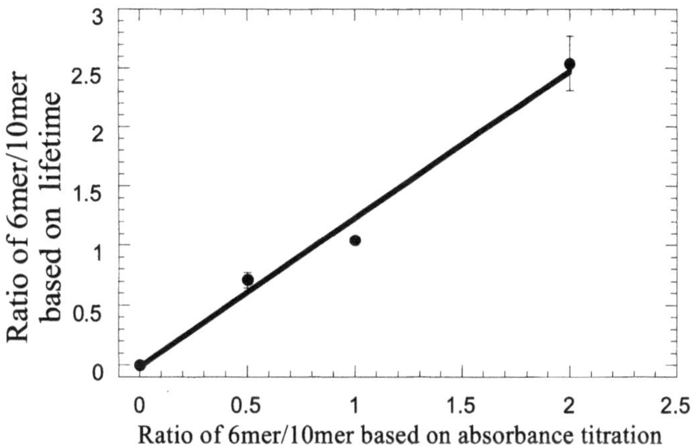

Figure 6.19. Two probes with the same emission color (520 nm) but different lifetimes can be independently identified and their populations quantified. Stock solutions of Tb-6-Fl and Tb-10-Fl were mixed at different ratios (determined by absorbance at 260 nm). The sensitized emission lifetime of the mixtures were fit to a sum of the Tb-6-Fl decay and Tb-10-Fl decay and the relative populations determined. The best fit line is: $y = -0.016 + 1.24x$; $r = 0.993$.

relying not just on conventional color discrimination but on both color and excited-state lifetime discrimination. If, for example, four colors and four lifetimes can be detected, a target DNA could potentially be labeled with 16 probes, each binding to a unique site and each probe/sequence uniquely identified by a combination of color and lifetime discrimination. These probes also extend fluorescence lifetime-based detection from the nanosecond range into the tens-to-hundreds of microseconds time domain. This time-scale is long enough to enable facile and nearly complete discrimination against short-lived background fluorescence and scattering, but short enough such that emission rate (saturation) is not a problem in many applications.

6.8. Conclusion

Lanthanides, as alternative probes to conventional fluorophores, can lead to enhanced sensitivity in applications where autofluorescence is a problem. Multiple labeling with lanthanide chelates have been shown to be an effective method for further increases in sensitivity. These probes, being nonisotopic, avoid the many practical problems associated with radioactive probes. Lanthanide probes are particularly advantageous in resonance energy transfer, whether for measuring distances in biocomplexes or when used to generate new lifetime tailored dyes.

References

1. Bunzli, J.-C. G. Luminescent probes. In *Lanthanide Probes in Life, Chemical and Earth Sciences, Theory and Practice*, J.-C. G. Bunzli and G. R. Choppin, eds. New York: Elsevier, 1989, pp. 219–293.
2. Diamandis, E. P. Time-resolved fluorometry in nucleic acid hybridization and Western blotting techniques. *Electrophoresis 14*, 866–875, 1993.
3. Diamandis, E. P., and Christopoulos, T. K. Europium chelate labels in time-resolved fluorescence immunoassays and DNA hybridization assays. *Anal. Chem. 62*, 1149A–1157A, 1990.
4. Dickson, E. F. G., Pollak, A., and Diamandis, E. P. Time-resolved detection of lanthanide luminescence for ultrasensitive bioanalytical assays. *J. Photochem. Photobiol. B: Biology 27*, 3–19, 1995.
5. Dickson, E. F. G., Pollak, A., and Diamandis, E. P. Ultrasensitive bioanalytical assays using time-resolved fluorescence detection. *Pharmac. Ther. 66*, 207–235, 1995.
6. Lövgren, T., and Iitia, A. Detection of lanthanide chelates by time-resolved fluorescence. In *Nonisotopic Probing, Blotting, and Sequencing*, L. J. Kricka, ed. San Diego, CA: Academic Press, 1995, pp. 331–376.
7. Sammes, P. G., and Yahioglu, G. Modern bioassays using metal chelates as luminescent probes. *Natural Products Reports 13*, 1–28, 1996.
8. Selvin, P. R. Principles and biophysical applications of luminescent lanthanide probes. *Annu. Rev. Biophys. Biomolec. Struct. 31*, 275–302, 2002.
9. Selvin, P. R. Luminescent lanthanide chelates for improved resonance energy transfer and applications to biology. In *Applied Fluorescence in Chemistry, Biology and Medicine*, W. Rettig, B. Strehmenl, S. Schrader, and H. Seifert, eds. New York: Springer Verlag, 1999, pp. 457–487.
10. Soini, E., and Lövgren, T. Time-resolved fluorescence of lanthanide probes and applications in biotechnology. *CRC Crit. Rev. Anal. Chem. 18*, 104–154, 1987.
11. Weissman, S. I. Intramolecular energy transfer: The fluorescence of complexes of europium. *J. Chem. Phys. 10*, 214, 1942.
12. Kwiatkowski, M., Samiotaki, M., Lamminmaki, U., Mukkala, V.-M., and Landegren, U. Solid-phase synthesis of chelate-labelled oligonucleotides: application in triple-color ligase-mediated gene analysis. *Nucleic Acids Res. 22*, 2604–2611, 1994
13. Samiotaki, M., Kwiatkowski, M., Ylitalo, N., and Landegren, U. Seven-color time-resolved fluorescence hybridization analysis of human papilloma virus types. *Anal. Biochem. 253*, 156–161, 1997.
14. Lehn, J. M. *Comprehensive Supramolecular Chemistry*. New York: Pergamon/Elsevier, 1996.
15. Alpha, B., Ballardini, R., Balzani, V., Lehn, J.-M., Perathoner, S., and Sabbatini, N. Antenna effect in luminescent lanthanide cryptates: A photophysical study. *Photochem. Photobiol. 52*, 299–306, 1990.
16. Mathis, G. Rare earth cryptates and homogeneous fluoroimmunoassays with human sera. *Clin. Chem. 39*, 1953–1959, 1993.
17. Mathis, G. Probing molecular interactions with homogeneous techniques based on rare earth cryptates and fluorescence energy transfer. *Clin. Chem. 41*, 1391–1397, 1995.
18. Mathis, G., Socquet, F., Viguier, M., and Darbouret, B. Homogeneous immunoassays using rare earth cryptates and time resolved fluorescence: Principles and specific advantages for tumor markers. *Anticancer Res. 17*, 3011–3014, 1997.
19. Prat, O., Lopez, E., and Mathis, G. Europium(III) cryptate: A fluorescent label for the detection of DNA hybrids on solid support. *Anal. Biochem. 195*, 283–289, 1991.
20. Saha, A. K., Kross, K., Kloszewski, E. D., Upson, D. A., Toner, J. L., Snow, R. A., Black, C. D. V., and Desai, V. C. Time-resolved fluorescence of a new europium chelate complex: Demonstra-

tion of highly sensitive detection of protein and DNA samples. *J. Am. Chem. Soc.* 115: 11032–11033, 1993.

21. Bailey, M. P., Rocks, B. F., and Riley, C. (1984) Terbium chelate for use as a label in fluorescent immunoassays. *Analyst 109*, 1449–1450, 1984.
22. Canfi, A., Bailey, M. P., and Rocks, B. F. Fluorescent terbium chelates derived from diethylenetriaminepentaacetic acid and heterocyclic compounds. *Analyst 114*, 1405–1406, 1989.
23. Li, M., and Selvin, P. R. Luminescent lanthanide polyaminocarboxylate chelates: The effect of chelate structure. *J. Am. Chem. Soc. 117*, 8132–8138, 1995.
24. Chen, J., and Selvin, P. R. Synthesis of 7-amino-4-trifluoromethyl-2-(1H)-quinolinone and its use as an antenna molecule for luminescent europium polyaminocarboxylate chelates. In press.
25. Savitsky, A. P., Chydinov, A. V., and Krilova, S. M. Novel fluorescent chelate for Eu. Presented at Advances in Fluorescence Sensing Technology II, San Jose, CA, 1995.
26. Hemmilä, I., Mukkala, V.-M., and Takalo, H. J. Development of luminescent lanthanide chelate labels for diagnostic assays. *J. Alloys Compounds* 249: 158–162, 1997.
27. Xiao, M., and Selvin, P. R. Quantum yields of luminescent lanthanide chelates and far-red dyes measured by resonance energy transfer. *J. Am. Chem. Soc. 123*, 7067–7073, 2001.
28. Yamada, S., Miyoshi, F., Kano, K., and Ogawa, T. Highly sensitive laser fluorimetry of europium(III) with 1,1,1-trifluoro-4-(2-thienyl)-2,4-butanedione. *Anal. Chim. Acta 127*, 195–198, 1981.
29. Siitari, H., Hemmila, I., Soini, E., Lövgren, T., and Koistinen, V. Detection of hepatitis B surface antigen using time-resolved fluoroimmunoassay. *Nature 301*, 258–260, 1983.
30. Xu, Y. Y., Pettersson, K., Blomberg, K., Hemmila, I., Mikola, H., and Lövgren, T. Simultaneous quadruple-label fluorometric immunoassay of thyroid-stimulating hormone, 17 alpha-hydroxyprogesterone, immunoreactive trypsin, and creatine kinase MM isoenzyme in dried blood spots. *Clin. Chem. 38*, 2038–2043, 1992.
31. Landegren, U., Kaiser, R., Caskey, andC. T., Hood, L. DNA diagnostics—Molecular techniques and automation. *Science 242*, 229–237, 1988.
32. Dahlén, P., Liukkonen, L., Kwiatkowski, M., Hurskainen, P., Iitiä, A., Siitari, H., Ylikoski, J., Mukkala, V. M., and Lövgren, T. Europium-labeled oligonucleotide hybridization probes—Preparation and properties. *Bioconj. Chem. 5*, 268–272, 1994.
33. Alpha-Bazin, B., Bazin, H., Preaudat, M., Trinquet, E., and Mathis, G. Rare earth cryptates and TRACE technology as tools for probing molecular interactions in biology. *New Trends Fluoresc. Spectrosc.: Appl. Chem. Life Sci. 1*, 439–455, 2001.
34. Takalo, H., Mukkala, V.-M., Mikola, H., Liitti, P., and Hemmila, I. Synthesis of europium(III) chelates suitable for labeling of bioactive molecules. *Bioconj. Chem. 5*, 278–282, 1994.
35. Sieving, P. F., Watson, A. D., and Rocklage, S. M. Preparation and characterization of paramagnetic polychelates and their protein conjugates. *Bioconj. Chem. 1*, 65–71, 1990.
36. Canfi, A., Bailey, M. P., and Rocks, B. F. Multiple labelling of immunoglobulin G, albumin and testosterone with a fluorescent terbium chelate for fluorescence immunoassays. *Analyst 114*, 1908–1911, 1989.
37. Moronne, M. M. Development of X-ray excitable luminescent probes for scanning X-ray microscopy. *Ultramicroscopy 77*, 23–36, 1999.
38. Lamture, J. B., and Wensel, T. G. Intensely luminescent immunoreactive conjugates of proteins and dipicolinate-based polymeric Tb (III) chelates. *Bioconj. Chem.* 6: 88–92, 1995.
39. Marriott, G., Heidecker, M., Diamandis, E. P., and Yan-Marriott, Y. Time-resolved delayed luminescence image microscopy using a europium ion chelate complex. *Biophys. J. 67*, 957–965, 1994.
40. de Haas, R. R., Verwoerd, N. P., van der Corput, M. P., van Gijlswijk, R. P., Siitari, H., and Tanke, H. J. The use of peroxidase-mediated deposition of biotin-tyramide in combination with time-resolved fluorescence imaging of europium chelate label in immunohistochemistry and in situ hybridization. *J. Histochem. Cytochem. 44*, 1091–1099, 1996.

41. Christopoulos, T. K., and Diamandis, E. P. Enzymatically amplified time-resolved fluorescence immunoassay with terbium chelates. *Anal. Chem. 64*, 342–346, 1992.
42. Chiu, N. H., Christopoulos, T. K., and Peltier, J. Sandwich-type deoxyribonucleic acid hybridization assays based on enzyme amplified time-resolved fluorometry. *Analyst 123*, 1315–1319, 1998.
43. Evangelista, R. A., Wong, H. E., Templeton, E. F., Granger, T., Allore, B., and Pollak, A. Alkyl- and aryl-substituted salicyl phosphates as detection reagents in enzyme-amplified fluorescence DNA hybridization assays on solid support. *Anal. Biochem. 203*, 218–26, 1992.
44. Templeton, E. F., Wong, H. E., Evangelista, R. A., Granger, T., and Pollak, A. Time-resolved fluorescence detection of enzyme-amplified lanthanide luminescence for nucleic acid hybridization assays. *Clin. Chem.* 37, 1506–1512, 1991.
45. Ioannou, P. C., and Christopoulos, T. K. Two-round enzymatic amplification combined with time-resolved fluorometry of Tb^{3+} chelates for enhanced sensitivity in DNA hybridization assays. *Anal. Chem. 70*, 698–702, 1998.
46. Galvan, B., Christopoulos, T. K., and Diamandis, E. P. Detection of prostate-specific antigen mRNA by reverse transcription polymerase chain reaction and time-resolved fluorometry. *Clin. Chem. 41*, 1705–1709, 1995.
47. Galvan, B., and Christopoulos, T. K. Quantitative reverse transcriptase-polymerase chain reaction for prostate-specific antigen mRNA. *Clin. Biochem. 30*, 391–397, 1997.
48. Halonen, P., Rocha, E., Hierholzer, J., Holloway, B., Hyypia, T., Hurskainen, P., and Pallansch, M. Detection of enteroviruses and rhinoviruses in clinical specimens by PCR and liquid-phase hybridization. *J. Clin. Microbiol. 33*, 648–653, 1995.
49. Dahlén, P., Iitiä, A., Mukkala, V. M., Hurskainen, P., and Kwiatkowski, M. The use of europium (Eu^{3+}) labelled primers in PCR amplification of specific target DNA. *Mol. Cell. Probes 5*, 143–149, 1991.
50. Ylikoski, A., Sjoroos, M., Lundwall, A., Karp, M., Lovgren, T., Lilja, H., and Iitia, A. Quantitative reverse transcription-PCR assay with an internal standard for the detection of prostate-specific antigen mRNA. *Clin. Chem. 45*, 1397–1407, 1999.
51. Iitia, A., Hogdall, E., Dahlen, P., Hurskainen, P., Vuust, J., and Siitari, H. Detection of mutation delta F508 in the cystic fibrosis gene using allele-specific PCR primers and time-resolved fluorometry. *PCR Methods Appl. 2*, 157–162, 1992.
52. Ried, T., Baldini, A., Rand, T. C., and Ward, D. C. Simultaneous visualization of seven different DNA probes by *in situ* hybridization using combinatorial fluorescence and digital imaging microscopy. *Proc. Natl. Acad. Sci. USA 89*, 1388–1392, 1992.
53. Speicher, M. R., Gwyn, B. S., and Ward, D. C. Karyotyping human chromosomes by combinatorial multi-fluor FISH. *Nature Genet. 12*, 368–375, 1996.
54. Sjoroos, M., Iitia, A., Ilonen, J., Reijonen, H., and Lovgren, T. Triple-label hybridization assay for type-1 diabetes-related HLA alleles. *Biotechniques 18*, 870–877, 1995.
55. Beverloo, H. B., van Schadewijk, A., Zijlmans, H. J., Verwoerd, N. P., Bonnett, J., Vrolijk, H., Tanke, and H. J. A comparison of the detection sensitivity of lymphocyte membrane antigens using fluorescein and phosphor immunoconjugates. *J. Histochem. Cytochem. 41*, 719–725, 1993.
56. Marriott, G., Clegg, R. M., Arndt-Jovin, D. J., and Jovin, T. M. Time resolved imaging microscopy. Phosphorescence and delayed fluorescence imaging. *Biophys. J. 60*, 1374–1387, 1991.
57. Tanke, H. J., De Haas, R. R., Sagner, G., Ganser, M., and van Gijlswijk, R. P. Use of platinum coproporphyrin and delayed luminescence imaging to extend the number of targets FISH karyotyping. *Cytometry 33*, 453–459. 1998.
58. Verwoerd, N. P., Hennink, E. J., Bonnet, J., Van der Geest, C. R. G., and Tanke, H. J. Use of ferroelectric liquid crystal shutters for time-resolved fluorescence microscopy. *Cytometry 16*, 113–117, 1994.
59. Hennink, E. J., de Haas, R., Verwoerd, N. P., and Tanke, H. J. Evaluation of a time-resolved fluorescence microscope using a phosphorescent Pt-porphine model system. *Cytometry 24*, 312–320, 1996.

60. Periasamy, A., Siadat-Pajouh, M., Wodnicki, P., Wang, X. F., and Herman, B. Time-gated fluorescence microscopy in clinical imaging. *Microsc. Anal. 11*, 33–35, 1995.
61. Seveus, L., Vaisala, M., Hemmila, I., Kojola, H., Roomans, G. M., and Soini, E. Use of fluorescent europium chelates as labels in microscopy allows glutaraldehyde fixation and permanent mounting and leads to reduced autofluorescence and good long-term stability. *Microsc. Res. Tech. 28*, 149–154, 1994.
62. Seveus, L., Vaisala, M., Syrjanen, S., Sandberg, M., Kuusisto, A., Harju, R., Salo, J., Hemmilä, I., Kojola, H., and Soini, E. Time-resolved fluorescence imaging of europium chelate label in immunohistochemistry and *in situ* hybridization. *Cytometry 13*, 329–338, 1992.
63. Bjartell, A., Laine, S., Pettersson, K., Nilsson, E., Lövgren, T., and Lilja, H. Time-resolved fluorescence in immunocytochemical detection of prostate-specific antigen in prostatic tissue sections. *Histochem. J. 31*, 45–52, 1999.
64. Mantrova, E. Y., Demcheva, M. V., and Savitsky, A. P. Universal phophorescence immunoassay. *Anal. Biochem. 219*, 109–114, 1994.
65. de Haas, R. R., van Gijlswijk, R. P., van der Tol, E. B., Zijlmans, H. J., Bakker-Schut, T., Bonnet, J., Verwoerd, N. P., and Tanke, H. J. Platinum porphyrins as phosphorescent label for time-resolved microscopy. *J. Histochem. Cytochem. 45*, 1279–1292, 1997.
66. Lakowicz, J. R. *Principles of Fluorescence Spectroscopy*. New York: Kluwer Academic/Plenum, 1999.
67. Cantor, C. R., and Schimmel, P. R. *Biophysical Chemistry*. San Francisco: W. H. Freeman, 1980.
68. Clegg, R. M. Fluorescence resonance energy transfer. *Curr. Opin. Biotech. 6*, 103–110, 1995.
69. Clegg, R. M. Fluorescence resonance energy transfer. In *Fluorescence Imaging Spectroscopy and Microscopy*, X. F. Wang and B. Herman, eds. New York: John Wiley & Sons, 1996, pp. 179–251.
70. van der Meer, B. W., Coker, G., III, and Chen, S. Y. *Resonance Energy Transfer: Theory and Data*. New York: VCH Publishers, 1994.
71. dos Remedios, C. G., and Moens, P. D. J. Fluorescence resonance energy transfer—Applications in protein chemistry. In *Resonance Energy Transfer*, D. L. Andrews and A. A. Demidov, eds. Chichester, UK: John Wiley and Sons, 1999, pp. 1–64.
72. Fairclough, R., H., and Cantor, C., R. The use of singlet–singlet energy transfer to study macromolecular assemblies. *Methods Enzymol. 48*, 347–379, 1978.
73. Herman, B. Resonance energy transfer microscopy. *Methods Cell. Biol. 30*, 219–243, 1989.
74. Selvin, P. R. Fluorescence resonance energy transfer. *Methods Enzymol. 246*, 300–334, 1995.
75. Stryer, L. Fluorescence energy transfer as a spectroscopic ruler. *Annu. Rev. Biochem. 47*, 819–846, 1978.
76. Blomberg, K., Hurskainen, P., and Hemmila, I. (1999) Terbium and rhodamine as labels in a homogeneous time-resolved fluorometric energy transfer assay of the B subunit of human chorionic gonadotropin in serum. *Clin. Chem. 45*, 855–861, 1999.
77. Burmeister-Getz, E., Cooke, R., and Selvin, P. R. Luminescence resonance energy transfer measurements in myosin. *Biophys. J. 75*, 2451–2458, 1998.
78. Root, D. D. *In situ* molecular association of dystrophin with actin revealed by sensitized emission immuno-resonance energy transfer. *Proc. Natl. Acad. Sci. USA 94*, 5685–5690, 1997.
79. Stenroos, K., Hurskainen, P., Eriksson, S., Hemmila, I., Blomberg, K., and Lindqvist, C. Homogeneous time-resolved IL-2-IL-2R alpha assay using fluorescence resonance energy transfer. *Cytokine 10*, 495–499, 1998.
80. Xiao, M., Li, H., Snyder, G. E., Cooke, R., G.Yount, R., and Selvin, P. R. Conformational changes between the active-site and regulatory light chain of myosin as determined by luminescence resonance energy transfer: The effect of nucleotides and actin. *Proc. Natl. Acad. Sci. USA* 95, 15309–15314, 1998.
81. Xu, J., and Root, D. D. Domain motion between the regulatory light chain and the nucleotide site in skeletal myosin. *J. Struct. Biol. 123*, 150–61, 1998.

82. Chen, J., and Selvin, P. R. Lifetime and color-tailored fluorophores in the micro- to milli-second time regime. *J. Am. Chem. Soc. 122*, 657–660, 2000.
83. Heyduk, E., and Heyduk, T. Thiol-reactive luminescent Europium chelates: Luminescence probes for resonance energy transfer distance measurements in biomolecules. *Anal. Biochem.* 248, 216–227, 1997.
84. Heyduk, E., and Heyduk, T. Architecture of a complex between the sigma70 subunit of *Escherichia coli* RNA polymerase and the nontemplate strand oligonucleotide. Luminescence resonance energy transfer study. *J. Biol. Chem. 274*, 3315–3322, 1999.
85. Heyduk, E., Heyduk, T., Claus, P., and Wisniewski, J. R. Conformational changes of DNA induced by binding of chironomus high mobility group protein 1a (cHMG1a). *J. Biol. Chem. 272*, 19763–19770, 1997.
86. Li, M., and Selvin, P. R. Amine-reactive forms of a luminescent DTPA chelate of terbium and europium: Attachment to DNA and energy transfer measurements. *Bioconj. Chem. 8*, 127–132, 1997.
87. Selvin, P. R., and Hearst, J. E. Luminescence energy transfer using a terbium chelate: Improvements on fluorescence energy transfer. *Proc. Natl. Acad. Sci. USA 91*, 10024–10028, 1994.
88. Selvin, P. R., Rana, T. M., and Hearst, J. E. Luminescence resonance energy transfer. *J. Am. Chem. Soc. 116*, 6029–6030, 1994.
89. Schobel, U., Egelhaaf, H.-J., Brecht, A., Oelkrug, D., and Gauglitz, G. New donor-acceptor pair for fluorescent immunoassays by energy transfer. *Bioconj. Chem.* 10, 1107–1114, 1999.
90. Horrocks, W. D., Jr., and Sudnick, D. R. Lanthanide ion probes of structure in biology. Laser-induced luminescence decay constants provide a direct measure of the number of metal-coordinated water molecules. *J. Am. Chem. Soc. 101*, 334–350, 1979.
91. Reifenberger, J. G., Snyder, G. E., and Selvin, P. R. Polarization of luminescent lanthanide chelates. *Biophys. J. 82*, 430a, 2002.
92. Vereb, G., Jares-Erijman, E., Selvin, P. R., and Jovin, T. M. Time and spectrally resolved imaging microscopy of lanthanide chelates. *Biophys. J. 75*, 2210–2222, 1998.
93. Xiao, M., and Selvin, P. R. An improved instrument for measuring time-resolved lanthanide emission and resonance energy transfer. *Rev. Sci. Inst. 70*, 3877–3881, 1999.
94. Callaci, S., Heyduk, E., and Heyduk, T. Core RNA polymerase from E. coli induces a major change in the domain arrangement of the sigma 70 subunit. *Mol. Cell. 3*, 229–238, 1999.
95. Helmann, J. D., and deHaseth, P. L. Protein-nucleic acid interactions during open complex formation investigated by systematic alteration of the protein and DNA binding partners. *Biochemistry 38*, 5959–5967, 1999.
96. Malhotra, A., Severinova, E., and Darst, S. A. Crystal structure of a sigma 70 subunit fragment from *E. coli* RNA polymerase. *Cell 87*, 127–136, 1996.
97. Gadella, T. W. J., Jovin, T. M., and Clegg, R. M. Fluorescence lifetime imaging microscopy (FLIM)—Spatial resolution of microstructures on the nanosecond time scale. *Biophys. Chem. 48*, 221–239, 1993.
98. Lakowicz, J. R. Long lifetime metal-ligand complexes as probes in biophysics and clinical chemistry. *Methods Enzymol. 278*, 295–321, 1997.
99. Benson, S. C., Mathies, R. A., and Glazer, A. N. Heterodimeric DNA-binding dyes designed for energy transfer: Stability and applications of the DNA complexes. *Nucleic Acids Res. 21*, 5720–5726, 1993.
100. Glazer, A., and Mathies, R. Energy-transfer fluorescent reagents for DNA analyses. *Curr. Opin. Biotechnol. 8*, 94–102, 1997.
101. Ju, J., Ruan, C., Fuller, C. W., Glazer, A. N., and Mathies, R. A. Fluorescence energy transfer dye-labeled primers for DNA sequencing and analysis. *Proc. Nat. Acad. Sci. USA 92*, 4347–4351, 1995.

7

DNA Arrays for Genetic Analyses and Medical Diagnosis

Sabato D'Auria, Mosè Rossi, Joanna Malicka, Zygmunt Gryczynski, and Ignacy Gryczynski

7.1. Introduction

During the past decade we have witnessed remarkable advances in molecular biology, the scientific discipline that studies the molecular basis of heredity, genetic variations, and the expression patterns of genes. While genetic engineering may appear to be a new science, the techniques that are in use today are the result of a series of discoveries made over a period spanning more than 125 years. In the "Genome Era" we are able to sequence whole genomes effectively and relatively quickly. Magazines and television routinely report advances in "gene therapy" and transgenics procedures, which would have been unimaginable even a few years ago. In 2001, Human Genome Project leaders announced the completion of a "working draft" DNA sequence of the entire human genome: the post-genomic era begins.[1–3]

The availability of DNA sequences from genomes of the several organisms,[4–7] and in particular from the human genome, has prompted the development of new techniques for tracing the links between genes and phenotypes, based on the variation in expression rather than in sequence. Alterations in gene expression patterns can have profound effects on biological functions of the organisms, and in fact are at the core of altered physiologic and pathologic processes. Biologists have long recognized the importance of gene regulation for a complete understanding of its functional role.[8–12] DNA sequencing has permitted the initial genomic surveys counting the occurrences of sequences representing each tran-

Sabato D'Auria, Joanna Malicka, Zygmunt Gryczynski, and Ignacy Gryczynski • Center for Fluorescence Spectroscopy, University of Maryland at Baltimore, Baltimore, Maryland 21201. *Mosè Rossi* • Institute of Protein Biochemistry and Enzymology, National Research Council, 801215 Naples, Italy. *Correspondence to*: Sabato D'Auria, Institute of Protein Biochemistry and Enzymology, National Research Council, 801215 Naples, Italy.

Topics in Fluorescence Spectroscopy, Volume 7: DNA Technology, edited by Joseph R. Lakowicz, Kluwer Academic / Plenum Publishers, New York, 2003

script in libraries obtained from different cell or tissue samples. However, even if major improvements in DNA sequencing are achieved, sequence-based methods still represent a slow and expensive approach to studying variation in gene expression patterns.[13,14] An important tool in this field has been the development of a new technique, DNA arrays.[15] These arrays consist of a highly ordered matrix of thousands of different DNA sequences that are used for detecting DNA or RNA variations in gene expression profiling, comparative genomics, genotyping, and so on.[16–19] For example, DNA arrays for assessing stress and damage caused by pollutants to microorganisms have wide-ranging applications in environmental toxicology. Application of DNA array technology to native organisms in the environment requires detection of families of genes with related but differing sequences. Consensus oligonucleotide probes based on conserved protein sequences are used to detect related genes.[20]

DNA microarrays typically consist of thousands of immobilized DNA sequences present on a miniaturized surface with a size of a business card or less, usually a glass slide. Fluorescently labeled RNA or DNA prepared from mRNA is hybridized to complementary DNA on the array and then detected by laser scanning. DNA arrays are used to analyze a sample for the presence of gene variations or mutations (genotyping), or for patterns of gene expression, performing the equivalent of ca. 5,000 to 10,000 individual "test tube" experiments in approximately two days of time. Microarrays are distinguished from macroarrays in that the DNA spot size is smaller, allowing the presence of thousands of DNA sequences instead of the hundreds present on macroarrays. In this chapter, we review the DNA arrays technology, and we will discuss its practical applications in genetics, medical, environmental and microbial systems.

7.2. Sample Preparation

Robotic technology is employed in the preparation of most arrays. Generally, the oligonucleotides are bound to a surface such as nylon membrane, glass slide, or silicon wafer at precisely defined location on a grid. There are two main methods for preparing DNA arrays samples: (1) the spotting method, and (2) the *in situ* syntheses method.

7.2.1. Spotting Method

This method allows us to covalently attach presynthesized DNA to the microchips (Figure 7.1). Two technologies are used for delivering oligonucleo-

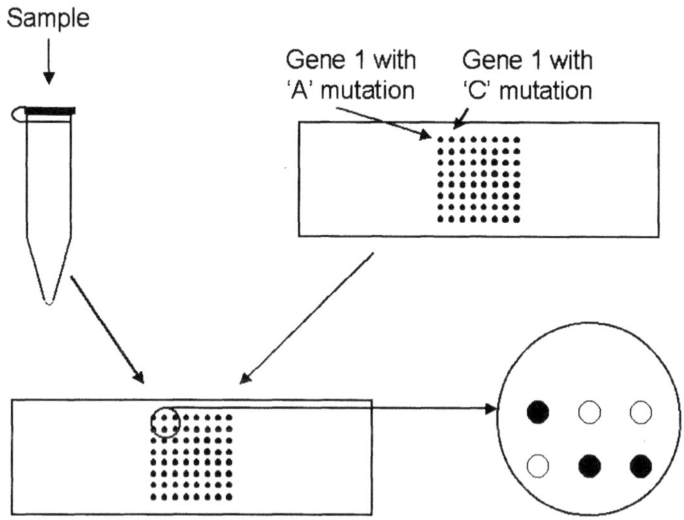

Figure 7.1. Mutation and gene expression analysis by cDNA microarrays.

tides to substrates: mechanical microspotting and ink jet ejection. The immobilization methods used to attach DNA molecules to surface include:

1. Adsorption of polyanionic probe molecules to polycationic poly-L-lysine-coated microscope slides[21] or its photocrosslinking after UV treatment.[15]
2. Covalent attachment of amino-linked (or thiol-linked) oligonucleotides to amino-functionalized (or thiol-functionalized) surface using homobifunctional molecule.[22]
3. Covalent attachment of DNA to self-assembled monolayers of amino-silanes using heterobifunctional crosslinkers.[23]
4. Tethering of DNA to glass slides via an epoxysilane-amine covalent linkage.[24]
5. Attraction of negatively charged DNA to a positively polarized electrode that is coated with gel containing functional groups for subsequent crosslinking.[25]
6. Electrochemically directed copolymerization of pyrrole and oligonucleotides bearing on their 5′-end a pyrrole moiety and deposition of this polymer on gold microelectrode.[26]

Okamoto and coworkers[27] have proposed a method for fabricating DNA microarrays based on the use of a Bubble Jet ink device to eject 5′-terminal-thiolated oligonucleotides onto a glass surface. The oligonucleotides are covalently attached to the glass surface by heterobifunctional crosslinkers that react with the amino group on the substrate and a thiol group on the oligonucleotide probes. By using this methodology, the authors developed DNA microarrays for analyzing the mutational hot spot of the p53 tumor-suppressor gene. Moreover, a simple procedure for manufacturing biochips in which oligonucleotides are immobilized within polyacrylamide gel pads has also been described.[28] In this procedure, the gel provides a stable support and presents a low fluorescence background.

7.2.2. *In Situ* Syntheses Method

The high-density DNA probe arrays are produced using technologies that combine photolithographic methods and combinatorial chemistry, and are referred as biochips or GeneChip™ (Figure 7.2). In this method DNA arrays are

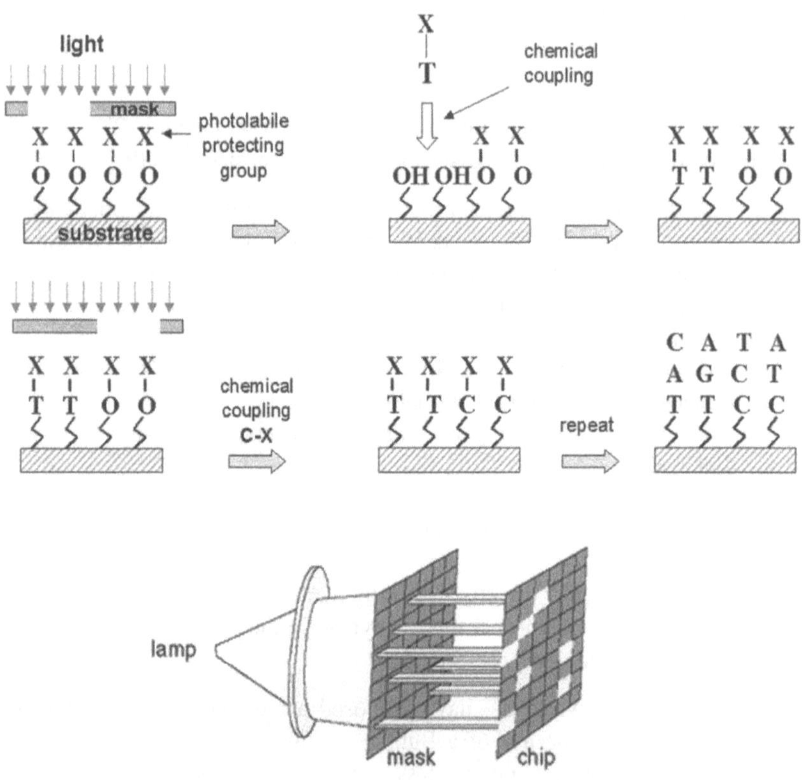

Figure 7.2. Photolithographic method for the synthesis of single-strand DNA on a chip.

usually synthesized using 5′-MeNPOC protected phosphoramidities.[29] In the first step, synthetic linkers modified with photochemically removable protecting group (for example MeNPOC-protected hexaethyleneglycol) are attached to a properly prepared glass substrate (silanated with bis(hydroxyethyl)aminopropyltriethoxy-silane). The substrate is then exposed to light (365 nm) through a lithographic mask that allows localized photodeprotection. Removal of the protecting groups (MeNPOC) results in the exposure of −OH groups that can be coupled to activated 5′-MeNPOC protected nucleoside phosphoramidities. After standard capping and oxidation steps, the light is directed to different regions of the substrate by a new mask and the process repeated. Additional cycles of light-directed deprotection and nucleoside coupling allow the efficient parallel synthesis of any desired combination of different oligonucleotides. Over 400,000 unique oligonucleotide probes are synthesized on each GeneChip representing in excess of 12,000 genes.[30] After the probe preparation the DNA arrays are placed in a temperature-controlled hybridization chamber. A fluorescently labeled nucleic acid sample injected into the chamber hybridizes to complementary oligonucleotides on the array. Fluorescence measurements give information about hybridization intensity.

In sequence analysis arrays, detection of mutations or polymorphisms is accomplished by using a four-probe interrogation strategy (Figure 7.3a). The probe that forms the most stable duplex provides the highest fluorescent signal among the four probes assigned to interrogate unknown base. A different set of four oligonucleotide probes is used for the identification of each nucleotide.

In gene expression monitoring, each gene is represented by 36–40 distinct oligonucleotide probe sequences, half of which are perfect match (PM) probes and the other half contain a single base mismatch (MM) and serve as controls for nonspecific binding (Figure 7.3b). This PM-MM system offers higher specificity in the presence of complex background signals and allows the optimal balance between sensitivity and specificity. Each probe sequence is located in a specific area of the array known as the probe cell and each probe cell contains more than 10 million copies of a given probe. Currently, Affymetrix is providing GeneChips that contain genes from mouse, rat, human, yeast, *E. coli*, and *Arabidopsis*. In addition, specialized chips to address mutations in HIV, cytochrome p450, P53, etc., are being generated. If one needs to monitor gene expression, instead of DNA the mRNA from cells or tissue is biotin-labeled and hybridized to chips containing the oligonucleotide arrays. The specificity of hybridization is measured by laser confocal fluorescence scanning. This technology allows massive, parallel data acquisition and analysis. Figure 7.4 shows the usual procedure used for the gene expression analysis. It is a combination of several interrelated steps: isolation of RNA, transcription to cDNA, reverse transcription and biotinylation, fragmentation, hybridization to the array, fluorescent labeling, and optical scanning of the fluorescent chip. The mRNA is first isolated from the target cells or tissue and

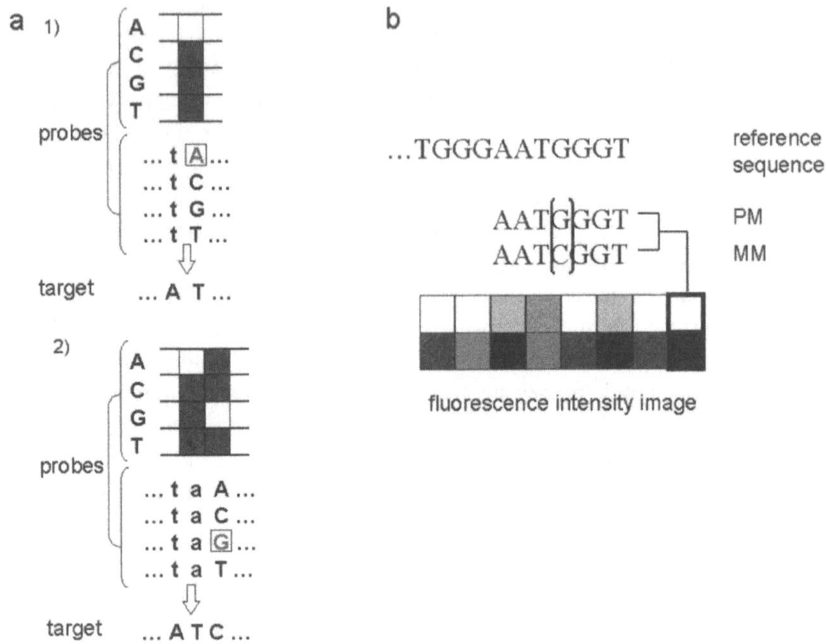

Figure 7.3. a) DNA sequence analysis by using four-probe interrogation strategy. b) Gene expression analysis by using perfect match (PM) and mismatch (MM) oligonucleotide probes.

transcribed to double strand cDNA using conventional protocols. The double strand cDNA is reverse transcribed to cRNA and at the same time is biotinylated using biotin-CTP and biotin-UTP. The biotinylated cRNA is then chemically fragmentized, mixed with control cRNA fragments, and hybridized to the oligonucleotide array. The biotinylated cRNA is fluorescently labeled by streptavidin and scanned to provide data for functional analysis. The hybridization information is obtained from the light emitted from the fluorescent reporter groups already incorporated into the target, which is now bound to the probe array. Probes that most clearly match the target generally produce stronger signals than those that have mismatches. Since the sequence and position of each probe on the array are known, by complementarity the identity of the target nucleic acid applied to the probe array can be determined. The preferred quantity of mRNA is about 1 mg although less is needed if single chips analyses are required.

The composition of the arrays is of three general types:

1. Oligonucleotides or DNA fragments, approximately 20–25 nucleotide bases. These arrays are frequently used in genotyping experiments. The sequences of alternate gene forms may be included for detection of mutations or normal variants (polymorphisms).

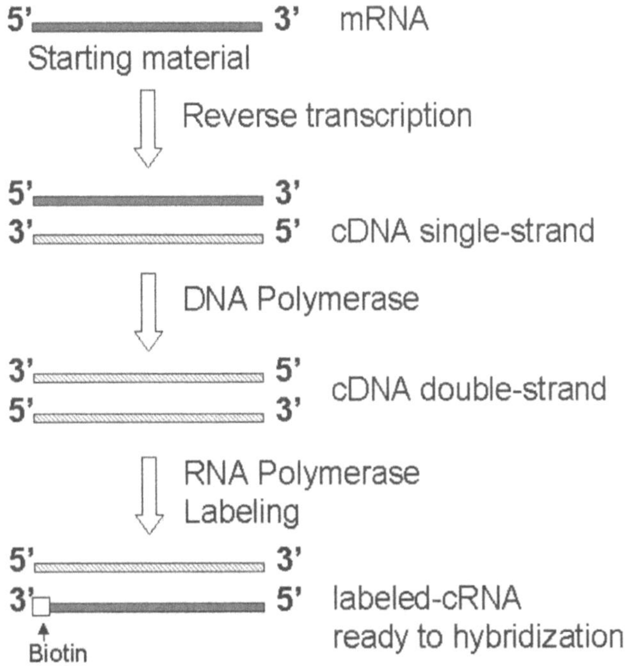

Figure 7.4. Schematic procedure for obtaining cRNA to use for arrays.

2. Complete or partial cDNA, approximately 500–5,000 nucleotide bases. These arrays are generally used for relative gene expression analysis of two or more samples; however, oligonucleotide-based arrays may also be used for these studies.
3. Messenger RNA are either directly or indirectly labeled with different fluorophores, hybridized to DNA on the glass slide and subsequently scanned to detect the fluorescence. These arrays are used for monitoring the gene expression.[16]

DNA/RNA samples are prepared from the cells or tissues of interest. For genotyping analysis, the sample is genomic DNA. For expression analysis, the sample is cDNA or RNA. The cDNA samples are tagged with a fluorescent label and applied to the array. Single-stranded DNA will bind to a complementary strand of DNA. At positions on the array where the immobilized DNA recognizes a complementary DNA in the sample, binding or hybridization occurs. The labeled sample DNA marks the exact positions on the array where binding occurs, allowing automatic detection. The output consists of a list of hybridization events, indicating the presence or the relative abundance of specific DNA sequences that are present in the sample.

Microarray technology is also useful for analyzing genome-wide patterns of mRNA expression. For mRNA expression studies, the ultimate goal is to develop arrays which contain every gene in a genome against which mRNA expression levels can be quantitatively assessed. The microarray technology is based on an approach where cDNA clone inserts are robotically printed onto a glass slide and subsequently hybridized to two differentially fluorescently labeled probes. The probes are pools of cDNAs that are generated after isolating mRNA from cells or tissues in two of the states that one wishes to compare. Resulting fluorescent intensities are produced using a laser confocal fluorescent microscope, and ratio information is obtained following image processing. Finally database development and design is critical for all aspects of this project.

7.3. Fluorescence Probes

The commonly used transduction mode for reading arrays is fluorescence, due to its superior limit of detection and the ability to monitor multiple reporter molecules simultaneously. In fact, in many applications it is necessary to screen for multiple probes that hybridize to a single target. In this case, it is common to use fluorescence wavelength-discrimination methods to identify whether the dyes hybridize to the target DNA found in a specific location in the array. Figure 7.5 shows some of the fluorescence probes that are usually utilized in DNA microarrays.

Two methods are used to determine the amount of duplex that is formed during DNA hybridization. In the first method, the single-strand DNA is labeled with a fluorophore before hybridization during the chemical solid-phase synthesis or the reverse transcription, or as a last step of these processes. In the second method, the DNA sample is first hybridized and then the solution with a fluorescence ligand capable of specifically binding to double-helical DNA is added.

In the first method (traditionally), the probes have been labeled by direct incorporation of labeled nucleotides into first-strand cDNA during synthesis.[31] Commercially available labeled nucleotides usually contain cytosine, thymine or uracyl (Amersham Pharmacia Biotech, Molecular Probes, Roche (Boehringer Mannheim)). Fluorophore labels are attached to the base moiety via aminoallyl or aminoalkynyl spacer at the position 5 of pyrimidine group (Figure 7.5a).[32] Recently, it has been shown that coumarin analogue of GTP, which contains coumarin fluorophore attached to the gamma phosphorus can be also used in the synthesis of labeled oligonucleotides.[33]

An alternative method, gaining in popularity, is the post-labeling method. A chemically reactive nucleotide analog (amino allyl-dUTP) is easily incorporated into the cDNA and then subsequently labeled with the reactive forms of fluoro-

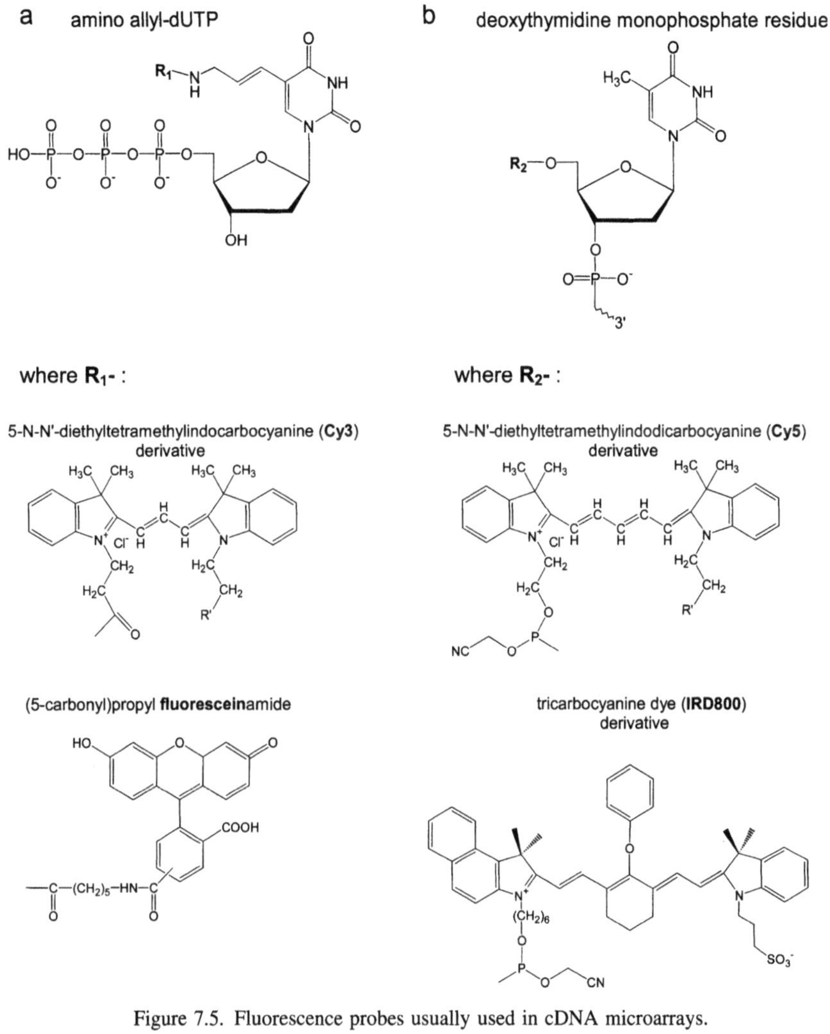

Figure 7.5. Fluorescence probes usually used in cDNA microarrays.

phore (for example N-hydroxysuccinimide ester of CyDye, or Alexa dyes).[34] The amino allyl nucleotide is incorporated more efficiently into the cDNA than bulky labeled nucleotide.

These two methods allow both internal and terminal oligonucleotide labeling. However, it is possible to label the 3′- or 5′-ends of oligonucleotides by coupling active form of the fluorophore (for example, iodoacetimidofluorescein) to the thiophosphate terminal group. Phosphoramidite derivatives of fluorophores (for example: 6-fluorescein phosphoramidite or Cy5 phosphoramidite) can be also used in chemical phosphorylation of the 3′- or 5′-ends of oligonucleotides (Figure 7.5b).[35] Moreover, incorporation of amino group at the 3′ terminal position of oligonucleotides gives the opportunity for coupling any amino reactive derivatives of dyes, which is used very often in DABCYL labeling.[36]

As an alternative to the covalently attached fluorophores, the ligand complexes with DNA can be used to identify the hybridization process. The homodimer dyes like TOTO and YOYO form stable and highly fluorescent complexes after their intercalation in double-stranded DNA, while in aqueous solution they are almost nonfluorescent. However, they can also bind to single-stranded DNA or RNA with concomitant increase in fluorescence quantum yield,[37] so they are less promising for the quantitation of DNA duplex formation in microchips. Hoechst 33258 shows no binding to ssDNA but binding to dsDNA with significant base pair specificity.[38] It is not widely used in DNA microarray, however, nor are the homodimeric monomethine cyanine dyes which were recently reported as able to distinguish between dsDNA and ssDNA.[39]

In the fluorescence-based DNA arrays the covalently attached fluorophores are usually employed. Their absolute intensity may be used to determine whether selective binding of the probe to the immobilized target occurred. The most commonly used fluorescence dyes include Cy3 (λ_{ex} = 514 nm) and Cy5 (λ_{ex} = 632 nm).[40–42] The choice of these dyes is due to their high photochemical stability, high quantum yields, and well-resolved emission profiles. Importantly, dUTP analogues of Cy3 and Cy5 are readily incorporated during the reverse transcription process. However, many other dyes are used in DNA arrays, for example: fluorescein derivatives (FITC, FAM, JOE), rhodamine derivatives (TAMRA, ROX, Texas Red, and some of Alexa dyes), phycoerythrin, phycocyanin, and some compounds from Cyanine Dyes family (Cy2, Cy3.5, Cy5.5, TOTO, YOYO, POPO-3, PO-PRO-3).[43,44] These dyes are often combined in two-, three-, and four-color applications.[42,45,46]

A few years ago, the use of fluorescence lifetimes for monitoring multiple reporter molecules simultaneously in DNA sequencing has been proposed as an alternative to the intensity measurements.[47,48] Recently Waddell and coworkers[49] have described the use of near-infrared time-resolved fluorescence as a detection method for reading DNA hybridization events on solid surfaces in DNA microarray applications. In their experiments they used two dyes, which were covalently attached to oligonucleotides: aluminum tetrasulfonated naphthalocyanine and tricarbocyanine dye (IRD800) (Figure 7.5b), with lifetimes 1.92 ns and 1.21 ns, respectively. The limit of detection for oligonucleotides containing a near-IR fluorescent reporter was found to be 0.38 molecules/μM^2. However, by setting up the time-gate, the authors were able to improve the resolution limit by a factor of 10. Although near-IR fluorescence is an attractive model for reading microarrays, the utilization of fluorophores with higher quantum yields and better photobleaching stability will further improve the sensitivity of the measurements.

7.4. Arrayers

Robots (arrayers) are required to place and/or array a large number of probes onto the slides. Several robotic instrumentations for arraying cDNA clones, as well as confocal scanning fluorescent microscopes or CCD systems to examine fluorescent intensities of microarrays have been developed. In this section we will briefly describe some of them. One example, the AECOM robot,[50] is designed to automatically collect samples from 96- or 384-well plates, with up 12 pens simultaneously (Figure 7.6). Each pen collects from 250 μl to 500 μl of solution and deposits 0.25–1 nl on each slide, creating spots ranging from 100–150 μm in diameter. The robot rests on an optical table, which allows many different configurations of microscope slides and plates. A typical full optical table supports 230 slides, 5 microtitre plates, up to three washing stations, and a dryer station. The wash stations are stationary containers of distilled water that is replaced after every two microtitre plates. When the pen tips are immersed, the robot shakes the pen assembly back and forth to enhance cleaning by computer-controlled water bath sonicator. The dryer is essentially a computer-controlled wet/dry vacuum cleaner with an adapter for inserting the pen tips. Drying is accomplished by the rapid air flow around the tips and the partial vacuum this creates.

A PC controls all of the other components and allows operator input of various parameters, such as number of probes, trays, and slides; spot spacing and

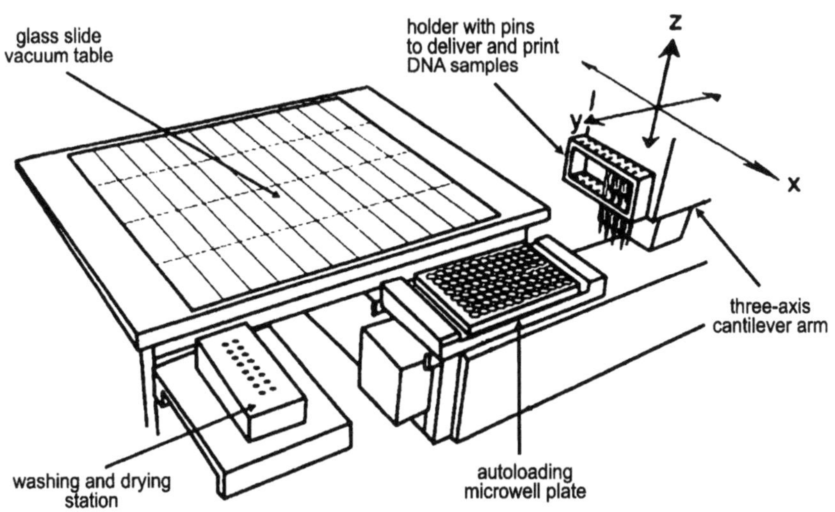

Figure 7.6. cDNA microarrays printing apparatus.

pattern on the slides; duration of each cycle component; and speeds and accelerations of the robot. A cabinet protects the operator and others in the lab from sudden movements of the robot and allows maintenance of clean and humid (~65%RH) conditions around the trays and slides. The air flow in the cabinet in the vicinity of the slides and trays is kept clean by a recirculating blower, HEPA filter, and ducting that keeps class 100 air circulating over the slides at about 100 cubic feet/minute. Humidity is controlled with a room humidifier controlled by an industrial humidistat. Usually, the arrayers are able to put down 16 spots on each of 48 slides, and to wash and dry the probes for the next set of cDNAs for the next tray, in about 70 seconds. Most of this time is taken up with the actual spotting, as the wash and dry cycles are about 2 seconds each and the loading is about 10 seconds. Thus, the contents of one 96-well tray can be spotted every 7 minutes, and 10,000 spots would take about 12 hours.

The reader is generally a computer-controlled inverted scanning fluorescent confocal microscope with a triple laser illumination system (Figure 7.7). The optical system is folded and arranged on an optical breadboard. The breadboard is hung with shock mounts in a vertical plane, inside a lightweight enclosure, which also protects the optics from laboratory dust and personnel from laser light. Illumination is provided by three air-cooled lasers: a 488 nm, 100 mW argon ion laser for exciting FITC; a 532 nm, 100 mW NdYAG for Cy3; and a 633 nm, 35 mW HeNe for Cy5. Any two lasers may be turned on simultaneously and their beams combined with dichroic mirrors and delivered to the specimen via a single dichroic and an objective lens (0.75 NA, 0.66 mm wd). The objective lens can be

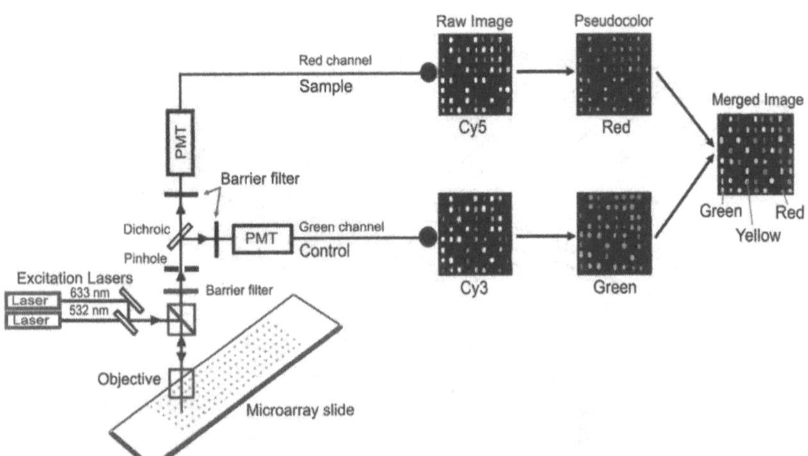

Figure 7.7. Confocal laser microscope scanner and image analysis.

remotely and reproducibly focused with a digital controller. The emitted light, after passing back through the objective and primary dichroic, is focused through a confocal pinhole and through a secondary dichroic onto two cooled PMTs, which operate in parallel for the two different wavelengths.

The stage is a standard computer controlled microscope stage capable of 100 mm/sec scans and 5 micron resolution. One or two standard 25 × 75 mm slides can be scanned at a time. Scanning is done in a comb pattern with data collected in both directions. Data is acquired with a custom integrator and standard 16 bit A/D card in a 133 MHz Pentium. The operator can set the gain, speed, pixel size, pattern position and size. All of the electronics and power supplies are mounted in the cabinet bay next to the optics. At 100mm/sec, with 20 micron pixels, a 50 × 20 spot array with spots on 400 micron centers involves 400 traverses each about 20 mm long and can be scanned in about 4 minutes. It is easily and reliably able to detect about 10 picograms/μl of each species of cDNA.

For microarray detection a CCD camera system has been also used.[51–53] CCD-based imaging often involves illumination and detection of a large portion of the substrate (1 cm^2) simultaneously, so in this case movable stages and optics is not required. Another difference between confocal scanners and CCD-based detection systems is that the latter often use a continuous wavelength light source such as arc lamps, thereby obviating the need for multiple lasers. Fastidious filtering of emission spectra in CCD-based systems minimizes optical crosstalk between different channels. However, it is noteworthy that the use of a confocal system results in a better resolution of the microarray image.

7.5. Image Analysis

The analysis of a microarray image consists of extracting probe intensities or ratios at each cDNA target location, and then crosslinking printed clone target information so that it becomes easy to interpret the outcomes for further high-level analysis. For the particular image example showed in Figure 7.7 an array of cDNAs was hybridized with a red fluor-tagged sample and green fluor-tagged sample (Figure 7.8), and then monochrome images of the fluorescent intensity were combined by placing each image into the appropriate color channel of the RGB color image. In this experiment, there were a total of 1,344 cDNA targets (printed from fourteen 96-well plates) using four print-tips. By defining the print-mode, the software can track the cDNA targets on this and the other array, and map them back to their original plate position.

Microarray image analysis can be divided into several parts: (1) array target segmentation, (2) background intensity extraction, (3) target detection, (4) target intensity extraction, and (5) ratio analysis.

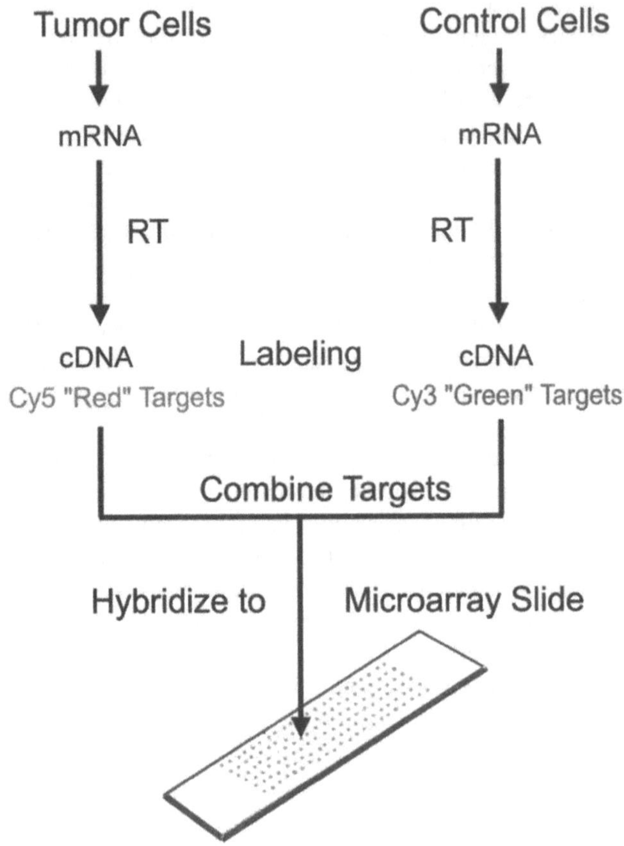

Figure 7.8. Gene expression in cancer and normal cells by cDNA microarrays.

7.5.1. Array Target Segmentation

Since each element of an array is printed automatically to a defined position, it is possible to assume that the final probe signals form a regular array, which can be automatically aligned to a predefined grid-overlay. The initial position of the grid can be either manually determined if no particular orientation markers (called "landing lights") are printed or there are no visible signals from the orientation markers, or automatically determined if the orientation markers are presented in the final image and the entire array has no obvious missing row or column signals. Usually the initial overlay grid does not precisely segment the targets. The refinement of the grid position is preferably done automatically. The usually utilized procedure is as follows: (1) detect strong targets, (2) find their centers (e.g., center of mass), and (3) regress four-corner coordinates of each subarray from these centers of strong signals. This last regression step is important since it avoids the problem of false centers due to noise blobs or miscalculated centers. The final grid-overlay precisely segments each target, which enables future processing tasks to concentrate on only one target. Also, the target position can be deconvoluted back to its original microplate position when the printing information is available on-line. Thus, the clone information can be attached to each target at this stage.

7.5.2. Background Intensity Extraction

The background of the microarray image is not uniform over the entire array and therefore it is necessary to extract local background intensity. The changes of fluorescent background across an array are usually gradual and smooth, and may be due to many technical reasons. Abrupt changes are rare; when they happen, the actual signal intensities of array elements near these changes are not reliable. Conventionally, pixels near the bounding box edge are taken to be background pixels, and thus the mean gray-level of these pixels provides an estimation of the local background intensity. This method becomes inaccurate when bonding box size is close to 10×10 pixels or the target fills the entire bounding box. Fluorescent background is typically modeled by a Gaussian process. For example, for a larger area (e.g., a 30×30 box centered at a particular target) the gray-level histogram within the box is usually unimodal since the majority of the background pixel values are similar while the target pixel values spread up to very high gray levels. The location of the mode of the histogram, therefore, provides the mean local background intensity and the left tail of the histogram will provide the spread (standard deviation) of the background intensity.

7.5.3. Target Detection

One of the difficult image processing tasks is to identify the target region within the bounding box. Each target is somewhat annular, resulting from both how the robot finger places the cDNA on the slide and how the slide is treated. However, the final image of the target may simply be a collection of sub-regions within the nominal circular target region due to the variability introduced by cDNA deposition or the hybridization process. It is important that the final signal intensity be measured over regions corresponding to the probe-hybridized-to-target area.

Conventionally a fixed thresholding method is used in image analysis. The threshold value T can be determined from the local background mean intensity m and its standard deviation s by the relationship $T = m + s$. However, the simple fixed thresholding method fails quite often due to variability of the background and the signal, particularly when the signal is weak (a frequent finding in cDNA array experiments). To avoid these problems, some sophisticated thresholding methods may be implemented. One of the methods that we utilize is the Mann-Whitney method, which takes sample pixels from the background and then performs a rank-sum hypothesis test on the target pixels. The rank-sum test is usually used to test the null-hypothesis that the two population distribution functions corresponding to the two random samples are identical against the alternative hypothesis that they differ. This approach allows users to specify the confidence level of target detection.

7.5.4. Target Intensity Extraction

After target regions are determined, intensity measurements are carried out. For a two-color system, the target region detected from the red channel will be overlaid to the target region detected from the green channel. The reason is simple: both probes were hybridized to the same target (Figure 7.8), so if we observe either one of them, the underlying region must belong to original target. The probe intensity measurement is chosen to be the average gray level within the target region. Keeping in mind that the final measurement is the ratio of two intensities (R/G), the average measurement will provide, to some degree, a data smoothing effect. The local background value is then subtracted from the reported probe intensities from the red channel (R) and the green channel (G); the ratio (R/G) is then calculated. Clearly, the ratio measurement is the ratio of two average intensity measurements. There are, of course, other choices for ratio measurement, including the average of ratios from every pixel location and the linear regression slope of R/G gray-values from every pixel location.

7.5.5. Ratio Analysis

In order to determine if gene expression differs significantly between the red and green channels, the expression ratio calculation is used. Such an approach is intuitive because two similar samples lead to a R/G ratio close to 1. Assume the ratio extracted from a microarray image satisfied the following conditions: normality, independence, being sufficiently positive, and having a constant-coefficient-of-variation. Then, the ratio distribution can be approximate as follows:

The distribution parameter can be estimated using a maximum likelihood method, and ratio calibration can be carried out by an iterative method (assuming that the two channels are not normalized). In order to satisfy the null hypothesis, which requires no intensity change from red to green, a set of "housekeeping" genes has been chosen. These genes have been selected and experimentally verified as being stable in most experiments (R/G = 1.0). While being referred to as a "housekeeping" gene set, their selection is based on biology as well as on their experimental calculation across thousands of observations under numerous experimental conditions. This selection of a "house-keeping" set has been modeled by labeling the same mRNA sample from the same cell line (UACC-903, a breast cancer cell line) by both Cy3 and Cy5. The ratio distribution estimated by the actual data from the housekeeping set closely predicts the spread of the ratio (by parameter c). Furthermore, the peak of the ratio distribution is correctly calibrated to 1, which is what we expected since the two samples should be essentially identical. The significance of basing our measurement on the analytical ratio distribution is that we can associate a confidence interval to each ratio measurement so that a significant difference can be easily detected. More importantly, this approach allows us to associate a p-value to each ratio measurement. Finally, it is possible to derive quality measurements from the above description.

7.5.6. Multiple Image Analysis

The value of microarray technology is not only that it enables a fast screening method for the expression of individual genes, but also that it enables the investigator to study gene interactions in parallel. Multiple images (slides) can be analyzed to explore the temporal expression pattern for a given gene, or to study the similarity between expression patterns from different samples (e.g., patterns of expression between stages of cancer progression). Applications where the scientist wishes to view the expression of genes across multiple slides (experiments) are numerous: for example, the clustering of genes based on their statistical behavior such that some functional relationship may be hypothesized, or the characteristics of expression patterns for all clusters may be extracted for fingerprinting purposes.

7.6. Applications

DNA array technology provides a method for rapid genotyping, facilitating the diagnosis of diseases for which a gene mutation has been identified (Figure 7.9). It also assists in the diagnosis of diseases for which known gene expression biomarkers of a pathologic state, or signature genes, exist. Signature genes are

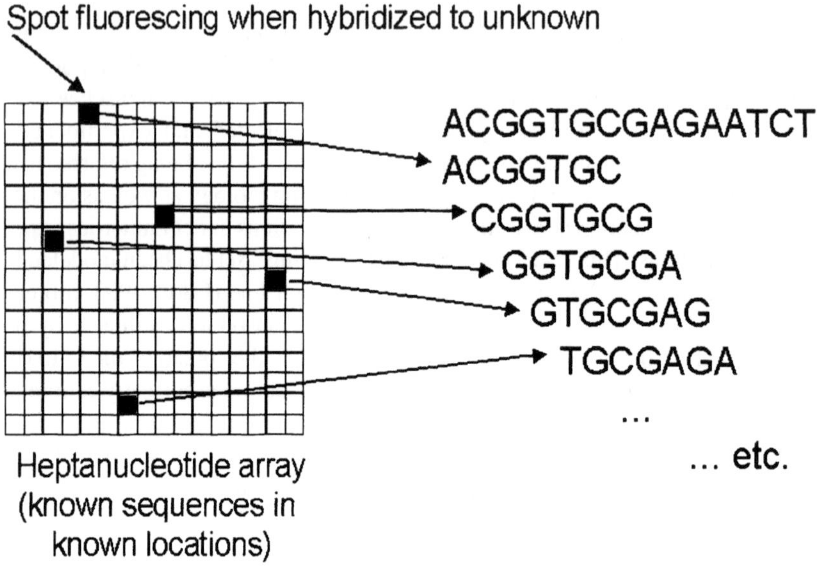

Figure 7.9. DNA-sequencing based on cDNA microarrays. The figure shows how to determine unknown nucleotide sequences by cDNA microarrays.

genes that are constitutively expressed in a normal or diseased cell or tissue that can serve as an identifier of that cell or tissue. We can say that microarrays of DNA and oligonucleotides are beginning to have the same impact on the biological sciences that integrated circuits produced on the physical sciences. In fact, as integrated circuits, microarrays can do many things in parallel, with very little material and with modest investment of labor.[54] Broadly speaking, the applications of microarrays fall into two categories: studies of genomic structure and studies of active-gene expression.

The applications of arrays to genomic studies primarily involve the search for single-nucleotide polymorphisms (SNPs), which have considerable importance regardless of whether they cause an apparent disease.[55] In fact, from these it is possible to use analytical techniques such as genetic-linkage mapping or association analysis to highlight genetic predisposition to disease, and classify them according to defect and the best treatment option.[56] Just a few years ago the people could think about this technology as a "Star Wars" project, but now it is almost obvious to imagine that, on a routine basis, all newborns might undergo 10,000 or more simultaneous genetic tests using only a single chip and a few drops of blood (Figure 7.10).

An additional interesting genomic application of microarrays is called "pharmacogenomics."[57] Since each individual has a different genetic makeup, each will

Figure 7.10. Image of cDNA microarrays from normal and prostate tumors patients.

have a unique set of polymorphism sites. Although these polymorphisms do not result in any disease, some of them may be particularly informative on how each individual responds to a particular drug. The presence of alternate gene forms or atypical expression of a gene involved in drug action or metabolism can manifest as resistance to therapy or an atypical response to therapy. Pharmacogenomic and toxicogenomic studies correlate the genetic profile of individual patients and the individual response to a drug or toxin, respectively. The information obtained from these studies can be used to design arrays that will assist in the selection of custom and rational drug therapy. Arrays assist in the identification of sentinel genes that demonstrate altered expression in a given cell or tissue type in response to drug or toxin exposure. Creating profiles of sentinel genes associated with drugs sharing a common mechanism allows potential new therapies to be rapidly screened for similar activities. This facilitates the selection of compounds for further investigation and may reduce the need for animal testing. An *in vitro* screen for potential toxicity has the potential to reduce drug-screening costs, prevent human suffering and reduce product liability. Other genomic applications are more evident, such as the identification of criminals, tissue typing in organ-donor-selection, studies of evolution, interspecies similarity.

As regards the use of microarrays in studying gene expression, there are numerous applications in this area. Arrays are used to study how cells respond to environmental changes and stress through changing mRNA patterns. For example, it is possible to monitor the toxicity caused by dioxin, or mercury, by looking for slight changes in gene expression.[58] In addition, the arrays can be used for evaluating the many products resulting from combinatorial chemistry. Such products might not cause visible changes in cellar appearance, but could induce change that shows up when we interrogate the mRNA by arrays. Of course, the array applications are not restricted to human cells; we can use them, for example, for studying the life cycle of a plant in deeper detail, or to figure out which specific insecticide that does not affect other species should be used.

There are many applications for this technology in the field of cancer.[59–61] Cancerous cells have a number of unique characteristics as loss of heterozygosity and fusion transcripts, and changes in tumor suppressor genes or oncogenes. Some of these changes are of phenotypic nature, but it is easier to detect them as change in genetic expression. In fact, it is well known that the activity of certain genes will be unregulated and others downregulated in actively dividing malignant cells. Recently, it has been shown that microarrays can be useful in distinguishing between acute lymphoblastic leukemias and acute myeloid leukemias.[62] The clinical diagnosis of leukemias and other cancers places a continuous spectrum of phenotypes into defined categories, and by using DNA array data it is now possible to generate "clinical categories" and reveal new groups of leukemias.[19] Microarray technology has also been applied to classifying different kinds of lymphomas.[63]

Critics of cancer cell lines as models of cancer argue that the cells have evolved and are no longer true genetic representations of tumors from which they were derived. There is, however, increasing evidence that cancer cell lines maintain the gene-expression pattern established during differentiation *in vivo*, with subtle adaptations to cope with the cell-culture environment. After cluster analysis of the expression profiles, Ross and coworkers[64] showed that the cell lines from many tissue types (renal, ovarian, central nervous system, melanoma, leukemia, and colon) were more closely related to lines from the same tissue rather than other tissues. Notable exceptions were the lines derived from breast and non-small cell lung cancer.

Elek and coworkers have applied the microarray method to identify prostate cancer-specific genes.[65] They built a microarray containing 588 known genes and analyzed the microarray by using cDNA probes derived from normal and three independent prostate tumors (Figure 7.10). The authors found that 19 of 588 genes were differentially expressed in the tumors in comparison to the normal tissue. In addition, the authors found that the gene glutathione-S-transferase theta 1 showed a correlation with the microarray results when analyzed by RT-PCR.

As last example of the high impact of array technology in oncology, we cite the work of Elkahloun and coworkers that studied the *in vivo* expression profile of human breast cancer progression.[66] In their work they monitored *in vivo* levels of gene expression in purified normal, invasive, and metastatic breast cell populations from a single patient. For this analysis, they used a combination of laser capture microdissection and high-throughput cDNA microarrays, and the obtained results were in agreement with those obtained by real-time quantitative PCR and immunohistochemistry.

Finally, we mention the application of DNA arrays technology for diabetics. Type 2 diabetes mellitus is an increasingly common disorder of carbohydrate and lipid metabolism. Approximately 16 million individuals in the United States have diabetes, and 800,000 new cases are identified each year. Two important characteristics of this disease are insulin resistance (the failure of peripheral tissues, including liver, muscle, and adipose tissue, to respond to physiologic doses of insulin) and failure of pancreatic beta-cells to properly secrete insulin in response to elevated blood glucose levels. Obesity is a significant risk factor for the development of type 2 diabetes mellitus. Nadler and Attie[67] analyzed the important role of adipose tissue dysfunction in the pathogenesis of diabetes by DNA microarrays. In particular, they studied the adipocyte gene expression in normal and diabetic mice. The expression level of over 11,000 transcripts was analyzed, and 214 transcripts showed significant differences between lean and obese mice, suggesting that a decrease in expression of genes normally involved in adipogenesis is associated with obesity.

7.7. Perspectives

As disease-related gene abnormalities are identified, the design of custom arrays will increase, tailoring sequence number and features to answer the type of question that is to be addressed. Custom designed DNA variation detection arrays will be used to scan the genome and detect single nucleotide polymorphisms (SNPs). The SNPs that are identified can be used in designing further genotyping chips for performing association. Reduction in DNA spot size and advances in scanner technology will allow larger portions of a genome to be present on an array. Decreasing sample size will permit analysis of small quantities of specialized cells acquired by techniques such as laser capture microdissection without the need for coupled RNA amplification.

DNA arrays will assist in the formation of information databases to assist in correlating genetic variation and gene expression with patient status, prognosis, and responsiveness to treatment. It is anticipated that tests will become quantitative so that information about gene expression changes can be measured. As quantitation approaches the limit of determining the exact number of copies of a gene expressed in a cell, the predictive ability of these tests will be maximized. DNA array analysis provides an excellent tool to enlarge our knowledge of genome function.

Acknowledgment

This work was supported by the American Diabetes Association, NIH National Center for Research Resources grant RR-08119, and by the Italian National Research Council.

References

1. Jang, W., Chen, H.-C., Sicotte, H., and Schuler, G. D. Making effective use of genomic sequence data. *Trends Genet. 15*, 284–286, 1999.
2. Venter, J. C., et al. The sequence of the human genome. *Science 291*, 5507, 1304–1351, 2001.
3. Lander, E. S., et al. Initial sequencing and analysis of the human genome. *Nature 409*, 860–921, 2001.
4. Richmond, C. S., Glasner, J. D., Mau, R., Jin, H., and Blattner, F. R. Genome-wide expression in *Escherichia coli* K-12. *Nucleic Acids Res. 27*, 3821–3835, 1999.
5. Lashkari, D. A., DeRisi, J. L., McCusker, J. H., Namath, A. F., Gentile, C., Hwang, S. Y., Brown, P. O., and Davis, R. W. Yeast microarrays for genome wide parallel genetic and gene expression analysis. *Proc. Natl. Acad. Sci. USA 94*, 13057–13062, 1997.
6. Wodicka, L., Dong, H., Mittmann, M., Ho, M-H., and Lockhart, D. J. Genome-wide expression monitoring in *Saccharomyces cerevisiae. Nature Biotechnol. 15*, 1359–1367, 1997.
7. Cho, R. J., Campbell, M. J., Winzeler, E. A., Steinmetz, L., Conway, A., Wodicka, L., Wolfsberg, T. G., Gabrielian, A. E., Landsman D., Lockhart, D. J., and Davis, R. W. A genome-wide transcriptional analysis of the mitotic cell cycle. *Mol. Cell. 2*, 65–73, 1998.
8. Detmer, J., Lagier, R., Flynn, J., Zayati, C., Kolberg, J., Collins, M., Urdea, M., and Sanchez-Pescador, R. Accurate quantification of hepatitis C virus (HCV) RNA from all HCV genotipes by using branched-DNA technology. *J. Clin. Microbiol. 34*, 901–907, 1996.
9. Horn, T., Chang, C. A., and Urdea, M. S. Chemical synthesis and characterization of branched oligodeoxyribonucleotides (bDNA) for use as signal amplifiers in nucleic acid quantification assays. *Nucleic Acids Res. 25*, 4842–4849, 1997.
10. Kopp, M. U., Mello, A. J., and Manz, A. Chemical amplification: continuous-flow PCR on a chip. *Science 280*, 1046–1048, 1998.

11. Ibrahim, M. S., Lofts, R. S., Jahrling, P. B., Henchal, E. A., Weed, W. V., Northrup, M. A., and Belgrader, P. Real-time microchip PCR for detecting single-base differences in viral and human DNA. *Anal. Chem. 70*, 2013–2017, 1998.
12. De Risi, J. L., Lyer, V. R., and Brown, P. O. Exploring the metabolic and genetic control of gene expression on a genomic scale. *Science 278*, 680–686, 1997.
13. Saiki, R. K., Walsh, P. S., Levenson, C. H., and Erlich, H. A. Genetic analysis of amplified DNA with immobilized sequence-specific oligonucleotide probes. *Proc. Natl. Acad. Sci. USA 86*, 6230–6234, 1989.
14. Holland, P. M., Abramson, R. D., Watson, R., and Gelfand, D. H. Detection of specific polymerase chain reaction product by utilizing the 5′ to 3′ exonuclease activity of *Thermus aquaticus* DNA polymerase. *Proc. Natl. Acad. Sci. USA 88*, 7276–7280, 1991.
15. Schena, M., Shalon, D., Davis, R. W., and Brown, P. O. Quantitative monitoring of gene expression patterns with a complementary DNA microarray. *Science 270*, 467–470, 1995.
16. Harrington, C. A., Rosenow, C., and Retief, J. Monitoring gene expression using DNA microarrays. *Curr. Opin. Microbiol. 3*, 285–291, 2000.
17. Ferea, T. L., Bolstein, D., Brown, P. O., and Rosenzweig, R. F. Systematic changes in gene expression patterns following adaptive evolution. *Proc. Natl. Acad. Sci. USA 96*, 9721–9726, 1999.
18. Graber, J. H., O'Donnel, M. J., Smith, C. L., and Cantor, C. R. Advances in DNA diagnostics. *Curr. Opin. Biotechnol. 9*, 14–18, 1998.
19. Golub, T. R., Slomin, D. K., Tamayo, P., Huard, C., Gaasenbeek, M., Mesirov, J. P., Coller, H., Loh, M. L., Downing, J. R., Caligiuri, M. A., Bloomfield, C. D., and Lander, E. S. Molecular classification of cancer: Class discovery and class prediction by gene expression monitoring. *Science 286*, 531–537, 1999.
20. Lee, P. T., and Lee, K. H. Genomic analysis. *Curr. Opin. Biotechnol. 11*, 171–175, 2000.
21. Maskos, U., and Southern, E. M., A novel method for the parallel analysis of multiple mutations in multiple samples. *Nucleic Acids Res. 21*, 2269–2270, 1993.
22. Cavic, B. A., McGovern, M. E., Nisman, R., and Thompson M. High surface density immobilization of oligonucleotide on silicon. *Analyst 126*, 485–490, 2001.
23. Chrisey, L. A., Lee, G. U., and O'Ferrall, C. E. Covalent attachment of synthetic DNA to self-assembled monolayer films. *Nucleic Acids Res. 24*, 3031–3039, 1996.
24. Beattie, W. G., Meng, L., Turner, S. L., Varma, R. S., Dao, D. D., and Beattie, K. L. Hybridization of DNA targets to glass-tethered oligonucleotide probes. *Mol. Biotechnol. 4*, 213–225, 1995.
25. Edman, C. F., Raymond, D. E., Wu, D. J., Tu, E., Sosnowski, R. G., Butler, W. F., Nerenberg, M., and Heller, M. J. Electric field directed nucleic acid hybridization on microchips. *Nucleic Acids Res. 25*, 4907–4914, 1997.
26. Bidan, G., Billon, M., Galasso, K., Livache, T., Mathis, G., Roget, A., Torres-Rodriguez, L. M., and Vieil, E. Electropolymerization as a versalite route for immobilizing biological species onto surface. *Appl. Biochem. Biotechnol. 89*, 183–193, 2000.
27. Okamoto, T., Suzuki, T., and Yamamoto, N. Microarray fabrication with covalent attachment of DNA using Bubble Jet technology. *Nature Biotechnol. 18*, 438–441, 2000.
28. Guschin, D., Yershov, G., Zaslavsky, A., Gemmel, A., Shick, V., Proudnikov, D., Arenkov, P., and Mirzabekov, A. Manual manufacturing of oligonucleotide, DNA, and protein microchips. *Anal. Biochem. 250*, 203–211, 1997.
29. Gentalen, E., and Chee, M. A novel method for determining linkage between DNA sequences: Hybridization to paired probe arrays. *Nucleic Acids Res. 27*, 1485–1491, 1999.
30. Lipshutz, R. J., Fodor, S. P. A., Gingeras, T. R., and Lockhart, D. J. High density synthetic oligonucleotides arrays. *Nature Genet. Suppl. 21*, 20–24, 1999.
31. Hedge, P., Qi, R., Abernathy, K., Gay, C., Dharap, S., Gaspard, R., Hughes, J. E., Snesrud, E., Lee, N., and Quackenbush, J. A concise guide to cDNA microarray analysis. *BioTechniques 29*, 548–562, 2000.

32. Henegariu, O., Bray-Ward, P., and Ward, D. C. Custom fluorescent-nucleotide synthesis as an alternative method for nucleic acid labeling. *Nature Biotechnol. 18*, 345–348, 2000.
33. Sastry, S. A fluorescence-based assay for transcription using a novel fluorescent GTP analogue. *Biophys. Chem. 91*, 191–208, 2001.
34. Lay, M. J., and Wittwer, C. T. Real-time fluorescence genotyping of factor V Leiden during rapid-cycle PCR. *Clin. Chem. 43*, 2262–2267, 1997.
35. Wittwer, C. T., Herrmann, M. G., Moss, A. A, and Rasmussen, R. P. Continuous fluorescence monitoring of rapid cycle DNA amplification. *BioTechniques 22*, 130–138, 1997.
36. Tyagi, S., Bratu, D. P., and Kramer, F. R. Multicolor molecular beacons discrimination. *Nature Biotechnol. 16*, 49–53, 1998.
37. Haugland, R. P. *Handbook of Fluorescent Probes and Research Chemicals*, 6th ed. Eugene, OR: Molecular Probes, 1996.
38. Drobyshev, A. L., Zasedatelev, A. S., Yershov, G. M., and Mirzabekov, A. D. Massive parallel analysis of DNA-Hoechst 33258 binding specificity with a generic oligodeoxyribonucleotide microchip. *Nucleic Acids Res. 27*, 4100–4105. 1999.
39. Timtcheva, I., Maximova, V., Deligeorgiev, T., Gadjev, N., Drexhage, K. H., and Petkova, I. Homodimeric monomethine cyanine dyes as fluorescent probes of biopolymers. *J. Photochem. Photobiol. B: Biol. 58*, 130–135, 2000.
40. Dugan, D. J., Bittner, M., Chen, Y., Meltzer, P., and Trent, J. M. Expression profiling using cDNA microarray. *Nature Genet. 21*, 10–14, 1999.
41. Khan, J., Saal, L. H., Bittner, M. L., Chen, Y., Trent, J. M., and Meltzer, P. S. Expression profiling in cancer using cDNA microarray. *Electrophoresis 20*, 223–229, 1999.
42. Yurov, Y. B., Soloviev, I. V., Vorsanova, S. G., Marcais, B., Roizes, G., and Lewis, R. High resolution multicolor fluorescence in situ hybridization using cyanine and fluorescein dyes: Rapid chromosome identification by directly fluorescently labeled alphoid DNA probes. *Hum. Genet. 97*, 390–398, 1996.
43. Kotova, E. Y., Kreindlin, E. Y., Barsky, V. E., and Mirzabekov A. D. Optical properties of fluorochromes promising for use in biological microchips. *Mol. Biol. 34*, 266–271, 2000.
44. Wessendorf, M. W., and Brelje, T. C. Which fluorophore is brightest? A comparison of the staining obtained using fluorescein, tetramethylrhodamine, lissamine rhodamine, Texas red, and cyanine 3.18. *Histochemistry 98*, 81–85, 1992.
45. Hacia, J. G., Edgemon, K., Sun, B., Stern, D., Fodor, S. P. A., and Collins, F. S. Two color hybridization analysis using high density oligonucleotide arrays and energy transfer dyes. *Nucleic Acids Res. 26*, 3865–3866, 1998.
46. Brelje, T. C., Wessendorf, M. W., and Sorenson, R. L. Multicolor laser scanning confocal immunofluorescence microscopy: practical application and limitations. *Methods Cell Biol. 38*, 97–181, 1993.
47. Bachteler, G., Drexhage, K. H., Arden-Jacob, A. J., Han, K. T., Kollner, M., Muller, R., Sauer, M., Seeger, S., and Wolfrum, J. Sensitive fluorescence detection in capillary gel electrophoresis using laser diodes and multiplex dyes. *J. Luminesc. 62*, 101–108, 1994.
48. Lassiter, S. J., Stryjewski, W., Legendre, B. L., Erdmann, R., Wahl, M., Wurm, J., Peterson, R., Middendorf, L., and Soper, S. A. Time-resolved fluorescence imaging of slab gels for lifetime base-calling in DNA sequencing applications. *Anal. Chem. 72*, 5373–5382, 2000.
49. Waddell, E., Wang, Y., Stryjewski, W., McWhorter, S., Henry, A. C., Evans, D., McCarley, R., and Soper, S. A. High-resolution near-infrared imaging of DNA microarrays with time-resolved acquisition of fluorescence lifetimes. *Anal. Chem. 72*, 5907–5917, 2000.
50. Cheung, V. G., Morley, M., Aguilar, F., Massimi, A., Kucherlapati, R., and Childs, G. Making and reading microarrays. *Nature Genet. 21*, 15–19, 1999.
51. Eggers, M., Hogan, M., Reich, R. K., Lamture, J., Ehrlich, D., Hollins, M., Kosicki, B., Powdrill, T., Beattie, K., Smith, S., Varma, R., Gangadharan, R., Mallik, A., Burke, B., and Wallace, D. A

microchip for quantitative detection of molecules utilizing luminescent and radioisotope reporter groups. *BioTechniques 17*, 516–524, 1994.
52. Chechetkin, V. R., Turygin, A. Y., Proudnikov, D. Y., Prokopenko, D. V., Kirillov, E. V., and Mirzabekov, A. D. Sequencing by hybridization with the generic 6-mer oligonucleotide microarray: an advanced scheme for data processing. *J. Biomol. Struct. Dyn. 18*, 83–101, 2000.
53. Brignac, S. J., Jr, Gangadharan, R., McMahon, M., Denman, J., Gonzales, R., Mendoza, L. G., and Eggers, M. A proximal CCD imaging system for high-throughput detection of microarray-based assays. *IEEE Eng. Med. Biol. Mag. 18*, 120–122, 1999.
54. Graves, D. J. Powerful tools for genetic analysis come of age. *TIBOTECH 17*, 127–134, 1999.
55. Drmanac, S., Kita, D., Labat, I., Hauser, B., Schmidt, C., Burczak, J. D., and Drmanac, R. Accurate sequencing by hybridization for DNA diagnostics and individual genomics. *Nature Biotechnol. 16*, 54–58, 1998.
56. Schafer, A. J., and Hawkins, J. R. DNA variation and the future of human genetics. *Nature Biotechnol. 16*, 33–39, 1999.
57. Marshall, A. Getting the right drug into the right patient. *Nature Biotechnol. 15*, 1249–1252, 1997.
58. Brown, P. O., and Hartwell, L. Genomics and human disease-variations on variation. *Nature Genet. 18*, 91–93, 1998.
59. DeRisi, J., Penland, L., Brown, P. O., Bittner, M. L., Meltzer, P. S., Ray, M., Chen, Y., Su, Y. A., and Trent, J. M. Use of a cDNA microarray to analyse gene expression patterns in human cancer. *Nature Genet. 14*, 457–460, 1996.
60. Hacia, J. G., Brody, L. C., Chee, M. S., Fodor, S. P., and Collins, F. S. Detection of heterozygous mutations in BRCA1 using high density oligonucleotide arrays and two-colour fluorescence analysis. *Nature Genet. 14*, 441–447, 1996.
61. Zhang, L., Zhou, W., Velculescu, V. E., Kern, S. E., Hruban, R. H., Hamilton, S. R., Vogelstein, B., and Kinzler, K. W., Gene expression profiles in normal and cancer cells. *Science 276*, 1268–1272, 1997.
62. Wooster, R. Cancer classification with DNA microarrays: Is less more? *TIG 16*, 327–329, 2000.
63. Alizadeh, A. A., et al. Distinct types of diffuse large B-cell lymphoma identified by gene expression profiling. *Nature 403*, 503–511, 2000.
64. Ross, D. T., Scherf, U., Eisen, M. B., Perou, C. M., Rees, C., Spellman, P., Iyer, V., Jeffrey, S. S., Van de Rijn, M., Waltham, M., Pergamenschikov, A., Lee, J. C., Lashkari, D., Shalon, D., Myers, T. G., Weinstein, J. N., Botstein, D., and Brown, P. O. Systematic variation in gene expression patterns in human cancer cell lines. *Nature Genet. 24*, 227–235, 2000.
65. Elek, J., Park, K. H., and Narayahan, R. Microarray-based expression in prostate tumors. *In Vivo 14*, 173–182, 2000.
66. Sgroi, D. N., Teng, S., Robinson, G., LeVangie, R., Hudson, J. R., Jr., and Elkahloun, A. G. In vivo gene expression profile analysis of human breast cancer progression. *Cancer Res. 59*, 5656–5661, 1999.
67. Nadler, S. T., and Attie, A. D. Please pass the chips: Genomic insights into obesity and diabetes. *J. Nutr. 8*, 2078–2081, 2001.

8

Flow Cytometric Sizing of DNA Fragments

W. Patrick Ambrose, Hong Cai, Peter M. Goodwin, James H. Jett, Robert C. Habbersett, Erica J. Larson, W. Kevin Grace, James H. Werner, and Richard A. Keller

8.1. Introduction

We have developed the capability to detect and identify single fluorescent molecules as they transit a focused laser beam. Single molecule detection allows one to measure properties of individual molecules that would be difficult or impossible with bulk measurements where properties of individual molecules are hidden in ensemble averages. This attribute is particularly important for analyzing the components of a heterogeneous mixture without separating the sample into its individual components. The use of this technology for sizing individual DNA fragments in a sample containing a mixture of DNA fragments of different sizes is an excellent example of the power of this approach. In this article, we summarize our approach to single molecule detection and the application of this technology to DNA fragment sizing.

8.2. Single Molecule Detection

Detection of single molecules labeled with 80–100 fluorescein fluorophores was reported by Hirschfeld in 1976.[1] Our quest for the detection of *single fluorophore* molecules began with the work of Dovichi in 1984[2] and progressed until

---

W. Patrick Ambrose, Hong Cai, Peter M. Goodwin, James H. Jett, Robert C. Habbersett, Erica J. Larson, W. Kevin Grace, James H. Werner, and Richard A. Keller • Bioscience Division, Los Alamos National Laboratory, Los Alamos, New Mexico 87545. *W. P. Ambrose's current address*: Lawrence Livermore National Laboratory, Livermore, California 94551. *E. J. Larson's current address*: Engineering Science and Applications Division, Los Alamos National Laboratory, Los Alamos, New Mexico 87545.

Topics in Fluorescence Spectroscopy, Volume 7: DNA Technology, edited by Joseph R. Lakowicz, Kluwer Academic / Plenum Publishers, New York, 2003

we demonstrated the detection of single fluorophore molecules in 1990.[3] Approximately a dozen workers around the world have now demonstrated single molecule detection and approximately two dozen highly fluorescent molecules have been detected at the single molecule level. The detection of single molecules in liquid solution was reviewed recently.[4–8] DNA fragment sizing is a spin-off of single fluorophore detection. We briefly review our recent work in single molecule detection as a lead-in to DNA fragment sizing.

The signature of a single molecule is the burst of photons emitted as the molecule traverses a focused laser beam. In a laser beam tuned to the absorption frequency of a highly fluorescent molecule, the molecule is continually cycled between its ground state and an excited electronic state with the emission of a photon (fluorescence) on most cycles. The maximum number of photons emitted by a molecule is limited by its fluorescence lifetime and is given by

$$N_{max} = \Phi_f(\tau_t/\tau_f) \tag{8.1}$$

N_{max} is the maximum number of photons that a molecule can emit, Φ_f is the fluorescence quantum yield, τ_t is the transit time across the laser beam, and τ_f is the fluorescence lifetime. For $\Phi_f = 1$, a transit time of 1 ms, and a fluorescence lifetime of 2 ns, N_{max} equals 500,000. Unfortunately, even relatively photostable molecules photobleach after absorbing 10^4–10^5 photons. It is customary to work at an irradiance such that <10% of the molecules photobleach while passing through the laser beam. An efficient detection system converts ~1% of the emitted photons into photoelectrons (pe) with the result that the burst of photoelectrons characteristic of the passage of a single molecule ranges between 10 and 200 pe. Hydrodynamic focusing[9] is used to confine the analyte stream to several microns in diameter. The excitation laser is focused to a waist of ~10 μm and an aperture in the detection path limits the detection length along the laser axis to ~10 μm. The detection volume, defined by the laser beam waist and the image of the aperture on the sample stream is approximately 1 pl (10 μm × 10 μm × 10 μm). Photon bursts associated with the passage of single molecules through the detection volume are detected and distinguished from the background associated with Raman scattering from the water solvent. This is achieved with pulsed excitation and time-gated detection.[3,10] A detailed description of our approach to single molecule detection is given in Goodwin *et al.*[11]

Efficient detection of analyte molecules can be realized if care is taken to ensure that the analyte molecules pass through the center of the detection volume such that all molecules receive similar irradiation and the optical collection efficiency is approximately constant across the analyte stream. One second of data from a dilute solution of tetramethylrhodamine isothiocyanate (TRITC) eluted from a 1 μm i.d. pipette is shown in Figure 8.1a. Each peak represents the passage of a single molecule. For comparison, data from a blank buffer solution is shown in Figure 8.1b. If a threshold were set at ~20 pe/ms, there would be few false

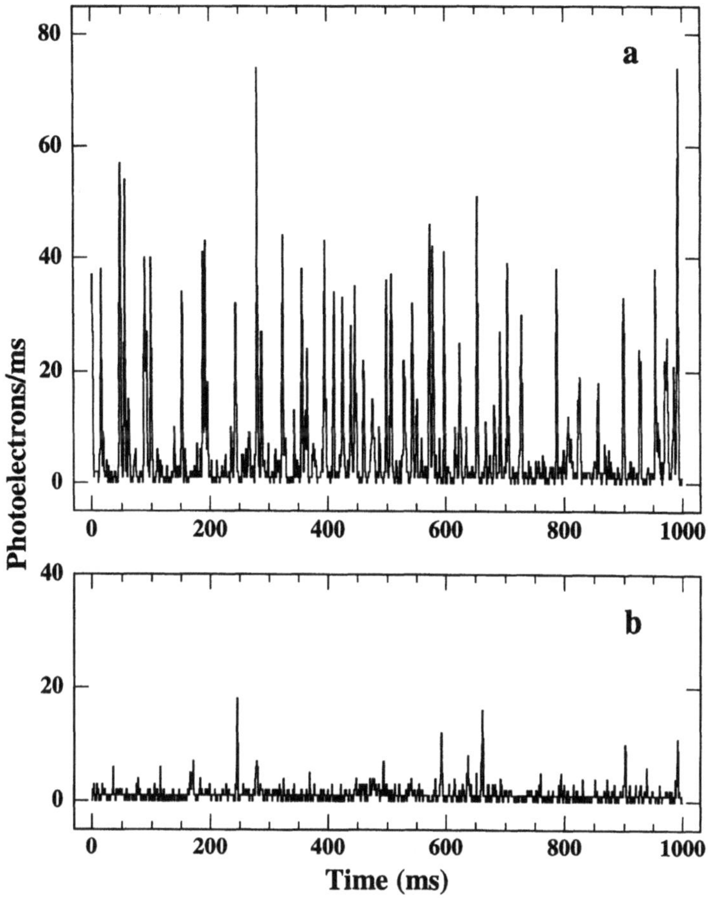

Figure 8.1. Detection of single TRITC molecules in dilute aqueous buffer. TRITC was introduced electrokinetically from a drawn capillary, i.d. ~1 μm at the tip. Conditions: excitation, 554 nm, mode-locked dye laser; average laser power, 20 mW; repetition rate, 82 MHz; beam waist, 15 μm ($1/e^2$); transit time, ~2ms. a. Sample stream on. b. Sample stream off.

positives from the blank and few of the TRITC molecules would be missed. A histogram of the burst sizes (total number of photoelectrons in a burst) is shown in Figure 8.2.[10] The peak at ~45 pe in the burst size distribution is the result of single molecules passing through the center of the detection volume. The secondary peak in the burst size distribution at ~85 pe is caused by the simultaneous presence of two analyte molecules in the probe volume. The probability of two molecules being in the probe volume at the same time depends upon the burst detection rate

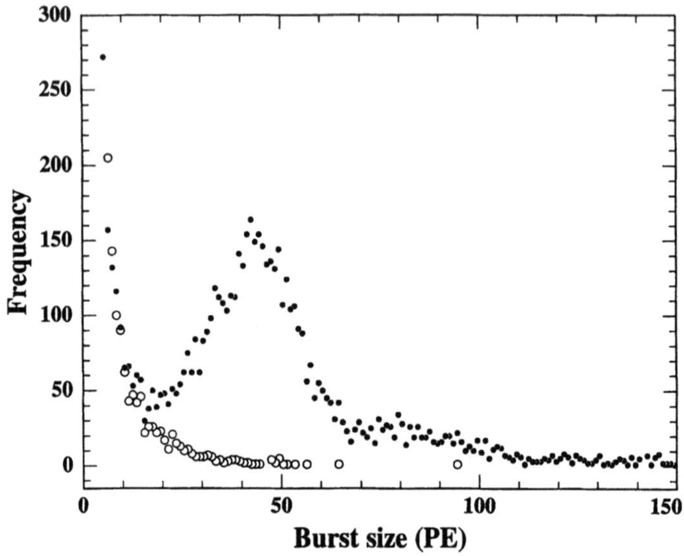

Figure 8.2. Efficient detection of single TRITC molecules. Histogram of fluorescence burst sizes from TRITC molecules obtained as in Figure 8.1. Solid circles are from buffer solution containing TRITC; open circles are from buffer alone. (From reference 12.)

and the transit time. When a burst threshold is set at 20 photoelectrons, over 90% of the molecules are detected.[12] At this threshold, there are few false positives from the background.

Single molecules can be identified by their fluorescence emission spectrum,[13] their fluorescence lifetime,[14,15] their burst size,[16] or a combination of burst size and fluorescence lifetime.[16] Distinguishing between Rhodamine 6G (R-6G) and Texas Red by their emission spectra is shown in Figure 8.3.[13] The measurement of the lifetime of a single molecule is shown in Figure 8.4.[17] An example of distinguishing between TRITC and R-6G by a combination of burst size and fluorescence lifetime is shown in Figure 8.5.[16] Burst size and fluorescence lifetime are measured simultaneously for each molecule. A two dimensional analysis is displayed. Burst size is plotted along one axis and fluorescence decay rate ($1/\tau_f$) along the other axis. The space is divided into four regions. Signatures from TRITC molecules appear mostly in Region I. A significant fraction of the R-6G molecules photobleach as they pass through the laser beam, which yields a broadened burst size distribution and results in signatures from R-6G being distributed between Regions II and III. Using this two dimensional analysis, TRITC and R-6G can be distinguished with greater than 80% confidence. The use of burst size and/or fluorescence lifetime for molecular identification has the advantage of requiring only one excitation wavelength and one detection channel.

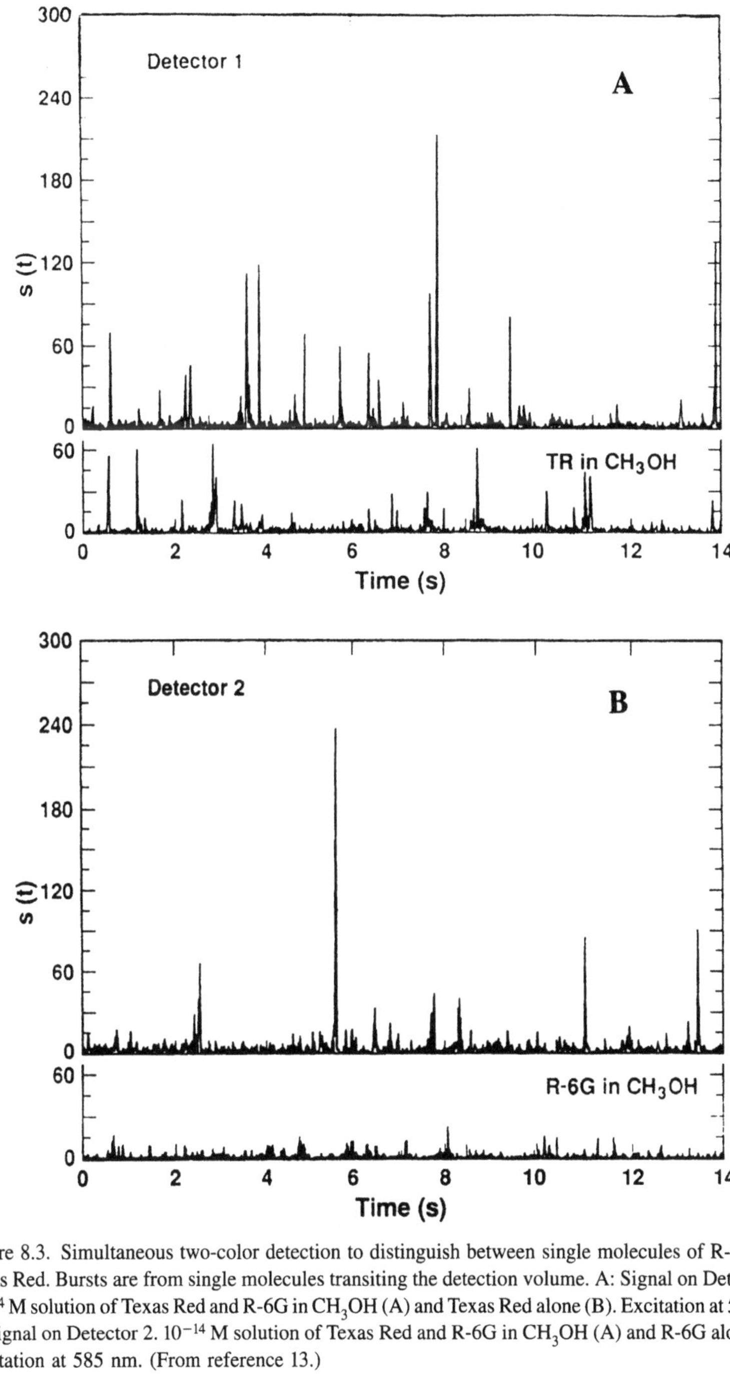

Figure 8.3. Simultaneous two-color detection to distinguish between single molecules of R-6G and Texas Red. Bursts are from single molecules transiting the detection volume. A: Signal on Detector 1. 10^{-14} M solution of Texas Red and R-6G in CH_3OH (A) and Texas Red alone (B). Excitation at 532 nm. B: Signal on Detector 2. 10^{-14} M solution of Texas Red and R-6G in CH_3OH (A) and R-6G alone (B). Excitation at 585 nm. (From reference 13.)

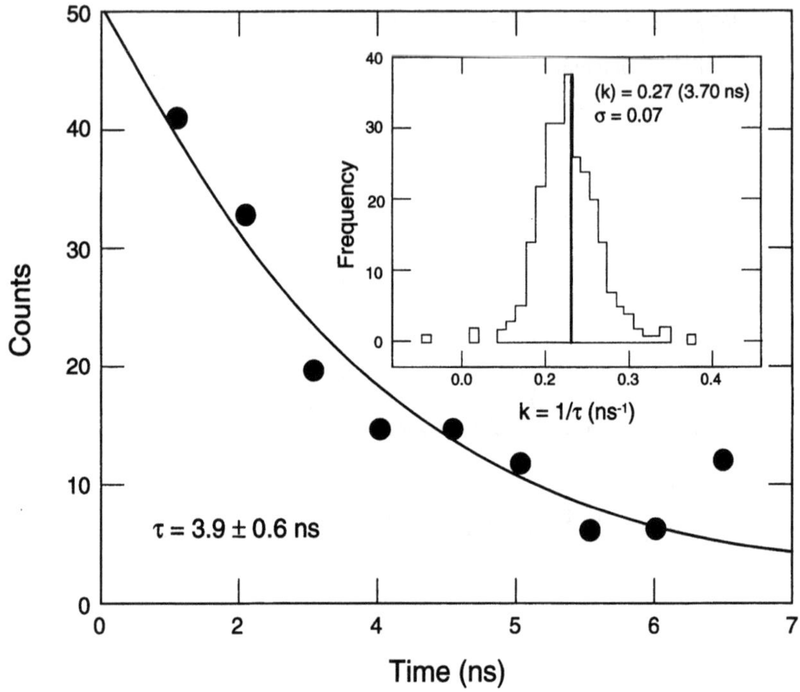

Figure 8.4. Fluorescence lifetime of a single rhodamine-110 molecule. Histogram of the arrival times of photons following pulsed excitation of single rhodamine-110 molecules in water. Conditions: excitation, 496 nm, mode-locked argon ion laser; average laser power, 10 mW; repetition rate, 82 MHz; pulse width, 200 ps ; beam waist, 13 μm ($1/e^2$). Insert: distribution of decay rates for 1408 single molecules. The weighted mean decay rate is indicated by a vertical line. (From reference 17.)

8.3. DNA Fragment Sizing

Our group[18–26] and others[27–33] have reported DNA single fragment analysis by flow measurements. Applications include: characterization of P1 artificial chromosome (PAC) clones for DNA sequencing; analysis of PCR fragments; and bacterial identification. These applications are described below. We have demonstrated that flow cytometric fragment sizing has significant advantages over pulsed-field gel electrophoresis (PFGE) for sizing large DNA fragments.[18] A cartoon describing our approach is shown in Figure 8.6. A DNA sample is stained with a fluorescent intercalating dye that binds stoichiometrically to the DNA such that the amount of dye incorporated is directly proportional to the fragment size [number of base pairs (bp)]. There is a large increase (~100×) in the fluorescence quantum yield of the intercalating dye upon binding to the DNA.[24,25,34] This makes it unnecessary to remove unbound dye from the solution before analysis.

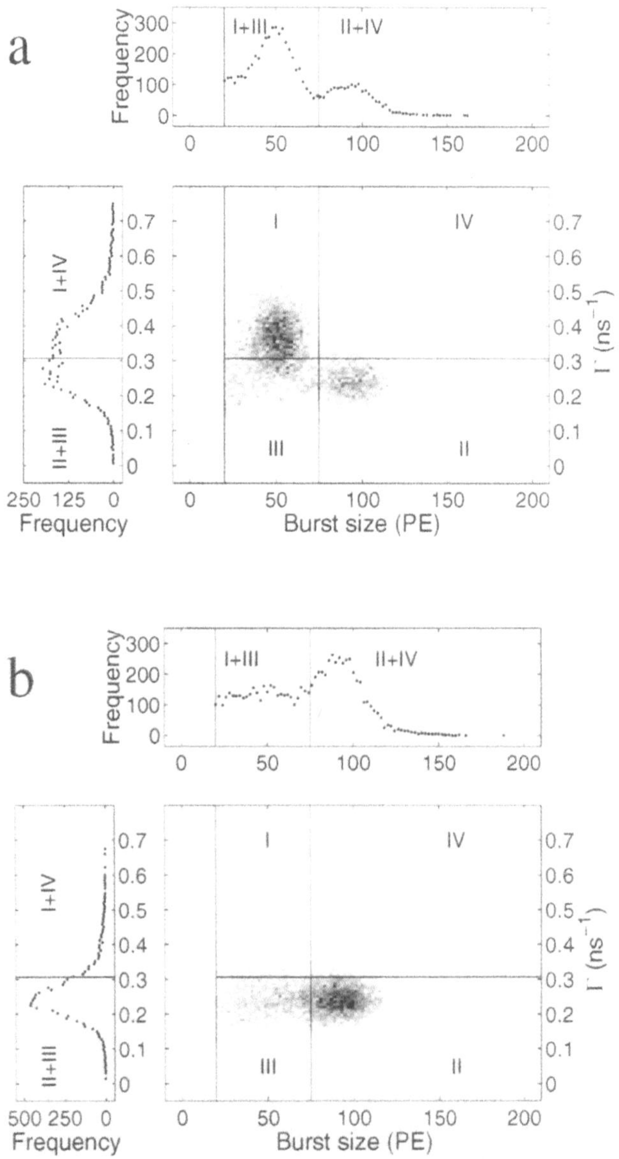

Figure 8.5. Distinguishing between TRITC and R-6G by fluorescence lifetime and/or fluorescence burst size. The data are presented as two dimensional histograms. The histograms were constructed from bursts >20 pe in size. a: distribution compiled from 6314 bursts found in 211 s of data collected from a dilute sample containing approximately equal amounts of TRITC and R-6G. b: distribution compiled from 7730 bursts found in 176 s of data collected from a dilute sample of R-6G. Conditions: sheath volumetric flow rate, 30 μL min^{-1}; average excitation laser power, 30 mW, 514.5 nm mode-locked argon ion laser; beam waist, 16 μm (1/e^2); sample transit time, 1.6 ms. (From reference 16.)

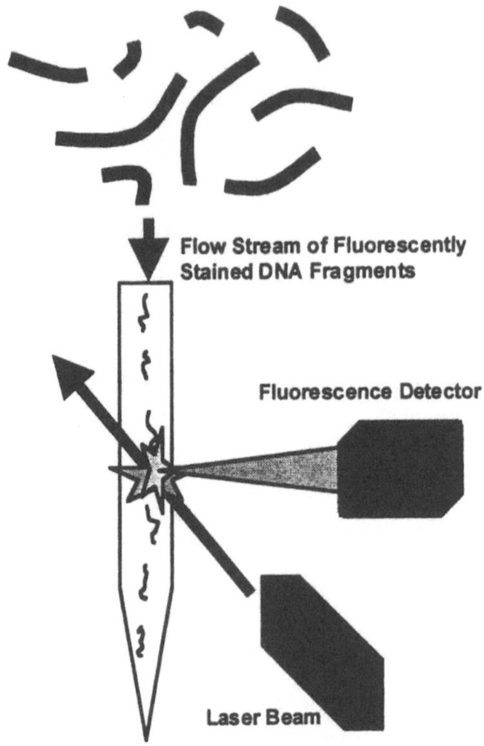

Figure 8.6. Cartoon describing fragment sizing. A restriction digest of DNA fragments is stained with an intercalating dye, diluted to a fragment concentration ~10^{-14} M, and passed through an ultrasensitive flow cytometer. Fluorescence from individual fragments is excited by a laser and detected by a sensitive detector. A histogram of the observed burst sizes is a DNA fingerprint.

The stained fragments are diluted to 10^{-12}–10^{-14} M and introduced into a single molecule detection apparatus developed in our laboratory. Fragments pass individually through the laser-illuminated detection region of the flow cytometer (~60 pl), each fragment producing a fluorescence burst as it transits the laser beam. Fluorescence bursts from individual fragments are detected and recorded. The burst size is a measure of the fragment size. A histogram of the burst sizes from a restriction digest of a DNA sample is a DNA fingerprint. Samples containing less than a picogram of DNA are analyzed in less than ten minutes. This represents an increase in sensitivity of approximately one million and a reduction in analysis time of approximately one thousand over conventional electrophoresis. Fragments ranging in size from 0.125 kbp to 425 kbp have been sized by this technique. The resolution exceeds that of gel electrophoresis for fragments larger than ~20 kbp.

In addition to high sensitivity and speed, our approach is quantitative (individual fragments are counted) and linear with fragment size. The analysis is conformation independent—linearization of circular and supercoiled DNA before analysis is not required.[20] In essence, flow cytometric sizing of DNA fragments is the inverse of gel electrophoresis. In gel electrophoresis, the smallest fragments give the largest signal (migrate farthest) while the largest fragments give the smallest signal (very large fragments show little or no movement on the gel). In contrast, in flow cytometric sizing the smallest fragments incorporate little dye and give small signals while large fragments incorporate a lot of dye and give large signals. The analysis of very large fragments by our technique is currently limited to <500 kbp by DNA fragmentation upon handling.

8.3.1. Apparatus

A schematic of the apparatus that we use to size DNA fragments is shown in Figure 8.7. Samples are introduced into a 250 μm × 250 μm, square bore flow cell from a 40 μm i.d., 240 μm o.d. capillary inserted into the flow cell. The fluid flow direction is into the sheet of paper. Hydrodynamic focusing is used to reduce the sample stream to a diameter of ~20 μm. Approximately 20 mW from a cw Ar^+/Kr^+ laser operated at 514.5 nm, polarized perpendicular to the flow axis, is focused to a diameter of ~50 μm ($1/e^2$) in the center of the flow cell. Fluorescence is collected at 90° to both the flow axis and excitation axis with a 40×, 0.85 NA

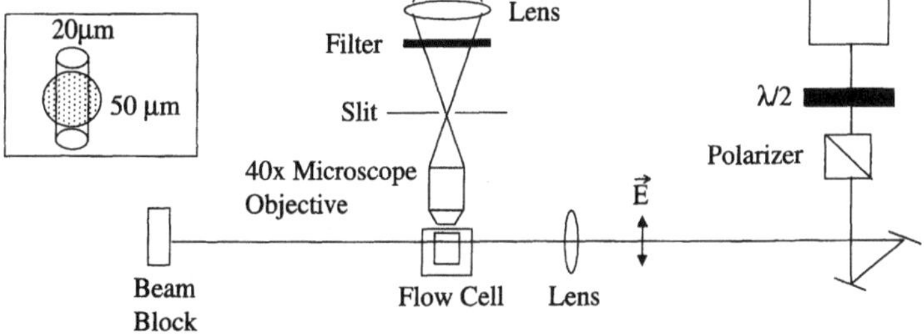

Figure 8.7. Apparatus for DNA fragment sizing. Details are described in the text.

microscope objective. The collected light is focused onto a 1.2 (parallel to the excitation beam) × 3.0 (parallel to the flow axis) mm aperture (spatial filter) located at the image plane of the microscope objective, passed through the appropriate optical filters, and focused onto the photocathode of a thermoelectrically cooled (−30°C) photomultiplier tube (PMT) (RCA, Model 31034A, Somerville, NJ). In some cases, a silicon avalanche photodiode (APD) (EG&G Optoelectronics, Vaudreuil, Québec, Canada) is used to increase the detection sensitivity. The detection volume, defined by the diameter of the laser beam and the image of the aperture in the sample stream, is ~60 pl. The transit time of the sample through the detection volume is controlled by the sheath flow rate (~30 μl/min) to yield a linear velocity of ~2 cm/sec at the center of the flow chamber, corresponding to a ~3 ms transit time through the laser beam. The sample volumetric flow rate is ~0.2 μl/min corresponding to ~40 fragments/s for a 2×10^{-14} M solution. At this sample rate, the probability of two fragments being in the detection volume at the same time is ~10%. Care is taken to position the sample stream in the center of the detection volume to ensure uniform excitation and detection of the DNA fragments.

The PMT photoelectron pulses are amplified, discriminated, and converted to fast NIM pulses (EG&G PAR, Model 1121A, Princeton, NJ). The output of the APD is a TTL-like pulse. A multichannel scaler (MCS) (Stanford Research Systems, Model SR430, Sunnyvale, CA) is used to sum the number of pulses in 40–100 μs bins.

At the flow rates used here, DNA fragments larger than ~10 kbp are stretched out and aligned parallel to the flow axis. Intercalated dye molecules are then aligned with their transition dipoles perpendicular to the flow axis. Care must be taken to avoid polarization effects in excitation and detection.[23,30] For our geometry, polarization effects are small but the coefficient of variation (CV) is a minimum when the polarization of the excitation laser is inclined 22° from perpendicular to the flow axis and both polarization components of the emission are detected.[23]

8.3.2. Data Analysis

Transit times of the analyte through the probe volume are determined by autocorrelation[35] of the first scan of the raw data. The background is determined by averaging the data below a level set near the maximum of the background noise. A burst is recorded when a series of points exceeds a threshold set above the average background. A typical background rate is 5 pe/MCS bin and a typical threshold is ~6 pe/bin. The criteria for choosing the proper threshold are discussed in detail elsewhere.[18–20] The bursts are integrated and the resultant sizes histogrammed to give a burst size distribution. Histograms are fit to a sum of Gaussians and either a fourth order polynomial or a decaying exponential to represent the background. The centroids are extracted from the fit and plotted against the known DNA fragment sizes.

8.3.3. Sample Preparation

A typical staining procedure for DNA fragments is described below. Stock DNA solutions are stored at 4°C. TOTO-1 (Molecular Probes, Eugene, OR) is stored as a 1 mM solution in DMSO at −20°C. A 1×10^{-5} M solution of TOTO-1 is prepared by diluting 1 μl of the stock solution in 99 μl of TE buffer (10 mM Tris-HCl, 1 mM EDTA, pH 8). To stain the DNA, TOTO-1 and DNA are added to TE buffer to give a total volume of 1 ml and concentrations of 1.2×10^{-7} M TOTO and 400 pg/μl DNA. These staining concentrations give a base pair:dye molecule ratio of 5:1. The resulting solution is incubated for 30–60 minutes at room temperature (22°C) in the dark and diluted in TE buffer to give a final total fragment concentration of 10^{-12}–10^{-14} M. The DNA/dye complex is very stable; narrow burst size distributions are obtained from samples stored at room temperature for several days.

8.3.4. Sizing of λ DNA, λ Digests, and λ Concatamers

A data stream from a mixture of λ phage DNA (48.502 kbp) and a *Hind* III digest of λ DNA (0.564 kbp; 2.027 kbp; 2.322 kbp; 4.361 kbp; 6.557 kbp; 9.416 kbp; and 23.130 kbp) stained with TOTO-1 is shown in Figure 8.8.[19] Care was taken to ensure that the sample stream passed through the center of the detection volume such that all fragments were excited with similar irradiance and detected with similar efficiency. The different burst heights are representative of the different fragment sizes in the mixture.

A histogram of the burst sizes from individual fragments in a mixture of λ DNA (48.502 kbp), a *Kpn* I digest of λ DNA (1.503 kbp; 17.053 kbp; and 29.946 kbp), and an *Apa* I digest of λ DNA (10.086 kbp; 38.416 kbp) stained with TOTO-1 is shown in Figure 8.9.[18] Bursts from the 1.503 kbp fragment are masked by scatter and impurity fluorescence backgrounds and are not visible in this data. Less than 1 pg of DNA was analyzed and the data were acquired in ~3 minutes. The histogram was fit to a sum of five Gaussians and an exponential function to represent the background. The reduced χ^2 of the fit is 0.999. The CV of the individual fragments ranges from 5% for the largest fragments to 14% for the smallest fragments. The larger CVs associated with the smallest fragments result from the smaller signal from these fragments. The CVs are within a factor of two of the shot noise limit

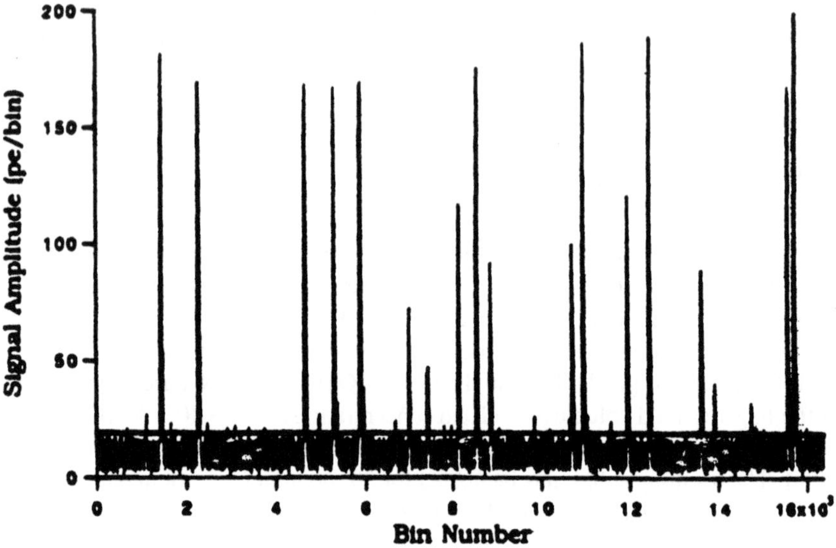

Figure 8.8. Approximately one second of data from a mixture of λ DNA and a *Hind* III digest of λ DNA stained with TOTO-1. Conditions: excitation—514.5 nm, 30 mW, cw; beam waist 46 μm ($1/e^2$); sample stream diameter, 20 μm; transit time, 10 ms; bin width, 82 μs. The different bursts heights are representative of the different fragment sizes in the mixture. (From reference 19.)

($\sqrt{\bar{X}_i}/\bar{X}_i$). Figure 8.10 shows a plot of the centroids of the burst size peaks versus the known fragment sizes. The data is fit to a straight line with a correlation coefficient, $r^2 = 0.9996$. Deviations of the measured values from the straight line are less than 2%. The sequence of λ DNA and the sizes of its restriction fragments are known, so the deviations from the straight line in Figure 8.10 are a good measure of the accuracy of sizing DNA fragments by this approach. For comparison, a similar sample was analyzed by conventional gel electrophoresis. A reproduction of the gel is shown in Figure 8.11. Approximately 300 ng of DNA were loaded onto the gel and the running time was 22 h. The 1.503 kbp fragment is visible on the gel but does not appear in the reproduction. The CV of the 48.5 kbp fragment is similar in the two approaches.

A histogram of the burst sizes from a *Hind* III digest of λ DNA stained with POPO-3 is shown in Figure 8.12.[36] In this case, the excitation laser was a doubled, diode pumped Nd:YAG laser (532 nm) and the detector was a silicon avalanche photodiode. Bursts associated with the 0.564 kbp fragment are visible clearly above background. Bursts from the two fragments at 2.027 and 2.322 kbp were not resolved. The CV at 23.1 kbp is ~5%.

The burst size distribution from a 0.212 kbp PCR fragment stained with

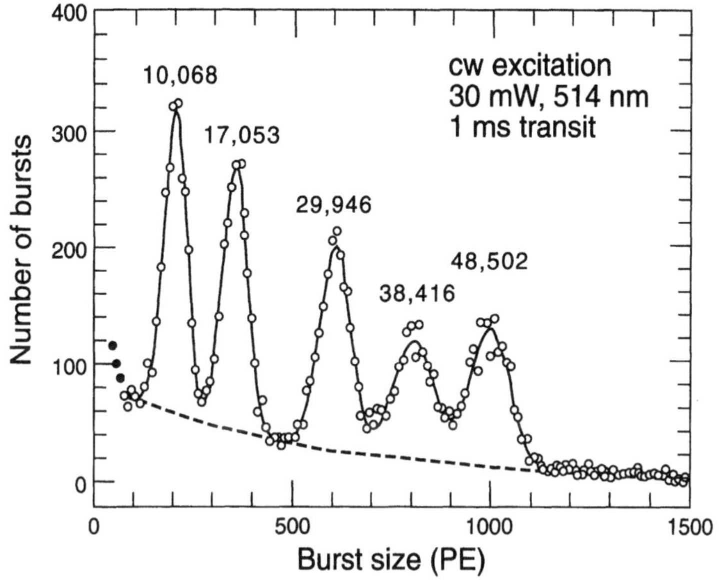

Figure 8.9. Histogram of burst sizes from a mixture of λ DNA, and *Apa* I, and *Kpn* I digests of λ DNA stained with TOTO-1. Approximately ~0.3 pg of DNA were analyzed in 164 s. The circles represent measured data and the solid line a fit to the data. The dashed line is an exponential fit to the background. Conditions: excitation 514.5 nm, 30 mW, cw; beam waist, 40 μm ($1/e^2$); sample stream diameter, 20 μm; transit time, 1 ms; detection rate, 60 fragments/s. (From reference 18.)

POPO-3 is shown in Figure 8.13. The smallest fragment that we have analyzed by flow cytometry is 125 bp.

λ DNA has a 12 basepair overhang at both ends (sticky ends). The base sequences at each end are complementary, allowing individual λ DNA fragments to hybridize, forming concatamers containing multiple units of λ DNA. Figure 8.14a shows a histogram of burst sizes from a mixture of λ DNA and λ concatamers stained with TOTO-1.[21] Spermine and spermidine were added to the solution to stabilize the large fragments. The solid lines represent data and the white lines represent a Gaussian fit to the data. This histogram is the result of analyzing 134 s of data obtained from approximately 10,000 DNA fragments. Concatamers containing up to seven units of λ DNA were observed and resolved. A plot of the centroids of the individual peaks versus the known fragment sizes is shown in Figure 8.14b. The plot is linear from 48.5 kbp to 339.5 kbp with a correlation coefficient of $r^2 = 0.99998$. The parameters determined from the fit are listed in Table 8.1. The average absolute deviation from the straight line is 0.4%. No point deviates from the line by as much as 2%, and most of the deviations are much less than 1%. Again, the CVs are determined mostly by shot noise. Since the sequence

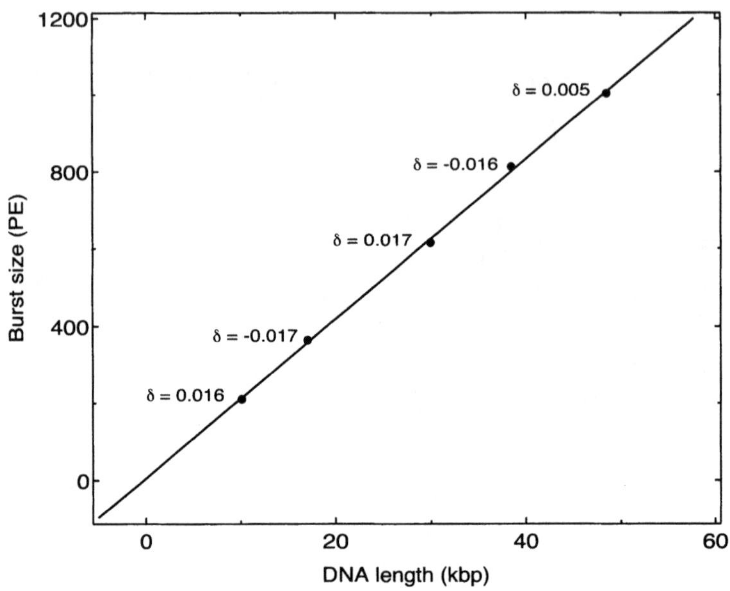

Figure 8.10. Correlation of the centroids of the burst sizes obtained from the fit in Figure 8.9 with the DNA fragment lengths. Deviations of the points from a straight line fit are shown in the figure. (From reference 18.)

of λ DNA is known, the number of base pairs in each concatamer is known exactly. Therefore, the deviations from the straight line are again a good measure of the accuracy of DNA fragment sizing by flow cytometry. The 7λ fragments contain 339.5 kbp. Sizing very large fragments by flow cytometry is limited by fragment stability. The largest fragments that we have sized by flow cytometry contain ~425 kbp.[22] We anticipate that the size range will increase as we learn how to stabilize very large fragments.

8.3.5. Applications

8.3.5.1. PAC Clones

The preparation and characterization of DNA libraries is the initial step in large scale DNA sequencing. Total genomic or chromosome specific DNA is digested by a restriction enzyme that cuts the DNA into 50–300 kb fragments with sticky ends. A cloning vector is prepared by cutting circular DNA with the same restriction enzyme to form linear DNA with the same sticky ends. The cloning

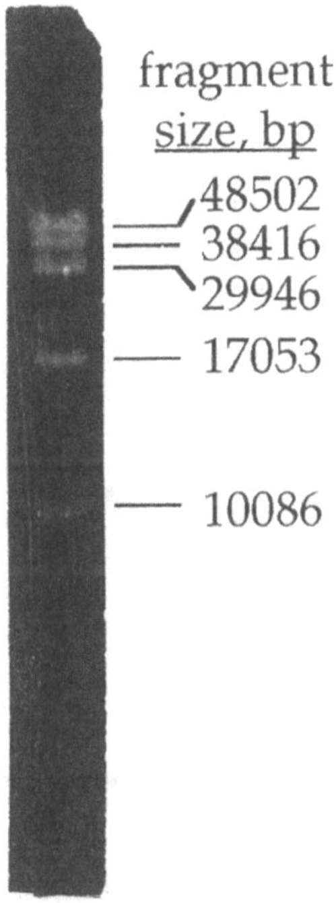

Figure 8.11. Gel electropherogram of a sample containing a mixture of λ DNA, and *Apa* I and *Kpn* I digests of λ DNA. The sample contained ~300 ng of DNA and the analysis time was 22 h. (From reference 18.)

vector is annealed to the restriction digest from the chromosome to form circular P1 artificial chromosomes (PACs). The PACs are inserted into *E. coli* cells at a very dilute concentration such that there is at the most one PAC in an *E. coli* cell. The bacteria replicates the vector plus the insert to form many copies of the PAC. The bacteria are then cultured into colonies (plaques) with each plaque containing only one type of insert. The plaques form a DNA library. The library components are ordered to form a physical map of the chromosome and are used as starting points in large scale DNA sequencing. It is important to characterize the PACs after replication to ensure that the replication went well. This characterization is normally done by PFGE—a process that requires micrograms of DNA and takes ~15 h.

We have demonstrated that flow cytometry can be used to characterize the PACs.[20] The PAC DNA extracted from lysed *E. coli* cells was stained with TOTO-1, diluted to approximately 10^{-14} M, and passed through our ultrasensitive flow cytometer. A histogram of burst sizes from a mixture of λ DNA, a *Kpn* I

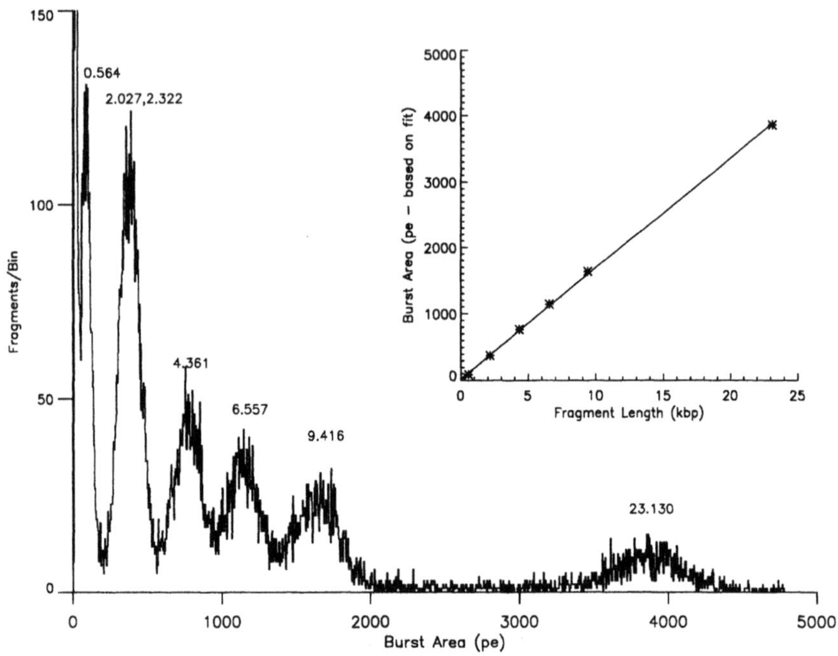

Figure 8.12. Histogram of burst sizes obtained from a *Hind* III digest of λ DNA stained with POPO-3. Conditions: excitation, 532 nm (doubled Nd:YAG), 10 mW, cw; beam waist, 11 μm high by 50 μm wide; detector, silicon avalanche photodiode; sample stream diameter, 3–5 μm; transit time, 1 ms; DNA analyzed, ~0.1 pg; analysis time, 5 min. A plot of the centroids of the burst sizes versus the DNA fragment lengths is shown in the insert.

digest of λ DNA, T4 DNA, and an uncut PAC clone is shown in Figure 8.15. The λ DNA, *Kpn* 1 digest, and T4 were used as internal size standards. The total amount of DNA analyzed was ~0.4 pg (~5000 fragments) and the analysis time was <3 min. The centroids of the standards from a fit to the data in Figure 8.15a were plotted versus the known fragment sizes in Figure 8.15b and fit to a straight line with a correlation coefficient of $r^2 = 0.9997$. The average absolute deviation of the points from the fitted line is 1.7%. The size of the intact supercoiled insert (including the cloning vector) determined from the fit was 88.9 ± 0.8 kbp. The uncut DNA clone was then digested with *Eag*I. This enzyme cuts at the vector and insert junction sites. The released vector and linearized insert were sized by flow cytometry to give fragment sizes of: insert—73.7 ± 0.4 kbp and vector—16.2 ± 0.2 kbp. Within experimental error, the size of the insert plus the vector is the same as the uncut DNA/vector adduct. The mixture of vector and linearized insert was also sized by PFGE. Approximately 0.2 μg of DNA was added to each well

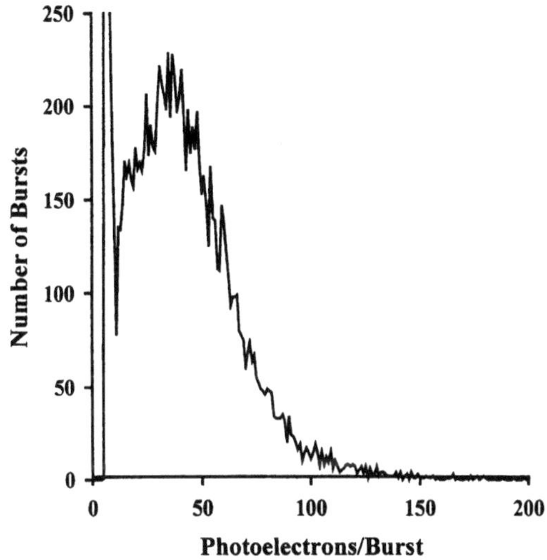

Figure 8.13. Analysis of small DNA fragments. A histogram of burst sizes from a sample containing 212 bp PCR fragments. Conditions: staining dye, POPO-3; excitation, 514.5 nm, 2 mW, cw; beam waist, 20 μm ($1/e^2$); detector, silicon avalanche photodiode; sample stream diameter, <10 μm; transit time, 1 ms; analysis time, 3 min.

and the pulsed-field gel was run for 15 h. The fragment sizes obtained from PFGE were: insert—71.0 ± 7.1 kbp and vector—16.7 ± 1.7 kbp. The uncertainty represents the typical 10% error associated with PFGE. The sizes determined by flow cytometry are within experimental uncertainties of the sizes determined by PFGE. Uncut clones do not migrate appreciably on a gel. The elimination of the need to linearize the DNA before sizing by flow cytometry removes a time consuming step associated with analysis of PAC clones by PFGE.

A comparison of the sizes obtained from six PAC clones obtained from a DNA library is shown in Table 8.2. Again, the sizes obtained by flow cytometry and PFGE are in good agreement. We attribute most of the differences to the 10% uncertainty associated with PFGE.

8.3.5.2. PCR Fragments

Polymerase chain reaction (PCR) is a valuable biochemical technique for amplifying the number of copies of DNA in a specific sequence region. This is useful for detection of a sequence in a sample containing very little DNA. One copy of a sequence may be amplified to 10^9 or more copies. In one variation of the

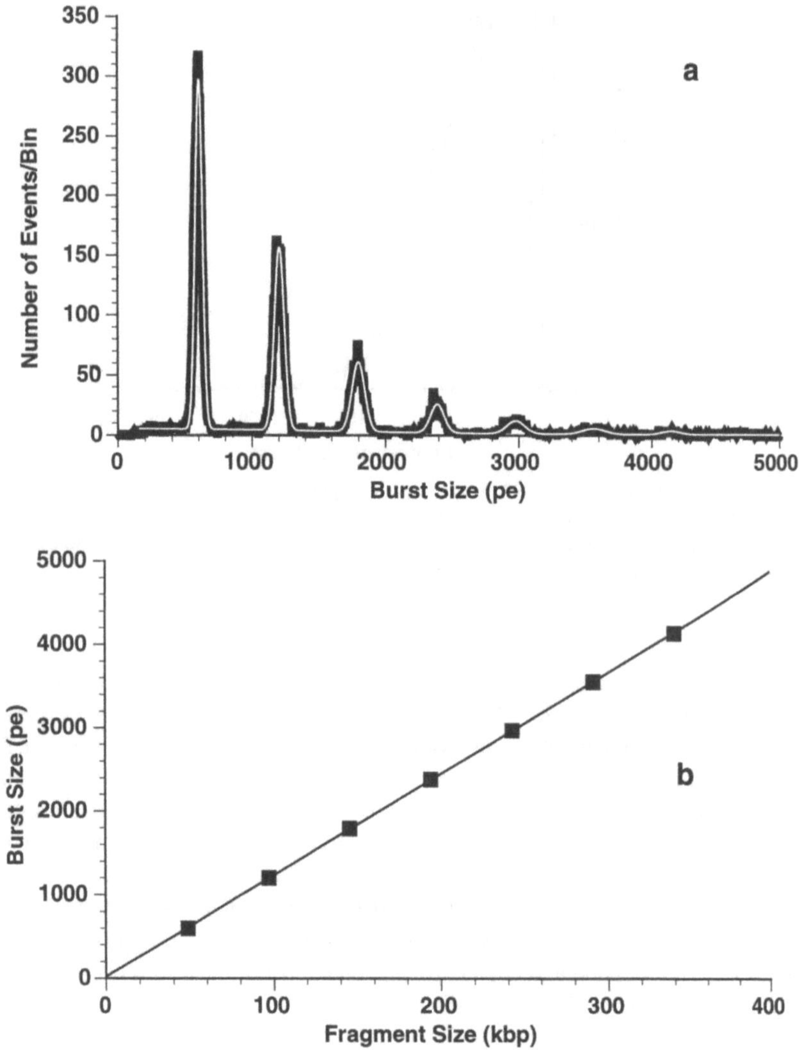

Figure 8.14. Sizing of λ DNA concatamers. a: Histogram of the burst sizes of λ concatamers. Conditions: staining dye, TOTO-1; excitation, 514.5 nm, 27 mW, cw; detector, PMT; transit time, 2 ms; DNA analyzed, ~0.4 pg (~10,000 fragments); analysis time, 134 s. The black line denotes data and the white line is the fit. b: Plot of burst size means from the Gaussian fit versus fragment lengths. (From reference 21.)

Table 8.1. Parameters Obtained from the Histogram of λ Concatamers (Figure 8.14)

λ concatamers (kbp)	Amplitude[a] (A_i)	Mean[a] ($\bar{X}_i$)	Standard deviation[a] (σ_i)	CV (%) ($\sigma_i/\bar{X}_i$)	Shot noise (%) ($\sqrt{\bar{X}_i}/\bar{X}_i$)	Deviations from linear fit (%)[b]
1λ (48.5)	292.9 ± 1.2	597.2 ± 0.1	27.5 ± 0.1	4.6	4.1	1.75
2λ (97.0)	152.6 ± 1.0	1199.1 ± 0.3	36.9 ± 0.3	3.1	2.9	−0.16
3λ (145.5)	57.8 ± 0.9	1792.6 ± 0.8	46.8 ± 0.9	2.6	2.4	−0.35
4λ (194.0)	23.7 ± 0.9	2384.3 ± 2.2	50.8 ± 2.3	2.1	2.0	−0.36
5λ (242.5)	10.7 ± 0.8	2971.3 ± 5.5	67.6 ± 6.0	2.3	1.8	−0.21
6λ (291.0)	4.7 ± 0.7	3555.6 ± 13.9	83.1 ± 15.1	2.3	1.7	−0.04
7λ (339.5)	2.8 ± 0.8	4135.0 ± 19.7	69.0 ± 22.7	1.7	1.6	0.21

[a]The uncertainties are the standard deviations of the fit parameters returned by the Marquardt-Levenberg curve fitting algorithm as implemented in SigmaPlot.
[b]The deviation of the data from the linear regression fit is defined as: 100 [(fitted burst size − measured burst size)/fitted burst size].

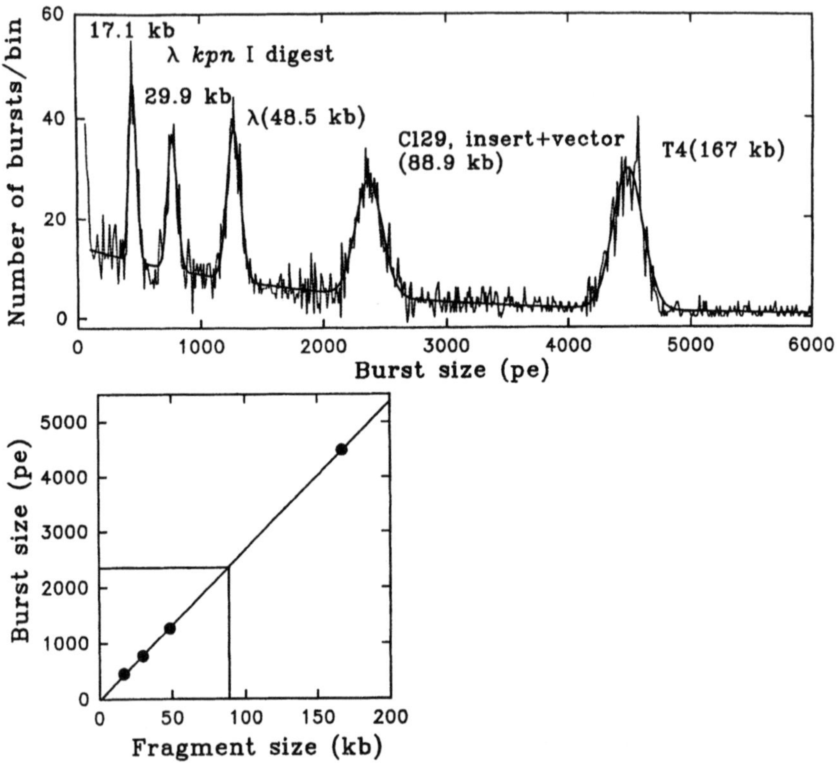

Figure 8.15. Sizing of PAC clones. a: Histogram of the burst sizes of an intact supercoiled clone. λ *Kpn* I digest, λ DNA, and T4 DNA were used as size standards. Conditions: staining dye, TOTO-1; excitation, 514.5 nm, 30 mW, cw; beam waist, 46 μm; detection volume, 10 pl; detector, PMT; transit time, 1.6 ms; DNA analyzed, <1 pg (~5,000 fragments); analysis time, 134 s. b: Plot of burst size means versus fragment lengths. The means of the DNA standard peaks were fit by linear regression. The size of the PAC clone was determined from the calibration line and its burst size mean. (From reference 21.)

technique, double-stranded DNA is denatured at high temperature, short DNA primers are hybridized to the complementary regions of the target sequence, and a polymerase enzyme produces new complementary strands in the region between the primers. This replication process is repeated many times, resulting in a geometric amplification of the target sequence. PCR fragments typically range in size from 0.5 to 5 kbp. For some applications it is necessary to know only that amplification has occurred. For example, one may want to know if a specific gene is present in a sample. In other applications, it is desirable to know the size of the target sequence, which will provide information on the particular polymorphism that has been amplified. The traditional method for determining the size of the PCR fragments is by gel electrophoresis. This analysis takes several hours, re-

Table 8.2. Comparison of the Sizes of PAC Clones Determined by Flow Cytometry and PFGE

Clone	Flow (kbp)	PFGE (kbp)
A	73.7 ± 0.4	71 ± 7
B	105.3 ± 0.9	107 ± 11
C	89.1 ± 1.7	97 ± 10
D	85.8 ± 1.8	93 ± 9
E	79.6 ± 0.9	81 ± 8
F	87.3 ± 1.9	94 ± 9

quires ~100 ng of DNA, and has a precision of approximately 0.1 kbp in the several kbp size range.

As an alternative to gel electrophoresis, we investigated the application of flow cytometric fragment sizing for the characterization of PCR products.[37] A set of eighteen PCR samples, each with a single fragment type in the 0.4 to 5 kbp size range, was analyzed. Internal standards were added and the samples were stained with POPO-3. For the small size fragments, we chose POPO-3 instead of TOTO-1 since POPO-3 is twice as bright as TOTO-1. One must be careful when using POPO-3, since the ranges of useful staining concentrations and dye:base-pair ratios are much narrower.[24] The size standards (0.684, 1.857, and 4.130 kbp) were DNA fragments purified from an enzymatic digestion of pBR322. The fragment sizes are known to single base-pair accuracy from sequence and restriction-site data.

Figure 8.16 shows burst size histograms for the eighteen PCR samples, each containing two of the size standards. Each histogram represents about 82 s of data from <0.1 pg of DNA (4,000–13,000 fragments). Three burst size histograms are overlaid for each sample to show the level of reproducibility. The vertical bars in the figure indicate the positions of the standards. The remaining peak in each trace is from a PCR fragment. Because there were variations in the absolute burst sizes due to alignment drifts and staining variations between samples, the ordinate in the figure was converted from burst size to fragment size for comparison using the known sizes of the internal standards. To assess the accuracy of the flow results in the PCR size range, these samples were sized on a standard gel, a CHEF (contour-clamped homogenous electric field) gel, and sequenced. Figure 8.17 shows the flow and gel results plotted against the sequence results. The flow and gel results are comparable in accuracy. The average absolute deviations from the sequence lengths are ~5%. The advantage that flow sizing has over all the other techniques is that a measurement on a single sample can be performed in one hundredth the time with one millionth the DNA with comparable accuracy in the 0.4 to 5 kbp range.

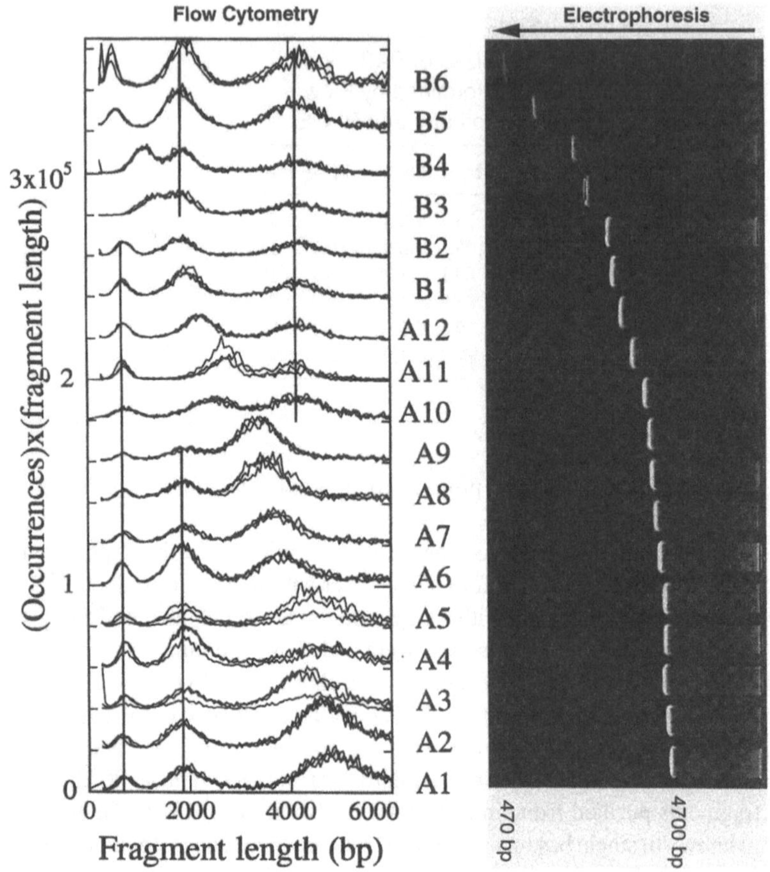

Figure 8.16. DNA fragment sizing of PCR products. Eighteen PCR samples were sized that had lengths from 0.4 to 5 kbp. Each sample was mixed with two internal standards and stained with POPO-3. Conditions: excitation, 514.5 nm, 8 mW, cw; beam waist, 10 μm high by 20 μm wide; detector, silicon avalanche photodiode; sample stream diameter, 10 μm; transit time, 1.5 ms; DNA analyzed, <1 pg (4,000–13,000 fragments); analysis time, 82 s/histogram. Three runs, each of 82 seconds of data, were obtained for each sample. Histograms of the burst sizes are shown on the left. The vertical bars denote the locations of the internal size standards (684, 1857, or 4130 bp in length). Results from gel electrophoresis (without internal standards) are shown on the right.

8.3.5.3. Bacterial Species and Strain Discrimination

There is considerable interest in the rapid identification of pathogenic bacteria for medical diagnostics, bioagent detection, monitoring hospital environments, and in the food processing industry. A commonly used technique is the analysis of

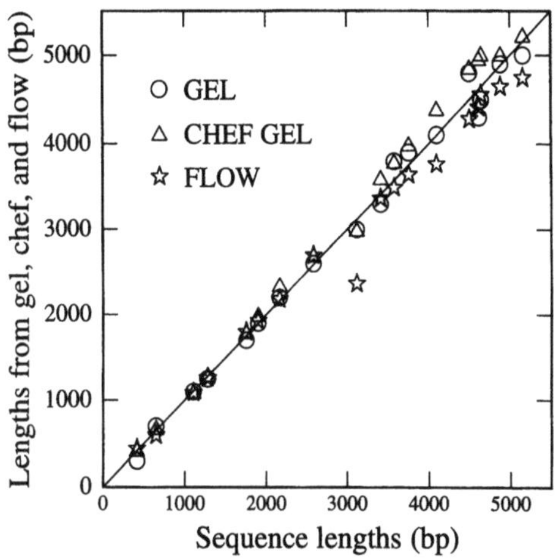

Figure 8.17. A comparison of the PCR fragment sizes from flow cytometry, gel electrophoresis, CHEF gel electrophoresis, and sequencing. The solid line is the sequence length. The open points show values obtained from flow and gels. The average absolute deviation for each technique from the sequence length is ~5%.

the fragment sizes produced by a restriction digest of the bacteria's chromosome. A restriction enzyme cuts the genome at sequences corresponding to the recognition site of the enzyme. For example, an eight-base cutter has a recognition site of eight bases. In general, the smaller the recognition site, the larger the number of restriction fragments obtained upon digestion of the DNA—a four-base cutter would yield many more (and smaller) fragments than an eight-base cutter. The number of base pairs between restriction sites is characteristic of the bacteria (fingerprint) and is used to identify bacteria species and strain. PFGE analysis of a restriction digest of the bacteria's genome is often used for bacterial identification. As stated above, PFGE requires micrograms of DNA, takes ~10–20 h for analysis, and has an uncertainty of ~10%. We have applied our flow cytometric approach for DNA fragment sizing to the discrimination of bacterial species[22] and strains.[28]

Approximately 10^6–10^8 bacteria cells are embedded into an agarose plug. The cells are lysed and RNAse and Proteinase K are added to degrade proteins and RNA. A restriction digest is carried out inside the agarose plug and the restriction fragments are removed from the plug by electroelution. Spermine and spermidine are added to the solution to reduce fragmentation during sample handling. This process currently takes several days. The electroeluted fragments are stained with the intercalating dye, TOTO-1, and diluted to a final fragment concentration of ~10^{-13} M. This solution is then introduced into our ultrasensitive flow cytometer for analysis.

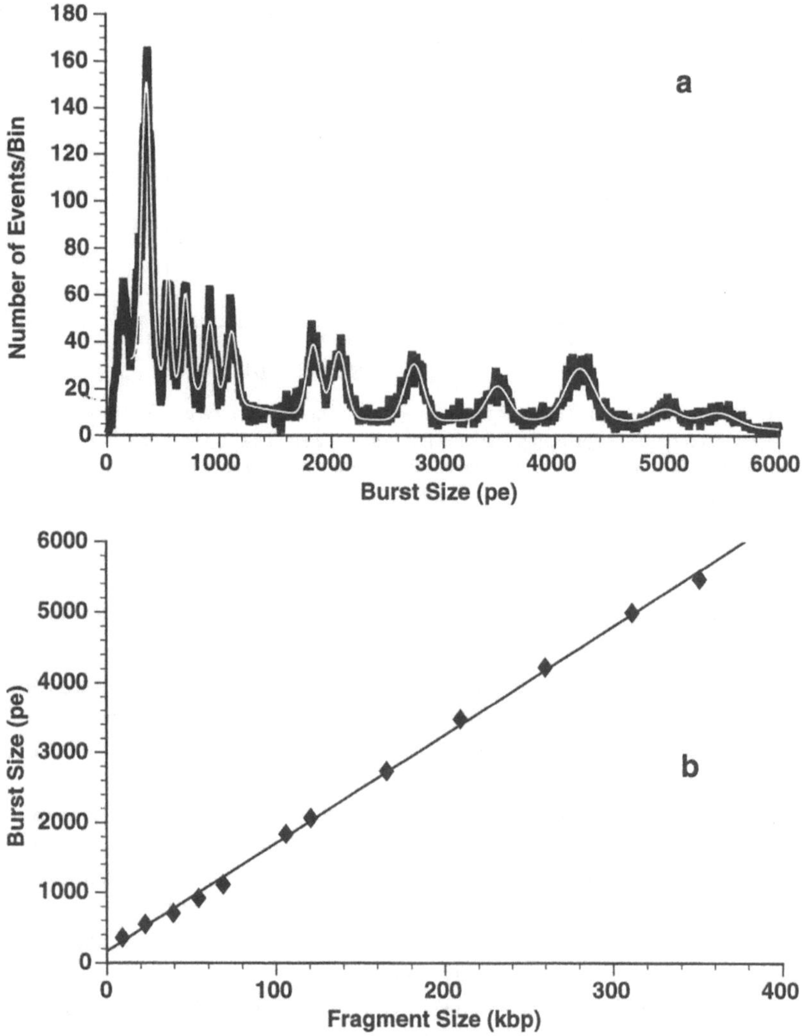

Figure 8.18. Sizing of a *Sma* I digestion of *S. aureus* DNA. a: Histogram of the fluorescence burst sizes. The histogram was fit with a sum of 12 Gaussians plus a fourth order background function. The black line denotes data; the white line is the fit. Conditions: staining dye, TOTO-1; excitation, 514.5 nm, 18 mW, cw; detector, PMT; transit time, 2.3 ms; DNA analyzed, <2 pg (~12,000 fragments from ~1000 bacteria); analysis time, 402 s. b: Plot of the burst size means from the Gaussian fit versus the fragment lengths determined by PFGE. (From reference 21.)

Figure 8.18 shows a histogram of burst sizes obtained from a *Sma* I restriction digest of the *S. aureus* genome.[21] The data were acquired in ~7 min from ~1 pg of DNA (~12,000 fragments from ~1000 bacteria). Twelve peaks were resolved. Bursts from the 670 kbp restriction fragment were not detected due to shear degradation. Also shown in Figure 8.18 is a plot of the centroids of the peaks extracted from the fit to the histogram versus the fragment sizes obtained from a PFGE analysis of the same sample. The plot is linear with a correlation coefficient of $r^2 = 0.99923$. The fragment sizes determined from the straight line are listed in Table 8.3. Also shown in Table 8.3 is a comparison of the fragment sizes determined by flow cytometry with the fragment sizes determined by PFGE. There is good agreement between the two approaches. The average absolute deviation is 5.2%. We attribute most of the deviations to the 10% uncertainty inherent in PFGE.

The applicability of this technology for bacteria species discrimination is shown in Figure 8.19.[22] Histograms of burst sizes obtained from *Not* I restriction digests of *E. coli*, *B. globigii*, and *E. herbicola* are displayed in this figure. DNA from ~1000 bacteria are sized in less than 10 min. The two largest fragments in the *E. coli* (576 and 629 kbp) and *B. globigii* (394 and 424 kbp) digests were not detected in flow. The peak patterns are unique for each bacterium. Also shown in Figure 8.19 are plots of the centroids of the peaks in the histogram versus the fragment sizes determined by PFGE. The plots are linear with a correlation

Table 8.3. Comparison of Fragment Sizes Obtained from a *Sma* I Restriction Digest of *S. aureus*: PFGE versus Flow Cytometry

PFGE (kbp)	Flow (kbp)	% difference[a]
9.0	12.7	−41.1
22.7	25.0	−10.1
39.2	35.0	10.7
54.2	48.9	9.8
68.5	61.2	10.7
105.8	108.4	−2.4
120.5	123.1	−2.2
165.5	166.1	−0.4
208.9	214.1	−2.5
259.8	261.9	−0.8
310.9	311.9	−0.3
350.7	342.5	2.3
670	—	

[a]100 [(PFGE − flow)/PFGE].

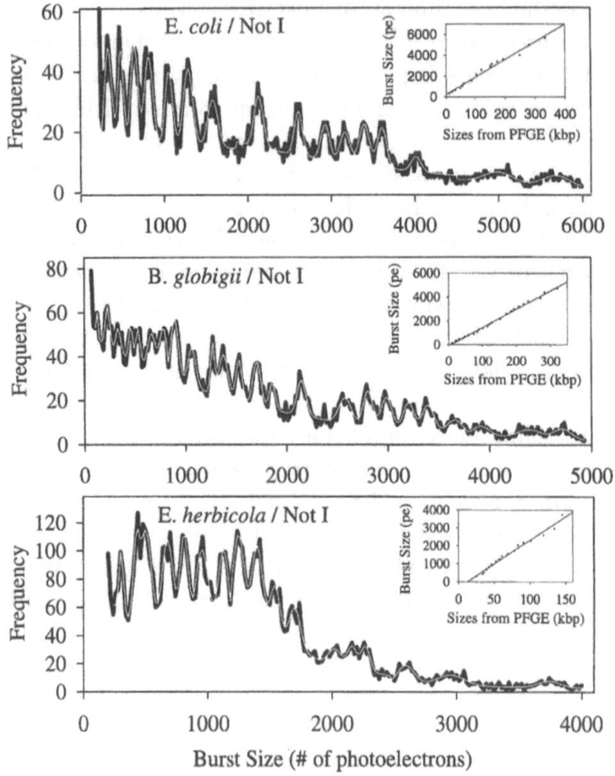

Figure 8.19. Histograms of burst sizes from a *Not* I restriction digest of *E. coli*, *B. globigii*, and *E. herbicola*. The black line is the data and the white line is the fit. Conditions: staining dye, TOTO-1; excitation, 514.5 nm, 25 mW, cw; detection volume, ~10 pl; detector, PMT; transit time, 2.1 ms; DNA analyzed, <1 pg; analysis time, ~300 s. A plot of the centroids versus the DNA fragment lengths determined by PFGE is shown in the insert. (From reference 22.)

coefficient $r^2 > 0.98$. Fragment sizes of *E. coli* determined by flow cytometry are compared to the PFGE results in Table 8.4. Again, within the 10% uncertainty attributed to PFGE, there is good agreement between the two approaches.

Histograms of the burst sizes for the *Not* I digestion of two different strains of *E. coli* (#15597 and #25922) are shown in Figure 8.20. It is clear that the patterns are different and the two strains are easily distinguishable. Also shown in Figure 8.20 are plots of the centroids of the peaks in the histogram versus the fragment sizes determined by PFGE.

The fragment sizes determined from the data for Figures 8.19 and 8.20 are based upon fragment sizes determined by PFGE. This is unsatisfactory because of

Table 8.4. Comparison of the Fragment Sizes of a *Not* I Digest of *E. coli* (#15597) Obtained from PFGE and Flow

PFGE (kbp)	Flow (kbp)	% difference[a]
—	17.5	
33.5	38.2	12.3
83.1	90.3	8.0
104.3	115.5	9.7
128.5	136.7	6.0
153.2	160.5	4.5
186.2	188.6	1.3
207.0	212.3	2.6
246.0	236.9	−3.8
274.8	258.1	−6.5
354.1	315.5	−12.2

[a]100 [(flow − PFGE)/flow].

the large uncertainty associated with PFGE analysis and because there is a need to size restriction digests of unknown bacteria when the PFGE data are not available. Absolute sizes can be determined by adding DNA fragments whose absolute sizes are known to the sample.[22] Figure 8.21 shows histograms of burst sizes from a *Not* I digest of *E. coli* (#15597) to which λ DNA was added as an internal standard. The sequence of λ DNA is known so that its size is known exactly. The concentration of λ DNA was large enough such that there was a 10% probability of two λ DNAs being in the detection volume at the same time. The peak in the histogram at burst size = 471 pe is from λ DNA (48.50 kbp) band the peak at burst size = 948 pe is from 2 λ (97.00 kbp). The straight line in Figure 8.21 was constructed to pass through the center of the points corresponding to λ and 2λ. The burst sizes measured for the bacteria fragments were placed on the line for fragment size determination. A comparison of the fragment sizes based upon PFGE with fragment sizes based on the internal λ standard is shown in Table 8.5.

8.4. Summary

Flow cytometry is an attractive alternative to pulsed-field gel electrophoresis for sizing of large DNA fragments. Samples containing ~1 pg of DNA are analyzed in less than 10 minutes with an uncertainty <2%. In contrast, PFGE requires micrograms of DNA, ~20 h for analysis and has an uncertainty of ~10%.

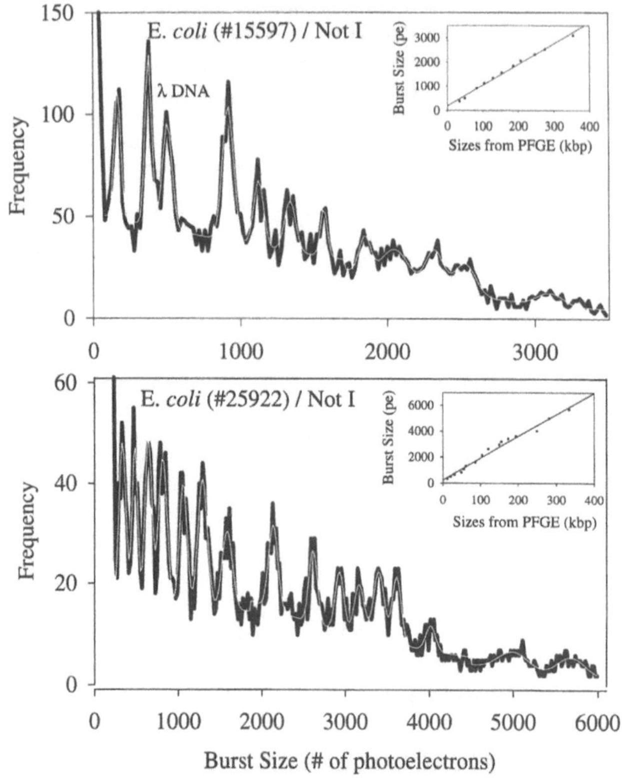

Figure 8.20. Histograms of burst sizes from a *Not* I restriction digest of *E. coli* (#15597) and *E. coli* (#25992). The black line is the data and the white line is the fit. Conditions: staining dye, TOTO-1; excitation, 514.5 nm, 25 mW, cw; detection volume, ~10 pl; detector, PMT; transit time, 2.1 ms; DNA analyzed, <1 pg; analysis time, ~300 s. A plot of the centroids versus the DNA fragment lengths determined by PFGE is shown in the insert.

Flow cytometric analysis is much more quantitative—individual fragments are counted. PFGE concentrations are estimated from the intensity of emission from the dye used to visualize the band. PFGE requires the DNA to be linearized before analysis while flow sizing is conformation independent—the same size is obtained for supercoiled, relaxed circular, and linear DNA.

Bacterial species and strains can be discriminated by the pattern of their restriction fragments (fingerprint). DNA from ~1000 bacteria containing fragments up to 425 kbp are sized in less than 10 minutes. Currently, the analysis time is dominated by sample preparation. In order to reduce shearing of large fragments, samples are handled inside of agarose plugs. Diffusion times of reagents

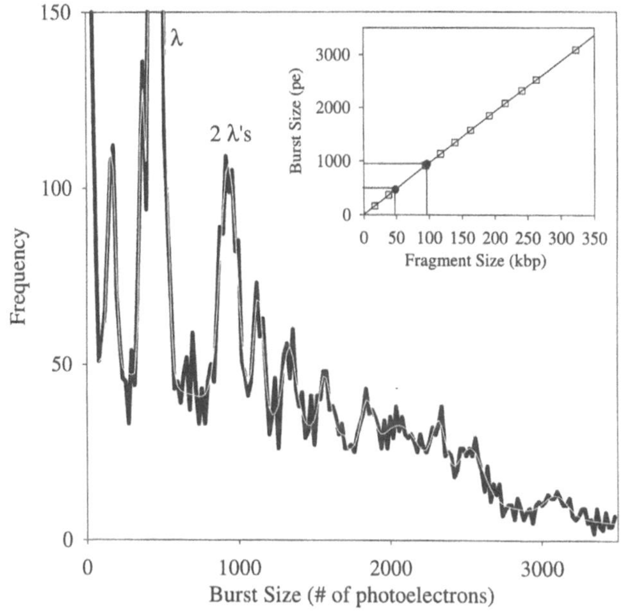

Figure 8.21. Two Point Calibration for *E. coli* (#15597) DNA digested by *Not* I. The peaks corresponding to one λ and two λ are labeled in the figure. The two-λ peak overlaps the bacteria peak at 90.3 kbp. Fragment sizes were determined from the calibration curve as described in the text. (From reference 22.)

Table 8.5. Comparison of Size Measurements for *E. coli* (#15597)/*Not* I Digests Obtained by PFGE and Flow with a Two Point Calibration

PFGE (kb)	Flow[a] (kb)	PFGE − flow (kb)	% difference[b]
—	16.7	—	—
33.5	38.2	−4.7	−14.1
83.1	90.3	−7.1	−8.6
104.3	115.5	−11.1	−10.7
128.5	136.7	−8.2	−6.4
153.2	160.5	−7.3	−4.7
186.2	188.6	−2.4	−1.3
207.0	212.3	−5.3	−2.6
246.0	236.9	9.1	3.7
274.8	258.1	16.6	6.1
354.1	315.5	38.6	10.9

[a]From two point calibration λ and 2λ ($y = 9.82x - 4.8$).
[b]100 [(PFGE − flow)/PFGE].

into the plug and restriction fragments out of the plug can take several days. We have recently reduced the sample preparation time from days to hours.[38]

In some cases it is desirable to size a small component of a complex DNA sample. Fragments from the major components would dominate the analysis and hide the signature of the desired components. In this case a two tag (color) assay could be used to identify fragments of interest. For example, a short, fluorescently tagged oligo, complementary to a sequence only on the fragments of interest, could be added to the sample. Simultaneous two color detection would be used—one color for the hybridization probe and one color for the intercalating dye. Only species containing both tags would be sized. See, for example, Agronskaia *et al.*[32]

Miniaturization of flow sizing of DNA restriction fragments was reported recently by Quake *et al.*[25] A microfabricated flow cell was constructed and sizing of DNA fragments from 2–200 kbp was demonstrated by measuring the pulse height as dye-stained fragments passed through the laser beam. The authors reported a resolution of 5–15%. See also Fouquet *et al.*[33] Our current analysis rate is ~100 fragments/s. This is too slow for many applications. Array detectors have the potential to increase the analysis rate to ~100,000 fragments/s.[39,40]

Acknowledgments

This work was supported by internal funding from Los Alamos National Laboratory, by the Department of Energy (NN20), and by the NIH funded National Flow Cytometry Resource (RR-01315). PCR samples were provided by and gel analysis of the samples was performed by Dr. Laurent Susini, Genset-Dpt Cartographie, Centre de Recherche Genomique-RN7, F-91030 Evry Cedex - France.

References

1. Hirschfeld, T. Optical microscopic observation of single small molecules. *Appl. Opt. 15*, 2965–2966, 1976.
2. Dovichi, N. J., Martin, J. C., Jett, J. H., M. Trkula, and Keller, R. A. Laser-induced fluorescence of flowing samples as an approach to single-molecule detection in liquids. *Anal. Chem. 56*, 348–354, 1984.
3. Shera, E. B., Seitzinger, N. K., Davis, L. M., Keller, R. A., and Soper, S. A. Detection of single molecules. *Chem. Phys. Lett. 174*, 553–557, 1990.
4. Barnes, M. D., Whitten, W. B., and Ramsey, J. M. Detecting single molecules in liquids. *Anal. Chem. 67*, A418–A423, 1995.
5. Keller, R. A., Ambrose, W. P., Goodwin, P. M., Jett, J. H., Martin, J. C., and Wu, M. Single-molecule fluorescence analysis in solution. *Appl. Spectrosc. 50*, 12A-32A, 1996.

6. Goodwin, P. M., Ambrose, W. P., and Keller, R. A. Single-molecule detection in liquids by laser-induced fluorescence. *Accounts Chem. Res. 29*, 607–613, 1996.
7. Keller, R. A., Ambrose, W. P., Arias, A. A., Cai, H., Emory, S. R., Goodwin, P. M., and Jett, J. H. Analytical applications of single molecule detection. *Anal. Chem. 74*, 317A–324A, 2002.
8. Zander, C., Enderlein, J., and Keller, R. A. *Single Molecule Detection in Solution: Methods and Applications*. Berlin: Wiley-VCH, 2002.
9. Kachel, V., Fellner-Feldegg, H., and Menke, E. Hydrodynamic properties of flow cytometry instruments. In *Flow Cytometry and Sorting, 2nd ed.*, Melamed, M. R., Lindmo, T., and Mendelsohn, M. L., eds. New York: Wiley-Liss, 1990, pp. 27–44.
10. Harris, T. D. and Lytle, F. E. Analytical applications of laser absorption and emission spectroscopy. In *Ultrasensitive Laser Spectroscopy*, Kliger, D. S., ed. New York: Academic Press, 1983, pp. 369–433.
11. Goodwin, P. M., Affleck, R. L., Ambrose, W. P., Jett, J. H., Johnson, M. E., Martin, J. C., Petty, J. T., Schecker, J. A., Wu, M., and Keller, R. A. Detection of single fluorescent molecules in flowing sample streams. In *Computer Assisted Analytical Spectroscopy*, Brown, S. D., ed. Chichester, UK: Wiley, 1996, pp. 61–80.
12. Goodwin, P. M., Affleck, R. L., Ambrose, W. P., Demas, J. N., Jett, J. H., Martin, J. C., Reha-Krantz, L. J. Semin, D. J., Schecker, J. A., Wu, M., and Keller, R. A. Progress toward DNA sequencing at the single molecule level. *Exp. Tech. Phys. 41*, 279–294, 1995.
13. Soper, S. A., Davis, L. M., and Shera, E. B. Detection and identification of single molecules in solution. *J. Opt. Soc. Am. B 9*, 1761–1769, 1992.
14. Zander, C., Sauer, M., Drexhage, K. H., Ko, D.-S., Schultz, A., Wolfrum, J., Brand, L., Eggeling, C., and Seidel, C. A. M. Detection and characterization of single molecules in aqueous solution. *Appl. Phys. B 63*, 517–523, 1996.
15. Enderlein, J., Goodwin, P. M., Van Orden, A., Ambrose, W. P., Erdmann, R., and Keller, R. A. A maximum likelihood estimator to distinguish single molecules by their fluorescence decays. *Chem. Phys. Lett. 270*, 464–470, 1997.
16. Van Orden, A., Machara, N. P., Goodwin, P. M., and Keller, R. A. Single molecule identification in flowing sample streams by fluorescence burst size and intraburst fluorescence decay rate. *Anal. Chem. 70*, 1444–1451, 1998.
17. C.W. Wilkerson, Jr., Goodwin, P. M., Ambrose, W. P., Martin, J. C., and Keller, R. A. Detection and lifetime measurement of single molecules in flowing sample streams by laser-induced fluorescence. *Appl. Phys. Lett. 62*, 2030–2032, 1993.
18. Goodwin, P. M., Johnson, M. E., Martin, J. C., Ambrose, W. P., Marrone, B. L., Jett, J. H., and Keller, R. A. Rapid sizing of individual fluorescently stained DNA fragments by flow cytometry. *Nucleic Acids Res. 21*, 803–806, 1993.
19. Petty, J. T., Johnson, M. E., Goodwin, P. M., Martin, J. C., Jett, J. H., and Keller, R. A. Characterization of DNA size determination of small fragments by flow cytometry. *Anal. Chem. 67*, 1755–1761, 1995.
20. Huang, Z., Petty, J. T., O'Quinn, B., Longmire, J. L., Brown, N. C., Jett, J. H., and Keller, R. A. Large DNA fragment sizing by flow cytometry: Application to the characterization of P1 artificial chromosome (PAC) clones. *Nucleic Acid Res. 24*, 4202–4209, 1996.
21. Huang, Z., Jett, J. H., and Keller, R. A. Bacteria genome fingerprinting by flow cytometry. *Cytometry, 35*, 169–175, 1999.
22. Kim, Y., Jett, J. H., Larson, E. J., Penttila, J. R., Marrone, B. L., and Keller, R. A. Bacterial fingerprinting by flow cytometry: Bacterial species discrimination. *Cytometry 36*, 324–332, 1999.
23. Werner, J. H., Larson, E. J., Goodwin, P. M., Ambrose, W. P., and Keller, R. A. Effects of fluorescence excitation geometry on the accuracy of DNA fragment sizing by flow cytometry. *Appl. Opt. 39*, 2831–2839, 2000.
24. Yan, X., Grace, W. K., Yoshida, T. M., Habbersett, R. C., Velappan, N., Jett, J. H., Keller, R. A.,

and Marrone, B. L. Characteristics of different nucleic acid staining dyes for DNA fragment sizing by flow cytometry. *Anal. Chem. 71*, 5470–5480, 1999.
25. Yan, X., Habbersett, R. C., Cordek, J. M., Nolan, J. P., Yoshida, T. M., Jett, J. H., and Marrone, B. L. Development of a mechanism-based, DNA staining protocol using SYTOX orange nucleic acid stain and DNA fragment sizing flow cytometry. *Anal. Biochem. 286*, 138–148, 2000.
26. Van Orden, A., Cai, H., Goodwin, P. M., and Keller, R. A. Efficient detection of single DNA fragments in flowing sample streams by two-photon fluorescence excitation. *Anal. Chem. 71*, 2108–2116, 1999.
27. Castro, A., Fairfield, F. R., and Shera, E. B. Fluorescence detection and size measurement of single DNA molecules. *Anal. Chem. 65*, 849–852, 1993.
28. Castro, A., and Shera, E. B. Single-molecule electrophoresis. *Anal. Chem. 67*, 3181–3186, 1995.
29. Haab, B. B., and Mathies, R. A. Optimization of single-molecule fluorescence burst detection of ds-DNA: Application to capillary electrophoresis separations of 100–1000 basepair fragments. *Appl. Spectrosc. 51*, 1579–1584, 1997.
30. Agronskaia, A., Schins, J. M., de Grooth, B. G., and Greve, J. Polarization effects in flow cytometric DNA sizing. *Appl. Opt. 38*, 714–719, 1999.
31. Chou, H-P., Spence, C., Scherer, A., and Quake, S. A microfabricated device for sizing and sorting DNA molecules. *Proc. Natl. Acad. Sci. USA 96*, 11–13, 1999.
32. Agronskaia, A., Schins, J. M., de Grooth, B. G., and Greve, J. Two-color fluorescence in flow cytometry DNA sizing: Identification of single-molecule fluorescent probes. *Anal. Chem. 71*, 4684–4689, 1999.
33. Foquet, M., Korlach, J., Zipfel, W., Webb, W. W., and Craighead, H. G. DNA fragment sizing by single molecule detection in submicrometer-sized closed fluidic channels. *Anal. Chem. 74*, 1415–1422, 2002.
34. Glazer, A. N., and Rye, H. S. Stable dye-DNA intercalation complexes as reagents for high-sensitivity fluorescence detection. *Nature 359*, 859–861, 1992.
35. Peck, K., Stryer, L., Glazer, A. N., and Mathies, R. A. Single-molecule fluorescence detection: Autocorrelation criterion and experimental realization with phycoerythrin. *Proc. Natl. Acad. Sci. USA 86*, 4087–4091, 1989.
36. Habbersett, R. C., Jett, J. H., and Keller, R. A. Molecular flow detection system. In *Cytometric Cellular Analysis, Emerging Tools for Cell Analysis: Advances in Optical Measurement*, Durack G., ed. New York: J. Wiley and Sons, 2000, pp. 115–137.
37. This was a collaborative project with Genset-Dpt Cartographie, Centre de Recherche Genomique-RN7, F-91030 Evry Cedex - France. PCR samples were prepared by Dr. Laurent Susini of Genset. Gel and sequence lengths were determined at Genset.
38. Larson, E. J., Hakovirta, J. R., Cai, H., Jett, J. H., Burde, S., Keller, R. A., and Marrone, B. L. Rapid DNA fingerprinting of pathogens by flow cytometry. *Cytometry 41*, 203–208, 2000.
39. Van Orden, A., Keller, R. A., and Ambrose, W. P. High-throughput flow cytometric DNA fragment sizing. *Anal. Chem. 72*, 37–41, 2000.
40. Ma, Y., Shortreed, M. R., and Yeung, E. S. High-throughput single molecule spectroscopy in free solution. *Anal. Chem. 72*, 4640–4645, 2000.

9

Fluorimetric DNA Biosensors

Paul A. E. Piunno and Ulrich J. Krull

9.1. Introduction

A biosensor is a device that physically combines a biologically active material to a transducer to convert a selective biochemical interaction into a measurable analytical signal[1,2] The advantages ideally offered by biosensors over other forms of analysis include ease of use (by nonexpert personnel), rapid response, low cost, ease of fabrication, small size, ruggedness, facile interfacing with computers, low detection limits, high sensitivity, selectivity, and reusability of the devices. Biosensors have been used to selectively detect cells, viruses, other biologically significant materials, biochemical reactions, and immunological reactions by using detection strategies that involve immobilization of enzymes, antibodies, or other selective proteins onto piezoelectric, optical, electrochemical, and surface plasmon resonance sensors[3,4] (see Table 9.1). Biosensors are not widely available from commercial sources due to problems associated with the long-term stability of the selective recognition elements when immobilized onto solid surfaces,[5,6] but some have already made significant impact in certain areas of analysis such as blood glucose determination.[7]

An approach that may potentially be used to create biosensors with long-term chemical stability takes advantage of the stability of DNA. With the advent of DNA probe technology, a number of selective oligomers which interact with the DNA of important biological species have been identified[8–11] These have been used to provide a new type of biorecognition element which is highly selective, stable, and can be easily synthesized in the laboratory.[12–14]

Paul A. E. Piunno • FONA Technologies, Inc., Waterloo, Ontario, Canada, N2V 2K1. *Ulrich J. Krull* • Chemical Sensors Group, Department of Chemistry, University of Toronto at Mississauga, Mississauga, Ontario, Canada, L5L 1C6.

Topics in Fluorescence Spectroscopy, Volume 7: DNA Technology, edited by Joseph R. Lakowicz, Kluwer Academic / Plenum Publishers, New York, 2003

Table 9.1. Examples of Biosensors for Detection of Nucleic Acids

	Electrochemical	
Cyclic voltammetry	Detection of redox active metallo-interaclators	Mikkelson[15–17]
Chronopotentiometry	Detection of electroactive intercalants for DNA, PNA, triplex formation, binding of pollutants and drugs	Wang[18–30]
Cyclic voltammetry and chronopotentiometry	Catalytic current of guanine in a target strand *via* reaction with a metallo-intercalator	Thorpe[31]
Peak current at constant potential	Polypyrrole molecular wiring to DNA with alterations in oxidation polypyrrole due to hybridization	Korri-Youssoufi[32]
Cyclic voltammetry	Ferrocene conjugated reporter oligonucleotide in sandwich-type assay	Ihara[33]
Peak current	Molecular writing in combination with electron transfer through fully complementary dsDNA	Meade[34]
	Acoustic wave	
AT-cut TSM	Network analysis for DNA hybridization and DNA-peptide interactions	Thompson[35,36]
AT-cut TSM	Frequency shifts for detection of hybridization with Au-thiol immobilization	Wang[37]
	Optical	
Surface plasmon resonance	Detection of optical mass changes due to hybridization, kinetics, triplex formation PCR assays, PNA	BIACore™[38–48]
Surface plasmon resonance	Improvements in immobilization and detection methods to increase sensitivity to try to match fluorescence detection (SPR reports LOD of 10^{11} molecules/cm^2)	Corn[49,50]
Interferometry	Analogue of Mach-Zender interferometer to detect change of refractive index due to hybridization	Gerdt[51]
Optical fiber	Evanescent-wave system for detection of hybridization of complement strands containing fluorescein	Squirrell[52]
Optical fiber	Automated system using acvidin-biotin immobilization of capture probes for both direct and competitive binding assays	Abel[53]
Planar monomodal waveguide	Detection limit of 6×10^7 molecules for target strands labeled with cyanine-based dye	Duveneck[54]
Tapered multimode fiber	Improved efficiency evancescent-wave device for DNA hybridization based on control of guided nodes in fiber	Ligler[56]
Tapered monomodal fiber	High efficiency evanescent-wave device using near IR dye labeled target strands	Pilevar[55]
Fiber bundle array	Microarray system where each fiber of a fiber bundle can be used as one element of an array with different selective chemistry	Walt[57–59]
Optical fiber	Detection of hybridization, triplex formation, self-contained reversible sensor	Krull[60–62]

9.2. Fluorimetric Fiber Optic Biosensors

Optical transduction techniques are being used for development of nucleic acid biosensors. The high sensitivity of fluorescence techniques has recently been highlighted by Weinfeld *et al.*[63] who have demonstrated zeptomole detection limits for the detection of radiation-induced DNA damage. Most efforts toward fluorimetric optical biosensor development incorporate optical fibers in the sensor design. The use of optical fibers in biosensing configurations is considered attractive due in part to their multiplex capability, since radiation of different wavelengths can propagate simultaneously in different directions along the fiber, in addition to their small size and potential use in applications of remote and continuous sensing.[64]

Surface plasmon resonance, acoustic wave, and interferometric sensors provide the advantage of operating in a label-free motif so that external reagent treatment to label the interaction pair is not required. This allows for these technologies to investigate a plethora of analytes and binding interactions, such as DNA-DNA duplex and triplex formation, DNA-protein, and DNA-drug interactions. As a consequence of the capability to detect ensemble alterations in interfacial structure, interference from nonspecific binding interactions is also embraced by these methods. Similar to the use of avidin films for oligonucleotide capture, the use of peptide nucleic acid recognition elements to boost the selectivity of the chemistry to permit single-base mismatch discrimination may also provide a surface that permits a multitude of nonspecific binding interactions. In order to avoid the problem of interferent binding leading to false-positive results, especially when keeping the goal of analyzing real-world samples (e.g., blood) in mind, a transduction strategy that is sensitive to the structure of the binding pair is preferred. This limitation may be overcome by using reporter molecules that associate selectively with the binding pair.

Fluorimetric detection of nucleic acid hybridization, where the nucleic acids are immobilized onto optical substrates, is the most commonly employed strategy. Fiber-optic devices are typically described in two configurations.[65] The extrinsic mode has the selective chemistry located at the distal terminus of an optical fiber in the form of a solution cell or as an immobilized layer. Such a system is easily constructed but in many cases, where thin films are desired, is limited by the quantity of reporter molecules that can be attracted to the end of a fiber, and by the propagation of excitation radiation beyond the terminus of the fiber into solution which often results in excitation of background fluorescence and scattering from targets that are not associated with the chemically selective layer.

An interesting approach to fiber-optic sensor development has been reported by Walt and coworkers[57,58] who have created a multiplexed extrinsic-mode fiber optic microarray biosensor. The sensor employed fiber optic bundles (2–3 feet in length) composed of many optical fibers each 200–350 μm in diameter. The microarray was fabricated by immobilizing a different oligonucleotide probe sequence onto the distal end of each fiber. Individual fibers within the bundle were made reactive for oligonucleotide immobilization by immersing the bundle in a solution of monomeric acrylamide functionalized with succinimidyl ester residues. Ultraviolet radiation (350 nm) was transmitted down the desired fiber to the distal terminus to polymerize acrylamide monomers, yielding an amine-reactive polymer matrix solely above the illuminated fiber. Subsequent oligonucleotide attachment was then done by incubating the terminus of the fiber bundle in a solution containing the desired 5-amino-functionalized oligonucleotide capture probe. Following incubation with the capture oligonucleotide probe, any remaining reactive sites were capped by treatment with ethanolamine. Hybridization experiments were done by dipping the functionalized fiber tips directly into a solution containing fluorescently-tagged target strands. Fluorescence detection was achieved by coupling the proximal end of the fiber bundle into a cooled CCD camera and measuring the intensity of fluorescence emission from the distal tips. Such a microarray biosensor arrangement provides the potential for remote multiplexed oligonucleotide sequence analysis with reported detection limits of 1.3 nM (in a 200 μl reaction volume to provide an absolute detection limit of 1.5×10^{11} molecules) and with a response time of ~7 minutes. Full regeneration was achieved in less than 10 s upon heating of the fiber bundle terminus to 65°C in hybridization buffer, and no discernible decrease in sensitivity was observed after three cycles of application. The unique aspect of this biosensor format is that in addition to intensity measurements, fluorescent images are also obtained by incorporating imaging fibers in the sensor design. Imaging fibers are simply bundled fibers that have their termini fully aligned at the distal and proximal ends. Thus each fiber behaves as a "pixel" to provide images with spatial resolutions on the order of micrometers.[59]

9.3. Evanescent Wave Biosensors

The intrinsic mode arrangement is more commonly employed to monitor fluorescence emission from thin films of selective material on waveguide substrates. Fluorophores present at the surface of the waveguide may be excited through the formation of a standing wave of electric field intensity which propagates normal to the surface of the fiber upon total internal reflection of radiation in the fiber. The amplitude of the electric field of the reflecting radiation decreases exponentially as a standing wave into the medium having the lower refractive index. This decaying radiation is referred to as an evanescent wave and can be used to excite fluorophores located near the boundary for TIR. The propagation intensity, I, of the evanescent wave depends on the reflection angle, θ, the wavelength of the transmitted radiation, λ, and a Fresnel transmission factor, T:

$$I = T(\theta)\exp(-2x/d_p) \tag{9.1}$$

where x represents distance normal to the boundary for TIR, and d_p is the penetration depth which is given by[66]

$$d_p = \frac{\lambda}{4\pi(n_1^2 \sin^2(\theta) - n_2^2)^{1/2}} \tag{9.2}$$

The penetration depth is defined as the distance at which the intensity of the evanescent field has decayed to $1/e$ of the intensity at the reflection boundary.

Typically, the evanescent wave propagates beyond the surface of a fiber with a penetration depth ranging from 200 nm to 400 nm for visible light. Light emitted from fluorophores (after evanescent or direct excitation) at the surface of the fiber is preferentially coupled back into the fiber and can be monitored by a PMT which is placed at an end of the fiber[67] Increasing the length of coated fiber results in a greater optical path length and improved sensitivity. This method is most suitable for the study of ultrathin films which have a film thickness on the order of 2–3 nm, where the intensity of the evanescent wave is near its maximum, thereby increasing the signal-to-noise ratio as interferences from the bulk environment are avoided. However, assay methods based on evanescent excitation are limited by the amount of signal that can be generated. For multimode waveguides, less than 0.01% of the optical radiation carried within the waveguide is exposed to the outer medium in the form of an evanescent wave[64] In the case where monomodal waveguides are used, ~10% of the radiation carried by the waveguide is exposed to the outer medium in the form of an evanescent wave.[51] In the classic total internal reflection fluorescence (TIRF) evanescent wave configuration, the critical angle (θ_c) for the waveguide/solution interface ($\theta_c^{W/S}$) is larger than θ_c for the waveguide/biological film interface ($\theta_c^{W/B}$). Only the evanescent component of the propagated radiation will enter the biological film. The principle of optical reciprocity states that light coupled back into a waveguide as a plane wave will be in the same way as the primary process when a plane wave generates an evanescent wave.[68] Thus, for the fluorophores excited by evanescent waves created from modes propagating at or near $\theta_c^{W/S}$, none of the fluorescence emission can be coupled back into the waveguide in the same propagation mode because $\theta_c^{W/S}$ would be >90.[67] Hence a large portion of the signal would be lost to the surroundings for systems in which fluorescence emission originates from thin films of a refractive index lower than that of the waveguide onto which they are immobilized. It has been shown by Love *et al.*[69] that under optimal conditions only 2% of the light emitted by the fluorophore in the medium of lower refractive index may be captured and guided by the fiber.

Alternatively, as the refractive index of a chemically selective monolayer of organic media ($n_{monolayer}$ = 1.46 to 1.5[70] can be very similar to that of the substrate on which it is immobilized ($n_{fused\ silica}$ = 1.46,[71] direct excitation of fluorophores in the membrane may occur as the boundary for TIR would be the membrane-solution interface. This concept will be explained further below.

Perhaps the first reported evanescent wave biosensor for detection of DNA hybridization was that of Squirrell *et al.*[52] In this work, single-stranded nucleic acid sequences ranging in length from 16-mer oligonucleotides to 204-base oligomers functionalized with an aminohexyl linker at the 5′ terminus were covalently attached to optical fibers functionalized with 3-aminopropyl triethoxysilane via a glutaraldehyde linkage. Investigations of nucleic acid hybridization were done by monitoring fluorescence intensity in an intrinsic mode configuration using complementary target strands that had been prelabeled with a fluorescein moiety. This yielded a reusable assay system in which signal generation was observed to occur within minutes and nanomolar detection limits were achieved.

Abel and coworkers[53] have reported an automated optical biosensor system. Their device utilizes 5′-biotinylated-16-mer oligonucleotide probes bound to an optical fiber functionalized with avidin to detect complementary oligonucleotides prelabeled with fluorescein moieties in a TIRF motif similar to that of Squirrell *et al.* Immobilization of nucleic acid probes onto the optical fiber substrate was achieved by functionalization of the surface with (3-aminopropyl)triethoxysilane (APTES) or mercaptomethyldimethylethoxysilane (MDS). A layer of biotin was coupled onto the short alkylsilane layer (by treatment of the aminosilane functionalized fibers with NHS-LC-biotin or by treatment of thiolsilane functionalized fibers with biotinylated BSA). The biotinylated fibers were then treated with avidin or streptavidin, followed by coupling of 5′-biotinylated oligonucleotide capture probes. Each assay consisted of a 3-min pre-equilibration, 15-min hybridization time, 10-min washing procedure, followed by a 5-min regeneration cycle (chemical or thermal). A chemical denaturation scheme using 50% aqueous urea was described as the preferred method for sensor regeneration, as exposure of the optical sensor to temperatures exceeding 52°C caused irreversible damage due to denaturation of the avidin used for immobilization. Assays for the detection of complementary target strands prelabeled with fluorescein showed a working range of almost five decades and a detection limit of 24 fmol (1.4×10^{10} molecules). Abel *et al.* employed a competitive binding assay to detect nucleic acids not prelabeled with fluorescein. Detection of the unlabeled analyte began with pretreatment of the sensor with fluorescein labeled "tracer-DNA." The experiments involved monitoring decreases of the fluorescence intensity from the sensor upon exposure to and subsequent displacement of the tracer-DNA by complementary nucleic acid. The dose-response curves reported by Abel *et al.* show a detection limit of 132 pmol (7.95×10^{13} molecules) with a working range of three decades for this detection strategy. Noncomplementary DNA was reported to produce a signal of only 1–2% of that for fully complementary material.

Various designs of fiber-optic devices have been investigated to achieve improved detection limits. The most significant developments stem from efforts that have been made to improve the intensity of power in the evanescent field, and to increase the capture of fluorescence into guided modes within optical fibers. Optical fibers that have a diameter that is substantially greater than the wavelength of light that is being carried by TIR will permit the light to travel through the fiber in a number of discrete pathways or modes. The sensitivity of a device is in part determined by the modes that can propagate in a fiber, which determine both the power of the evanescent field that is created by a portion of light that travels in the core and the efficiency of coupling of fluorescence from an area outside of a fiber into guided modes. Marcuse[72] has described a model that is useful for evaluation of the efficiency of coupling by considering an estimate of the number of modes in a multimode fiber as a "normalized frequency":

$$\text{Freq} = \frac{2\pi}{\lambda r \sqrt{n_1^2 - n_2^2}} \tag{2.3}$$

where r is the radius of the core of the fiber. Analysis of models and practical experiments[72] indicates that the evanescent field of multimode fibers often represents less than 2% of the total power in a fiber because many of the lower-order modes are confined to the center of the core region. Monomodal fibers, where the core diameter is of a dimension approximately the same as the wavelength of the light being carried, have much higher evanescent field powers, representing much better than 10% of the total power in a fiber in many cases.[69]

Duveneck *et al.*[54] have investigated the use of planar waveguides as a platform for an optical nucleic acid biosensor. The device consisted of a single-mode planar waveguide composed of a tantalum pentoxide film of ± 100 nm thickness deposited on a glass substrate. The waveguide material was chosen so as to provide a high refractive index ($n = 2.2$) so as to maximize evanescent field strength in the biorecognition layer immobilized above the waveguide. Coupling of excitation radiation into the waveguide was accomplished by use of a grating coupler that was etched into the glass substrate onto which the waveguide was deposited. Detection was done in a "volume collection mode" wherein luminescence that was evanescently excited and isotropically emitted in the half-sphere outside of the waveguide was collected and quantitatively measured. Collection and guiding of the emitted light to a detector was achieved by use of a high numerical aperture lens and intensity measurements were done by use of either a photodiode coupled to a high gain amplifier or a photomultiplier tube in combination with a photon-counting unit. Immobilization of capture oligonucleotides onto the waveguide surface was done by a direct immobilization scheme wherein the waveguide surface was functionalized with glycidyloxypropyltrimethoxy silane followed by extension of the substrate linker with hexaethylene glycol and assembly of the capture sequence by automated solid-phase oligonucleotide synthesis. Hybridization experiments were done *in situ* and revealed full analytical signal evolution in a time of ~6 minutes with a detection limit of 100 attomoles (6×10^7 molecules) for target strands labeled with red cyanine-based fluorophores. Experiments were done in which one sensor was used for 20 consecutive hybridization assays, with regeneration being achieved by treatment with 50% urea. After the first two cycles, no discernible change in sensitivity was observed with each complete hybridization/regeneration cycle requiring ~40 minutes using an unoptimized fluidics system and flow cell geometry.

Ligler and Thompson[64] have described the preparation of tapered fibers of many centimeters length that are produced by an acid-etching process. The tapering geometry provided for significant increases in evanescent field strength and also reduction of modes that were scattered and lost at the transition of a fiber from an aqueous environment (immersion into sample solution) to air.

A similar physical design was used by Pilevar *et al.*[56] to prepare an adiabatically tapered fiber. The method of fiber preparation involved heating an optical fiber with a CO_2 laser and then drawing the fiber with a micropipet puller to reproducibly create tapered and approximately cylindrical fibers of 2–3 cm length. The fiber were described as being single-mode along the taper length, with a diameter of 8–10 μm. It was estimated that more than 50% of the total guided power within the fibers was present in the evanescent field. Pilevar[56] went on to use the tapered fibers to detect hybridization of nucleic acids by use of long excitation wavelength fluorescent dyes and a pulsed laser diode source. The use of fluorophores which could emit at longer wavelengths (750–850 nm range) was intended to reduce background signals that are commonly acquired from natural intrinsic fluorescence of biological samples. Single-stranded oligomers of 20-mer length were immobilizd in one step by two methods, one involving aminopropyltriethoxysilane and glutaraldehyde, and the other a 3-mercaptotrimethoxysilane and γ-maleimidobutyric acid N-hydroxysuccinimide ester. Complementary sequences labeled with a near-IR fluorescence dye were detected at picomolar concentrations, and sandwich assays to detect rRNA from *Helicobacter pylori* again showed selective detection at picomolar concentrations. The time course for hybridization reactions showed completion of signal development within one to three minutes.

The most sensitive optical biosensor device reported to date is that of Bier *et al.*[73] The intrinsic mode fiber optic sensor provided a detection limit of 3.2 attomoles (1.9×10^6 molecules) of non-labeled fully complementary nucleic acid target sequence in the presence of a double-strand selective fluorochrome with full regenerability for over 60 cycles of application without discernible loss of sensitivity. Capture oligonucleotides were immobilized onto the optical fiber substrate by either an avidin-biotinylated oligomer approach or by covalent attachment of the capture oligomer by carbonyldiimidazole activation. The avidin-biotin method involved adsorption of avidin onto the surfaces of fused silica optical fibers, which was subsequently crosslinked via treatment with glutaraldehyde. 5′-biotinylated oligonucleotides were then coupled onto the immobilized avidin layer. In the covalent coupling strategy, clean optical fibers were incubated with carbonyldiimidazole overnight in dry acetone, washed, and then treated with nucleic acid probes activated with 1-ethyl-3-(3-dimethylaminopropyl)-carbodiimide. Following a washing procedure, the fibers were used immediately or stored for up to three months at $-18°C$. Hybridization experiments were done using a protocol involving exposure of the sensor situated in a continuous flow apparatus [100 $\mu l\ min^{-1}$ flow rate, 100 μl volume exposed to the fiber-optic sensor to the target nucleic acid in hybridization buffer for 60–180 s, followed by a 120–180 s wash with buffer, 45–60 s treatment with a solution of fluorochrome (either Pico-Green or YOYO-1), 30 s wash with buffer (at 2.0 ml min^{-1}), and finally immersion in buffer for 210–255 s (steady state for data collection)]. Experiments done involving the use of biotin-avidin immobilized oligonucleotides provided a hybridization efficiency of 55%, however, the sensor could not be regenerated owing to the high stability of the double-stranded oligonucleotide fluorochrome complex. Chemical denaturation (e.g., treatment with NaOH) did not regenerate the sensor. Thermal regeneration of the sensor at high temperature (90°C) could denature the nucleic acid fluorochrome complex (T_m = 88°C for Pico-Green/double-stranded DNA (dsDNA), T_m > 90°C for YOYO-1/dsDNA, T_m = 45°C for dsDNA alone) and regenerate the sensor, however, the avidin-biotin linkage of the oligonucleotide to the waveguide substrate would also be compromised. This led Bier *et al.* to abandon the avidin-biotin immobilization protocol for covalent attachment methods, where complete regeneration was possible using thermal denaturation via a stepwise heating to 90°C following hybridization and staining. Single-base mismatch discrimination was also investigated. It was demonstrated by Bier *et al.* that under appropriate conditions of temperature, pH, and ionic strength a high degree of selectivity for detection of a central single-base (T-T) mismatch could be achieved for a tridecanucleotide.

9.4. Nonevanescent Intrinsic Mode Biosensors

Efforts in our research laboratory have also been directed toward the development of fiber-optic nucleic acid biosensors.[60–62,74] In general, we have chosen to pursue a sensing scheme based on covalent immobilization of oligonucleotides via long-chain polyether substrate linker molecules. Following substrate functionalization with these linker molecules terminated with dimethoxytrityl protecting groups, assembly of oligonucleotide capture probes is done by automated solid-phase oligonucleotide synthesis.

We have switched to the use of polyether substrate linkers as these types of tethers (in contrast to alkyl chains) have been shown to reduce nonselective binding of peptides or proteins that could be present as contaminants in real-world samples[75] Immobilization of a linker such as hexaethylene glycol (HEG) can be

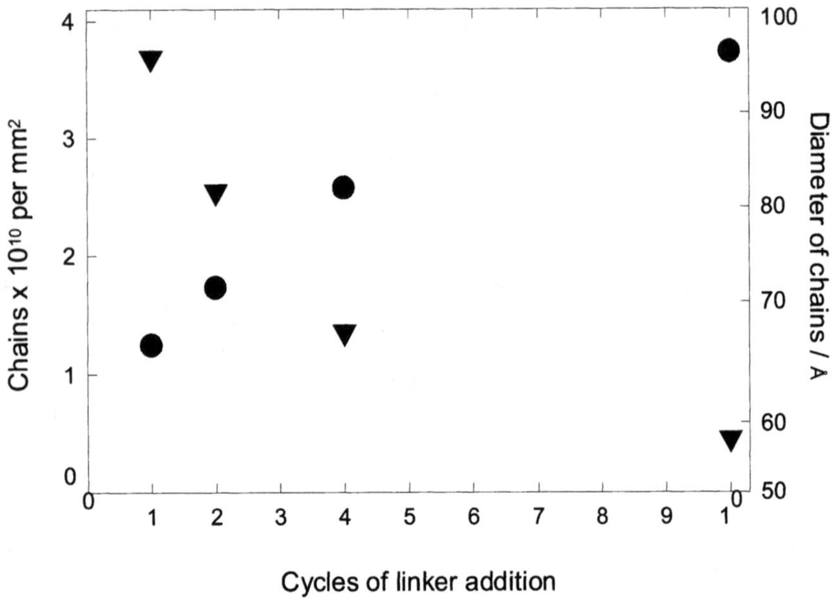

Figure 9.1. Packing density of immobilized oligomers as a function of cycles of application of HEG linker molecules onto activated silica substrates. Circles: chains of oligonucleotide-linker conjugates immobilized. Triangles: effective diameter available to each oligonucleotide-linker conjugate on the silica surface.[76]

done in cycles so as to control the density of HEG, and therefore ultimately the density of covalently immobilized oligomer on a surface. Figure 9.1 indicates the degree of control that can be achieved in terms of design of oligomer packing density based on the number of cycles of additions of HEG.

Much effort has been put into engineering surfaces with controlled density of immobilized oligonucleotides in order to provide control of the thermodynamics and kinetics of interfacial nucleic acid hybridization and to provide a means to overcome the limitations imparted by evanescent based signal generation. Maxwell-Garnet theory suggests that the ensemble refractive index of an immobilized film of nucleic acid molecules will be governed by the packing density of the oligonucleotides on the waveguide surface.[77] By use of angularly dependent light scattering techniques,[78] we have determined that oligonucleotide films of refractive index equal to or greater than that of the underlying fused silica substrate may be created. This suggests incorporation of the selective chemistry into the waveguide so that the boundary for total internal reflection for the optical sensors developed in our laboratories exists at the oligonucleotide membrane/solution interface as opposed to the fused silica/nucleic acid membrane interface as experienced in evanescent mode experiments. In this way, all of the light guided by the waveguide may be experienced by the film of immobilized oligonucleotides and light emitted from fluorophores associated with the oligonucleotide molecules will be done from within the waveguide and should preferentially be launched into a guided mode to provide efficient signal generation and recovery (est. $>10^3$ relative to evanescent wave methods in multimodal waveguides of large diameter).

Hybridization of the immobilized capture oligonucleotides to target oligonucleotides was done by treating the optical sensor immersed in hybridization buffer with various quantities of complementary and noncomplementary nucleic acids in a hybridization buffer (1.0M NaCl, 50 mM phosphate, pH 7.0). Detection of hybrid formation was done by addition of ethidium bromide to the reaction vessel at various concentrations and times to affect *in situ* staining of the immobilized hybrids with the intercalant fluorophore. Detection of stained hybrids was done by excitation of fluorophores along the length of the optical sensor, followed by recovery and quantitative measurement of guided fluorescence by use of a photomultiplier tube. Figure 9.2 shows a schematic representation of an instrument that permits rapid replacement of fibers and which is suitable for automated multisample analysis. Using this strategy we have been able to detect as few as 10^8 molecules of fully complementary hybrid molecules with linear calibration over a three-decade concentration range using ethidium bromide for transduction of hybridization. The speed of response to samples of approximately 100 λ volume in the reaction vessel is very fast, and full signal is usually achieved within one minute. Reversibility, reproducibility, and response to a noncomplementary sequence is shown in Figure 9.3.

Our research group has also investigated detection of triple-helix formation.[62] Elucidation of triple-strand hybridization in Hoogsteen and reverse-Hoogsteen motifs between an immobilized single-stranded probe oligonucleotide and linear or branched complementary oligonucleotides was done by monitoring alterations in the fluorescence temperature coefficient of intercalated ethidium bromide. Ligand exclusion concomitantly occurred with the onset of triple-strand formation upon decreasing the temperature of the system below the T_m for triplex formation, causing the temperature coefficient of the fluorescence signal to switch from negative to positive. For most systems of triplex forming oligonucleotides, the inversion of the fluorescence temperature coefficient could be used to determine the melting temperature of the triple-stranded complex (Figure 9.4).

9.5. Selectivity and Calibration Issues

We have begun to use our fiber-optic devices to investigate whether the density of DNA that is immobilized and the extent of hybridization can directly influence thermodynamic selectivity and quantitative calibration. The density of

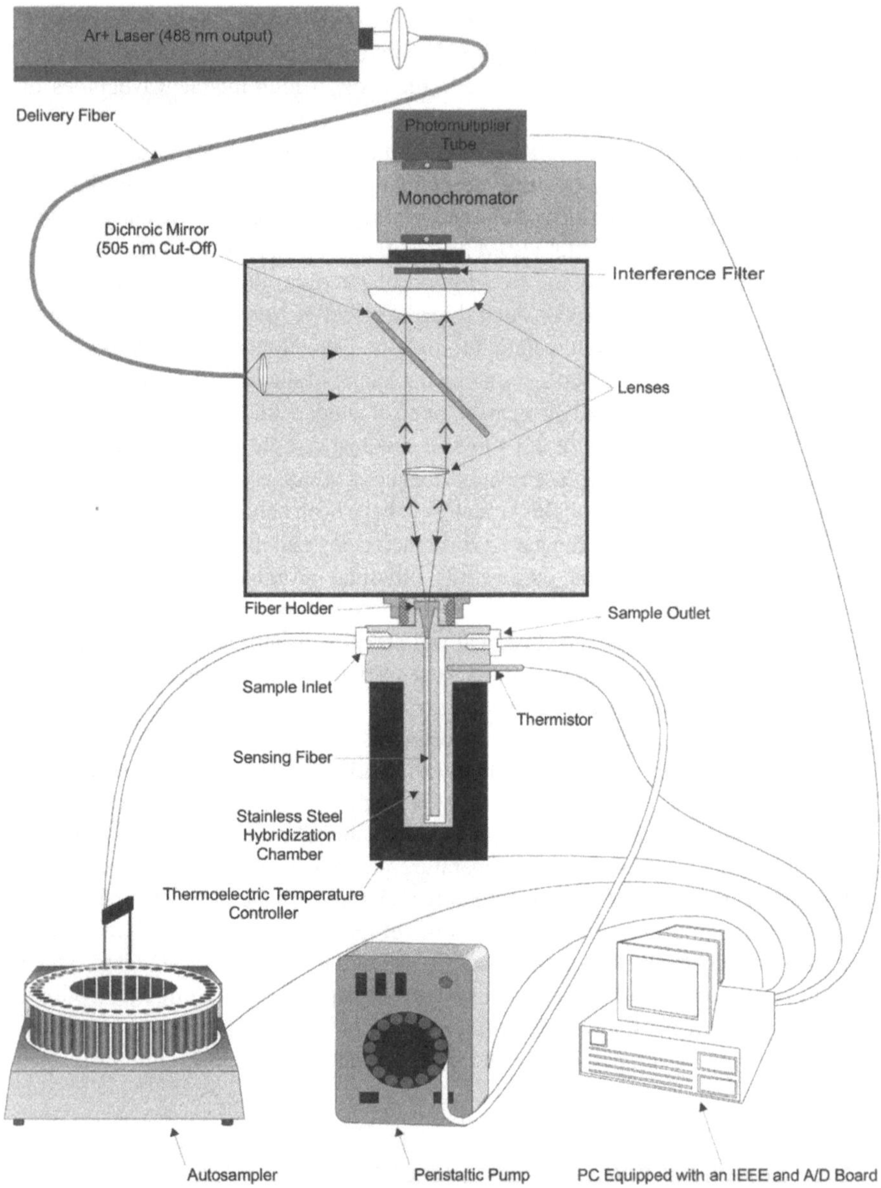

Figure 9.2. Schematic illustration of the instrument used for fluorimetric investigations of interfacial nucleic acid hybridization on from optical fiber sensors.

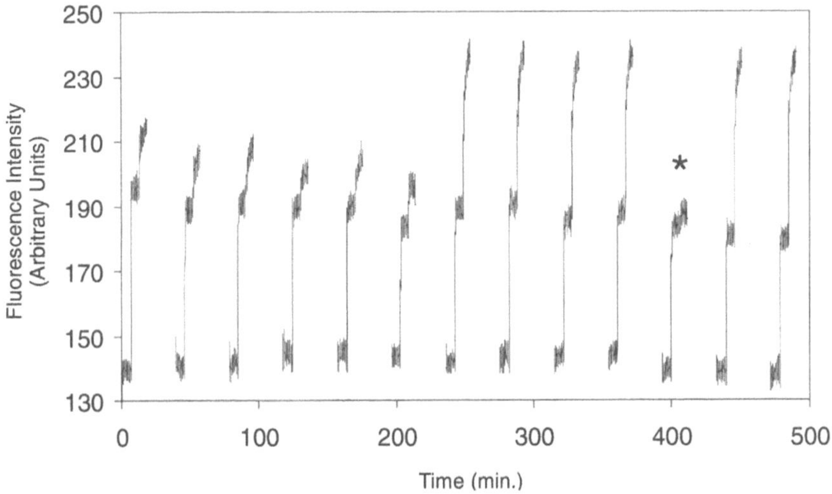

Figure 9.3. Response of an optical sensor for *Candida albicans* to complementary and noncomplementary nucleic acid sequences. Each analysis curve is comprised of the sensor challenged to hybridization buffer (1 × PBS: 1.0 M NaCl, 50 mM PO_4^{-n}, pH 7.0), 1 × PBS containing 2.5×10^{-8} M ethidium bromide, and 1 × PBS containing 2.5×10^{-8} M ethidium bromide and nucleic acid. Response curves 1–6 are for the sensor challenged with 10^{13} molecules of complementary DNA, response curves 7–10 and 12 and 13 are for the sensor challenged with 10^{14} molecules of complementary DNA and response curve 11 (*) is for the sensor challenged with 10^{14} molecules of a noncomplementary sequence.

single-stranded DNA (ssDNA) on a surface will determine the charge density due to ionizable phosphate groups. Such a negatively charged interface will attract positive counterions from solution, which may result in a local ionic strength, pH, and dielectric constant at the surface that is substantially different from that in bulk electrolyte solution. It is the local conditions that influence the thermodynamics of hybridization, and these can be studied by investigations of thermal denaturation profiles of immobilized oligonucleotide duplexes. Experimental work and theoretical models have been used to examine whether hybridization reactions at a surface can cause dynamic changes of local charge density, and therefore changes in selectivity and drift in calibration for quantitative analysis.[80] Fiber-optic biosensors based on fused silica that was coated with DNA were used in a total internal reflection fluorescence instrument to determine T_m from the dissociation of duplexes of fluorescein-labeled dA_{20}:dT_{20}. The experimental results suggest that the thermodynamic stability of duplexes that are immobilized on a surface is dependent on the density of immobilized DNA and on the extent of hybridization of DNA. The experimental results show that the thermodynamic stability of immo-

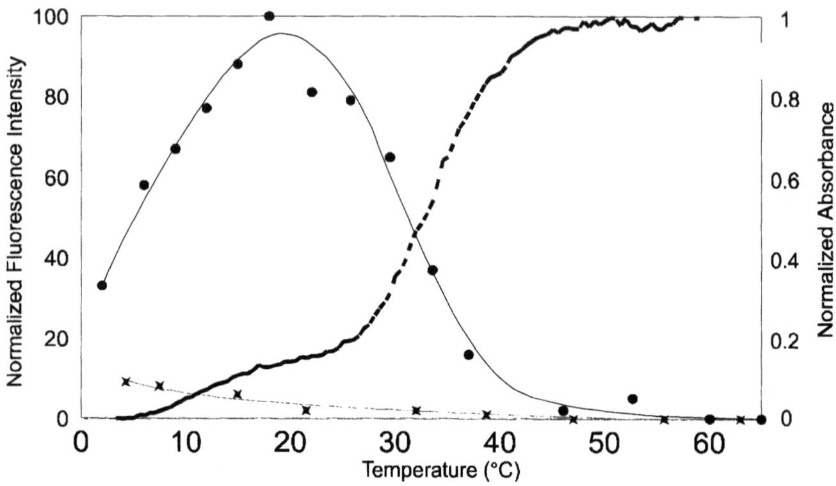

Figure 9.4. Fluorescent intensity as a function of temperature for dT_{10} using reversed orientation 3′-dA_{10}-5′-fiber derivatized sensors. Response of the optical sensor to 40 pmol of dT_{10} in the presence of 2.5×10^{-8} M ethidium bromide (circles) and to the 2.5×10^{-8} M ethidium bromide solution alone (pinwheels). Cooling profile of the same nucleic acid system in bulk solution by measurement of absorbance at 260 nm (broken line).[62]

bilized dsDNA is significantly different from that of dsDNA in bulk solution, and include observations of variation of enthalpy at different ionic strengths, asymmetry in the melt curves, and the possibility of a reduced dielectric constant within the layer of immobilized DNA relative to that in bulk solution. Thermodynamic considerations are often used to evaluate selectivity, and it is clear from this perspective that selectivity is not just a function of the nucleic acid sequence that is used to define a "receptor." The thermodynamic stability of dsDNA is dependent on nearest-neighbor interactions, including the extent of surface occupancy by ssDNA and dsDNA. This has consequences in terms of both selectivity and quantitative binding (equilibrium partitioning based on thermodynamic stability), and each can change as a result of the extent of formation of hybrids during an analytical experiment.

9.6. A Reagentless Biosensor

By associating a selective transduction element with the biorecognition element, the device may function without the need for external reagent treatment and obviate the need to collect and dispose of hazardous waste. Such a technology readily lends itself to automated and in-line analysis, and precludes the need for skilled technicians to partake in the analysis procedure or disposal of waste (provided the sample itself is not biohazardous). Our research work has led to development of a reagentless sensor that incorporates the transduction element into the sensor, thereby fulfilling the true definition of a biosensor (Figure 9.5). The use of a fluorescent double-stranded nucleic acid binding ligand tethered onto the terminus of the immobilized strand provides a small quantity of background fluorescence when in the presence of ssDNA and exposed to solution. This provides a means for internal calibration of the device via normalization of the background signal in terms of monitoring photobleaching and drift in the output of the excitation source and detector gain. The reporter in the presence of single-stranded nucleic acid may provide a unique baseline signal to which all signals can be referenced, hence providing meaningful analytical data and perhaps insight into the physical properties of the interaction surface. Also, the useful lifetime of the device can be determined from alterations in the background signal from the reporter molecules over time. Therefore, by including a selective reporter, an internal reference marker and diagnostic tool for the device status is included as an integral part of the biosensor. A fluorescent reporter molecule that selectively binds to double-stranded nucleic acids and provides increased quantum efficiency when bound to the double-stranded target is desirous as a tethered transducer. Our

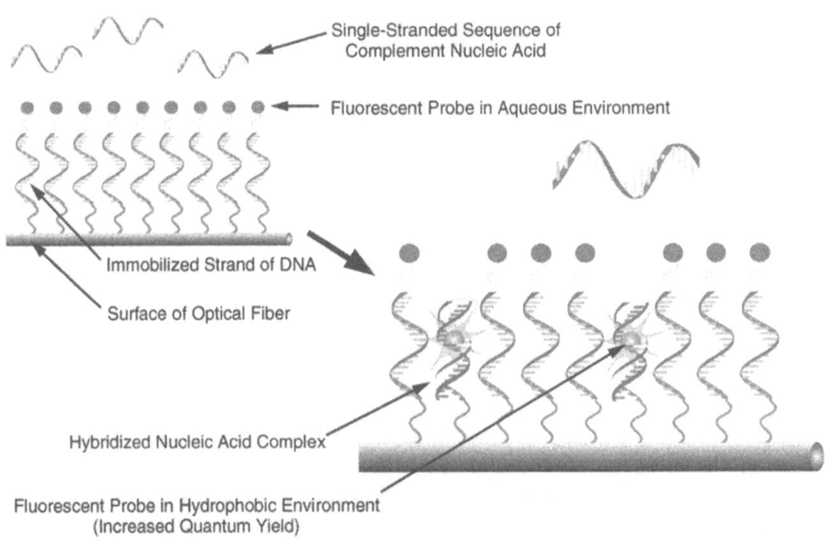

Figure 9.5. Schematic of the response mechanism of optical sensors employing the tethered dye transduction system.[74]

initial investigations have lead to the creation of a tethered analogue of the ethidium fluorophore that was attached to the end of a 5′-amino functionalized oligonucleotide via a 19-carbon molecular tether.

9.7. Future

It is interesting to see that the evolution of biosensors is beginning to follow the trend of a number of instrumental analytical techniques where such techniques are combined to increase informing power. A recent contribution by Asanov *et al.*[81] is a good example of such an approach, and describes the combination of TIRF with a transparent electrode for simultaneous electrochemical and optical investigations of immunochemical interactions. Some detailed contributions by Heller *et al.*[82,83] have reported the use of electrochemistry to manipulate the migration of ssDNA and the stability of dsDNA in a configuration that used fluorescent dyes to detect the presence of hybrids. Fabrication using conventional microelectronic technology yielded a 1 cm^2 device with 25 central 80 μm diameter electrodes in a 1 mm^2 test area. The surface of the electrodes was coated with low-stress silicon nitride onto which an ~1 μm thick permeation layer of streptavidin-doped agarose was then formed via a spin-casting method. This layer served to provide a means of ssDNA attachment, in addition to preventing electrochemical DNA damage while maintaining ion movement. Studies employing ssDNA bis-functionalized with biotin and the fluorescent probe Bodipy Texas Red were done and showed that immobilization occurred only at those sites (to within the detection limit of the experiment) which were held at positive potential. Hybridization investigations were carried out and revealed that hybrids formed 25 times faster at positive electrical potential sites than at neutral sites, indicating that kinetic limitations of passive, diffusion-limited hybridization on solid supports may be overcome. The device was demonstrated to be capable of distinguishing single-base mismatches as revealed by a decrease in fluorescence signal of approximately 90% after 50 s of application of an alternating field of negative bias (0.6 μA, 0.1 s on, 0.2 s off, repeated for 150 cycles) to the sites containing the mismatch hybrid. A more detailed analysis of the physical properties of the system was undertaken and reported in a subsequent publication by researchers at Nanogen.[83]

In general, once nucleic acid sequences of known composition and length are selected, the energetics underlying hybrid formation resulting from the annealing of two complementary nucleic acids is governed by several independent variables, including temperature, pH, and ionic strength. Alteration of any one of these parameters will affect hybrid stability and hence stringency. The concept of electric field assisted hybridization introduces another variable affecting hybrid stability, namely, electric field strength, which now adds a new dimension of complexity to the system as it is interdependent with the aforementioned key variables.

Our experiments are now combining optically transparent indium-tin oxide (ITO) electrodes with fiber-optic techniques for detection and manipulation of nucleic acid hybridization. Fused silica optical fibers are being coated with ITO, and ssDNA is being assembled on these surfaces by automated synthesis. The electrochemical properties of these optical fibers will allow manipulation of the kinetics of hybridization and the stability of hybrids, while providing for an efficient and sensitive fluorescent method for detection of the presence and extent of hybridization.

References

1. Thompson, M., and Krull, U. J. *Trends Anal. Chem. 3*, 173–178, 1984.
2. Guilbault, G. G., *Curr. Opin. Biotechnol. 2*, 3–8, 1991.
3. Buck, R. P., Hatfield, W. E., Umana, M., and Bowden, E. F., eds. *Biosensor Technology: Fundamentals and Applications*. New York: Marcel Dekker, 1990.
4. Wise, D. L., ed. *Bioinstrumentation: Research, Developments and Applications*, Stoneham, MA: Butterworth, 1990.
5. Kallury, K., Lee, W. E., and Thompson, M. *Anal. Chem. 64*, 1062–1068, 1992.
6. Krull, U. J., Brown, R. S., Vandenberg, E. T., Heckl, W. M. *J. Electron Microsc. Tech. 18*, 212–222, 1991.
7. Ramsay, G., ed. *Commercial Biosensors*, New York: Wiley Interscience, Vol. 148, 1998.
8. Symons, R. H. *Nucleic Acid Probes*. Boca Raton, FL: CRC Press, 1989.
9. Bock, L. C., Griffin, L. C., Latham, J. A., Vermaas, E. H., and Toole, J. J., *Nature 355*, 564, 1992.
10. Tay, F., Liu, Y. B., Flynn, M. J., and Slots, J. *Oral Microbiol. Immunol. 7*, 344, 1992.
11. Sherman, M. I., Bertelsen, A. H., and Cook, A. F., *Bioorg. Medic. Chem. Lett. 3*, 469, 1993.
12. Letsinger, R. L., and Lunsford, W. B. *J. Am. Chem. Soc. 98*, 3655, 1976.
13. Beaucage, S. L., and Caruthers, M. H. *Tetrahedron Lett. 22*, 1859–1862, 1981.
14. Alvarado-Urbina, G., Sathe, G. M., Liu, W-C., Gillen, M. F., Duck, P. D., Bender, R., and Ogilvie, K. K. *Science 214*, 270, 1981.
15. Millan, K. M., Saraullo, A., and Mikkelsen, S. R. *Anal. Chem. 66*, 2943–2948, 1994.
16. Millan, K.M., Spurmanis, A. J., and Mikkelson, S. R. *Electroanalysis 4*, 929–932, 1992.
17. Millan, K. M., and Mikkelson, S. R. *Anal. Chem. 65*, 2317–2323, 1993.
18. Wang, J., Cai, X., Gustavo, R., Shiraishi, H., Farias, P. A. M., and Dontha, N. *Anal. Chem. 68*, 2629–2634, 1996.
19. Wang, J., Cai, X., and Jonsson, C. *Anal. Chem. 67*, 4065, 1995.
20. Wang, J., Rivas, G., Cai, X., Dontha, N., Shiraishi, H., Luo, D., and Valera, F. S. *Anal. Chim. Acta 337*, 41–48, 1997.
21. Wang, J., Rivas, G., Ozsoz, M., Grant, D. H., Cai, X., and Parrado, C. *Anal. Chem. 69*, 1457–1460, 1997.
22. Wang, J., Rivas, G., Cai, X., Parrado, C., M. Chicharro, D. Grant, and Ozsoz, M. *Electrochem. Soc. Proc. 19*, 727–740, 1997.
23. Wang, J., Cai, X., Rivas, G., Shiraishi, H., and Dontha, N. *Biosensors Bioelectron. 12*, 587–599, 1997.

24. Wang, J., Rivas, G., Cai, X., M. Chicharro, Parrado, C., Dontha, N., Begleiter, A., Mowat, M., Palecek, E., and Nielsen, P. E. *Anal. Chim. Acta 344*, 111–118, 1997.
25. Wang, J., Rivas, G., and Cai, X. *Electroanalysis 9*, 395–398, 1997.
26. Wang, J., Rivas, G., Parrado, C., Cai, X., and Flair, M. N. *Talanta 44*, 2003–2010, 1997.
27. Cai, X., G. Rivera, Shiraishi, H., P. Farias, Wang, J., Tomschik, M., Jelen, F., Palecek, E. *Anal. Chim. Acta 344*, 65–76, 1997.
28. Wang, J., Palecek, E., Nielsen, P. E., Rivas, G., Cai, X., Shiraishi, H., Dontha, N., Luo, D., and Farias, P. A. M. *J. Am. Chem. Soc. 118*, 7667–7670, 1996.
29. Wang, J., Chicharro, M., Rivas, G., Cai, X., Dontha, N., Farias, P. A. M., and Shiraishi, H. *Anal. Chem. 68*, 2251–2254, 1996.
30. Wang, J., Rivas, G., Cai, X., Palecek, E., Nielsen, P., Shiraishi, H., Dontha, N., Luo, D., Parrado, C., Chicharro, M., Farias, P. A. M., Valera, F. S., Grant, D. H., Ozsoz, M., and Flair, M. N. *Anal. Chim. Acta 347*, 1–8, 1997.
31. Napier, M. E., Loomis, C. R., Sistare, M. F., Kim, J., Eckhardt, A. E., and Thorp, H. H. *Bioconj. Chem. 8*, 906–913, 1997.
32. Kori-Youssoufi, H., Garnier, F., Srivastava, P., Godillot, P., and Yassar, A. *J. Am. Chem. Soc. 119*, 7388–7389, 1997.
33. Ihara, T., Nakayama, M., Murata, M., Nakano, K., and Meada, M. *Chem. Commun.*, 1609–1610, 1997.
34. Wilson, E. K. *Chem. Eng. News 5*, 47–49, 1998.
35. Su., H., Kallury, K. M. R. and Thompson, M. *Anal. Chem. 66*, 769, 1994.
36. Su., H., Chong, S., and Thompson, M. *Biosensors Bioelectron. 12*, 161–173, 1997.
37. Wang, J., Neilsen, P. E., Jiang, M., Cai, X., Fernandes, J. R., Grant, D. H., Ozsoz, M., A. Beglieter, and Mowat, M. *Anal. Chem. 69*, 5200–5202, 1997.
38. Biacore AB, Rapsgatan 7, S-754 50 Uppsala, Sweden.
39. Griffiths, D., and Hall, G. *Trends Biotechnol. 11*, 122, 1993.
40. Bianchi, N., Rutigliano, C., Tomassetti, M., Feriotto, G., Zorato, F. and Gambari, R. *Clin. Diagn. Virol. 8*, 199–208, 1997.
41. Rutigliano, C., Bianchi, N., Tomassetti, M., Pippo, L, Mischiati, C., Feriotto, G., and Gambari, R. *Int. J. Oncol. 12*, 337–343, 1998.
42. Thiel, A. J., Frutos, A. G., Jordan, C. E., Corn, R. M., and Smith, L. M. *Anal. Chem. 69*, 4948–4956, 1997.
43. Jordan, C. E., Frutos, A. G., Thiel, A. J., and Corn, R. M. *Anal. Chem. 69*, 4939–4947, 1997.
44. Kruchinin, A. A., and Vlasov, Y. G. *Sensors and Actators B 30*, 77–80, 1996.
45. Jensen, K. K., Orum, H., Nielsen, P. E., and Nordén, B. *Biochemistry 36*, 5072–5077, 1997.
46. Persson, B., Stenhag, K., Nilsson, P., Larsson, A., Uhlén, M., and Nygren, P. *Anal. Biochem. 246*, 34–44, 1997.
47. Bates, P. J., Dosanjh, H. S., Kumar, S., Jenkins, T. C., Laughton, C. A., and Neidle, S. *Nucleic Acids Res. 23*, 3627–3632, 1995.
48. Gotoh, M., Hasegawa, Y., Shinohara, Y., Shimizu, M., and Tosu, M. *DNA Res. 2*, 285–293, 1995.
49. Thiel, A. J., Frutos, A. G., Jordan, C. E., Corn, R. M., and Smith, L. M. *Anal. Chem. 69*, 4948–4956, 1997.
50. Jordan, C. E., Frutos, A. G., Thiel, A. J., and Corn, R. M. *Anal. Chem. 69*, 4939–4947, 1997.
51. Gerdt, D. W., and Herr, J. C. Fiber Optic Evanescent Wave Sensor for Immunoassay. United States Patent 5,494,798.
52. Graham, C. R., Leslie, D., and Squirrell, D. J. *Biosensors Bioelectron. 7*, 487–493, 1991.
53. Abel, A. P., Weller, M. G., Duveneck, G. L., Ehrat, M., and Widmer, H. M. *Anal. Chem. 68*, 2905–2912, 1996.
54. Duveneck, G. L., Pawlak, M., Neuschäfer, D., Bär, E., Budach, W., Pieles, U., and Ehrat, M. *Sensors and Actuators B 38–39*, 88–95, 1997.

55. Anderson, G. P., Golden, J. P., Lynn, P. C., and Ligler, F. *IEEE Eng. Med. Biol. 13*, 358–363, 1994.
56. Pilevar, S., Davis, C. C., and Portugal, F. *Anal. Chem. 70*, 2031–2037, 1998.
57. Ferguson, J. A., Boles, T. C., Adams, C. P., and Walt, D. R. *Nature Biotechnol. 14*, 1681, 1996.
58. Healey, B. G., Matson, R. S., and Walt, D. R. *Anal. Biochem. 251*, 270–279, 1997.
59. P. Pantano, and Walt, D. R. *Anal. Chem. 67*, 481A, 1995.
60. Piunno, P. A. E., Krull, U. J., Hudson, R. H. E., Damha, M. J., and Cohen, H. *Anal. Chim. Acta 288*, 205, 1994.
61. Piunno, P. A. E., Krull, U. J., Hudson, R. H. E., Damha, M. J., and Cohen, H. *Anal. Chem. 67*, 2635–2643, 1995.
62. Uddin, A. H., Piunno, P. A. E., Hudson, R. H. E., Damha, M. J., and Krull, U. J. *Nucleic Acids Res. 25*, 4139–4146, 1997.
63. Le, Z. C., Xing, J. Z., Lee, J., Leadon, S. A., and Weinfeld, M. *Science 280*, 1066–1069, 1998.
64. Thompson, R. B., and Ligler, F. S. Chemistry and technology of evanescent wave biosensors. In *Biosensors with Fiberoptics*, Wise, D. L. and Wingard, L. B., eds. Clifton, NJ: Humana, 1991, pp. 111–138.
65. Wolfbeis, O. S., ed. *Fiber Optic Chemical Sensors and Biosensors*, Vol I. Boca Raton, FL: CRC Press, 1991.
66. Krull, U. J., Brown, R. S., Hougham, B. D., and Brock, I. *Talanta 37*, 801–807, 1990.
67. Krull, U. J., Brown, R. S., and Vandenberg, E. T. Fiber optic chemoreception. In *Fiber Optic Chemical Sensors and Biosensors*, Vol II., O.S. Wolfbeis, ed. Boca Raton, FL: CRC Press, 1991, pp. 315–340.
68. Sutherland, R. M., Bromley, P., and Gentile, B. Analytical Method for Detecting and Measuring Specifically Sequenced Nucleic Acid. European Patent Application Number 87810274.8, Publication Number 0 245 206 A1, date of filing: 30 April 1987, p. 13.
69. Love, W. F., Button, L. J., and Slovacek, R. E. Optical characteristics of fiberoptic evanescent wave sensors: Theory and experiment. In *Biosensors with Fiberoptics*, Wise, D. L., and Wingard, L. B., Clifton, NJ: Humana, 1991, pp. 139–180.
70. Ducharme, D., Max, J. J., Salsee, C., and LeBlanc, R. M. *J. Phys. Chem. 94*, 1925, 1990.
71. Ohanian, H. C., *Physics*. New York: W.W. Norton, 1985, p. 835.
72. Marcuse, D. J. *Lightwave Technol. 6*, 1273–1279, 1988.
73. Kleinjung, F., Bier, F. F., Warsinke, A., and Scheller, F. W. *Anal. Chim. Acta 350*, 51–58, 1997.
74. Krull, U. J., Piunno, P. A. E., Wust, C., Li, A., Gee, A., and Cohen, H. A Fiber Optic DNA Sensor for Rapid Detection of Environmental *E. coli*. In *Biosensors for Direct Monitoring of Environmental Pollutants in Field*, Nikolelis, D. P., *et al.*, eds. Dordrecht, The Netherlands: Kluwer Academic, 1998, pp. 67–77.
75. Lee, J. H., Kopecek, J., and Andrade, J. D. *J. Biomed. Mater. Res. 23*, 351–368, 1989.
76. Sojka, B., Piunno, P. A. E., and Andrade, J. D. *Appl. Biochem. Biotechnol. 89*, 85–103, 2000.
77. Azzam, R. M. A., and Bashara, N. M. *Ellipsometry and Polarized Light*. Amsterdam: North-Holland Personal Library, 1987, p. 359.
78. Henke, L., Piunno, P. A. E., Nagy, N., and Krull, U. J. *Anal. Chim. Acta 433*, 31–45, 2001.
79. Hanafi-Bagby, D., Piunno, P. A. E., Wust, C. C., and Krull, U. J. *Anal. Chim. Acta 411*, 19–30, 2000.
80. Piunno, P. A. E., Watterson, J., Wust, C. C., and Krull, U. J, *Anal. Chim. Acta 400*, 73–89, 1999.
81. Asanov, A. N., Wilson, W. W., and Oldham, P. B. *Anal. Chem. 70*, 1156–1163, 1998.
82. Sosnowski, R. G., Tu, W., Butler, W. F., O'Connell, J. P., and Heller, M. J. *Proc. Natl. Acad. Sci. USA 94*, 1119–1123, 1997.
83. Edman, C. F., Raymond, D. E., Wu, D. J., Tu, E., Sosnowski, R. G., Butler, W. F., Nerenberg, M., and Heller, M. J. *Nucleic Acids Res. 25*, 4907–4914, 1997.

10

Technicolor Genome Analysis

Michael J. Difilippantonio and Thomas Ried

10.1. Introduction

Our understanding of the organization of complex genomes has progressed rapidly in the last fifty years. This has been the result of intense and dedicated work not only in the biological sciences but also chemistry and physics. Our current knowledge about genomes owes a debt of gratitude to the interplay between these three disciplines. The desire for a deeper understanding of the complex mechanisms of life has posed numerous technical challenges. Among these was the need for increased optical resolution. One of the best examples of the *merging* of the three sciences is the development of fluorescence *in situ* hybridization (FISH) and its application to other newly developed techniques for the analysis of genomes. In this chapter we will discuss the combined application of chemically modified *nucleosides* and advances in microscopy, optics, digital imaging devices, and image analysis software.

10.1.1. Historical Perspective

The first published use of a nucleic acid probe for the *in situ* detection of its complimentary sequence came from two different groups. Both Gall and Pardue[1] and John, Birnstiel, and Jones[2] hybridized tritium-labeled *Xenopus* rRNA to cell squashes of ovarian tissue fixed on microscope slides. The specificity of the hybridization reaction was demonstrated through the use of such controls as: treatment of the slide with DNase; not denaturing the cellular DNA prior to hybridization; treatment with RNase A to demonstrate the stability of the DNA/RNA complex; and competing the hybridization reaction with unlabeled rRNA. Despite the undesirable use of radioactivity and the long exposure times of 5–22 days, this was the first technique that could be used to "... assign the chromosomal

Michael J. Difilippantonio and Thomas Ried • Center for Cancer Research, National Cancer Institute, National Institutes of Health, Bethesda, Maryland 20892-8010.

Topics in Fluorescence Spectroscopy, Volume 7: DNA Technology, edited by Joseph R. Lakowicz, Kluwer Academic / Plenum Publishers, New York, 2003

location of a given DNA segment by means other than conventional cytogenetic and genetic analysis. Thus chromosomal mapping of sequences detectable by molecular hybridization is limited to a small number of genetically favorable organisms, or to rare viable deletion mutants, or to comparisons between the sex chromosomes of some organisms."[2]

Around the same time that *in situ* hybridization was being developed, chemists and biochemists had begun experimenting with different nucleoside analogs in the search for agents with antiviral and anticancer activity. One approach was to identify agents that would interfere with the biosynthesis or function of the viral nucleic acids without causing host toxicity. This could involve inhibition of viral genome replication or transcription, viral mRNA translation, viral enzyme catalysis, or virus maturation. Perhaps differences in the bioavailability of the nucleoside analogs and differences in incorporation rates would result in sensitivity to these chemically modified DNA precursors. It was from these same chemists that attempts were made to improve *in situ* hybridization.

One of the first reported attempts to improve *in situ* protocols took advantage of the strong affinity between biotin and avidin.[3] The removal of radioisotopes from the reaction increased the spatial resolution of the signal and indefinitely extended the shelf-life of the probe. Manning *et al.*[3] covalently coupled biotin (vitamin H) to the rRNA probe via a cytochrome-c bridge. The efficiency of this reaction resulted in about one biotin-cyt-c label for every 100–200 ribonucleotides. This ratio, and the use of the small molecule biotin remain important factors in trying to keep steric hindrance of the probe to a minimum, thereby maintaining the hybridization kinetics. The synthesis reaction first linked the cytochrome-c and biotin molecules. This moeity was then coupled to the rRNA by reaction with formaldehyde. It is important to note that this labeling reaction was not specific for any nucleotide in particular. The *in situ* hybridization protocol followed that of Gall and Pardue[1] with a few minor modifications. To reduce nonspecific background, the slides were blocked with cytochrome-c prior to detection with avidin-labeled polymethacrylate spheres. The positively charged cytochrome-c effectively bonded to the negatively charged sites on the microscope slide, thus preventing interaction with the positively charged avidin. The avidin-spheres, each of which on average contained a few avidin molecules, were then directly visualized by scanning electron microscopy, abrogating the need for long photographic emulsion exposure times as previously required. Variations on this protocol included the replacement of avidin-spheres with the electron-opaque protein ferritin covalently coupled to avidin and labeling of the RNA probe with biotin via a diamine bridge instead of the polyamine cytochrome-c.[4]

Researchers at Yale University were also working on the development of nucleoside analogs with antiviral activity. They used their antiviral approach of synthesizing nucleosides that were efficient substrates for polymerase to improve upon the above *in situ* hybridization procedure. As such, they covalently coupled a number of different potential probe determinants directly to either the purine or pyrimidine rings of different nucleotides.[5] Once synthesized, these labeled nucleotides were then used in a *nick-translation* protocol. Briefly, the nucleic acid probe of choice was incubated in the presence of DNase I, DNA polymerase, and free nucleotides. The nucleotide cocktail contained all four nucleotides (dATP, dCTP, dGTP, and dTTP), plus a lower molar concentration of Bio-dUTP. DNase I causes single-strand breaks, or nicks, in the DNA backbone which are then repaired by the polymerase. DNA polymerase I has both $5' \rightarrow 3'$ and $3' \rightarrow 5'$ exonuclease activity, which allows it to remove bases from both sides of the nick site. It is during the replacement of these bases by the DNA polymerase activity of the enzyme that the biotin-labeled nucleotide, as well as the others in the reaction, are incorporated. At the ratios of DNase I to polymerase used, not all of the nicks are repaired, thus the size of the DNA probe is decreased, resulting in a smear on an agarose gel.

The size of the probe after labeling is important for both the hybridization kinetics as well as accessibility to the genomic target DNA. There is also a mixture of dTTP and dUTP in the reaction. Since dUTP is not a normal substrate for DNA polymerase, the incorporation efficiency of dUTP is less than for the other nucleotides and therefore not every thymidine residue is replaced with a labeled nucleotide. This is important for maintaining the proper hybridization kinetics as well as an efficient nick translation reaction.

These hapten-labeled nucleotide probes were then used for detecting and localizing specific sequences in cell, tissue, and chromosome preparations as well as membrane bound nucleic acids.[6,7] Biotin-labeled probes were initially detected using rabbit anti-biotin antibodies. These were either directly conjugated to a fluorescent molecule like fluorescein isothyocyanate (FITC) or were themselves detected with a peroxidase-conjugated sheep anti-rabbit IgG.[6] There are currently many different directly-conjugated nucleotides available. Some are conjugated to fluorochromes such as FITC, Spectrum Green, TRITC, Rhodamine, Spectrum Orange, Texas Red, Cy5, while others are conjugated to nonfluorescent molecules like biotin, dintrophenol (DNP) or digoxigenin that are detected with fluorescence-conjugated avidin or antibodies directed against them.

FISH has revolutionized the way in which geneticists are able to analyze and visualize the genomes of various organisms. Research applications include the mapping of newly isolated DNA or RNA clones to their chromosomal locus and determination of the relative order of genes or clones in a contig with respect to each other. The fluorescence detection of nucleic acid probes has added to the repertoire of techniques already available for the analysis of genes and genomes, such as southern blot hybridization and cytogenetic banding. Microscopic analysis of chromosomes hybridized directly with fluorescent DNA probes allows rapid mapping of genes and clones to their genomic chromosomal locus, as well as sometimes identifying other loci containing related sequences. The positional relationship between different clones along the chromosome can also be readily determined. By labeling each clone with a different fluorescent molecule their simultaneous hybridization to either condense chromosomes or elongated chromatin fibers allows determination of order. The advent of different fluorescent dyes, suitable optical filters, and imaging technology has increased the amount of information that can be obtained from a single hybridization. The culmination of these advances can be found in the technique known as spectral karyotyping (SKY), in which 24 different chromosomes can be distinguished by their spectral signature after *in situ* hybridization.

10.2. Principles behind FISH

In the general sense of the term, FISH involves the hybridization of fluorescently labeled nucleic acid probes to genomic DNA preparations on a microscope slide. Since there are many different types of specimens and probes, other names have been given to the more specific applications of this technique. FISH is used primarily to refer to the use of probes shorter than a few megabases and chromosome painting probes to metaphase chromosome preparations. The typical application of FISH is for the genomic localization of particular probes (i.e., gene-specific plasmids, cDNA, ESTs, YACs, BACs, and PACs) by physical mapping onto metaphase chromosomes (Figure 10.1A, B, E). Interphase FISH, sometimes referred to as interphase cytogenetics, involves the visualization of chromosome aberrations in cell nuclei by hybridization with nucleic acid probes. This technique

---

Figure 10.1. A: Fluorescence *in situ* hybridization of painting probes specific for human chromosomes 5 (green), 7 (purple), and 12 (yellow) and gene-specific probes for the mammalian bitter taste receptors (M. Difilippantonio and N. J. P. Ryba). B: Mouse tumor metaphase chromosomes hybridized with probes specific for the immunoglobulin heavy chain locus (red) and the c-myc gene (green). A translocation between the chromosomes on which these genes reside has also resulted in amplification of the genes on the derivative chromosome (arrowhead). This is also evident in the adjacent interphase nucleus where this DNA leaked out of the nuclear membrane during the processing of the slide (arrow). The other loci appear as discrete dots in the nucleus (M. Difilippantonio and A. Nussenzweig). C: Interphase mapping of three gene-specific probes reveals that the chromosomal position of the gene detected with the FITC-labeled probe (green) is between the two genes detected with the TRITC-labeled probes (red) (courtesy of R. Yonescu). D: Biotin (green) and digoxigenin (red) labeled probes specific for two halves of exon 45 in the Duchenne's muscular dystrophy gene can be spatially resolved on chromatin fibers using DNA fiber FISH extracted nuclei (M. Difilippantonio and T. Ried). E: High resolution FISH mapping of two BAC clones corresponding to adjacent regions of DNA separated by approximately 1 Mb (courtesy of R. Yonescu). F: Hybridization of a human chromosome 4 painting probe (yellow) to metaphase chromosomes from a human-hamster hybrid cell line reveals the presence of three copies of the human chromosome among the hamster genome (M. Difilippantonio and L.A. Doucette-Stamm).

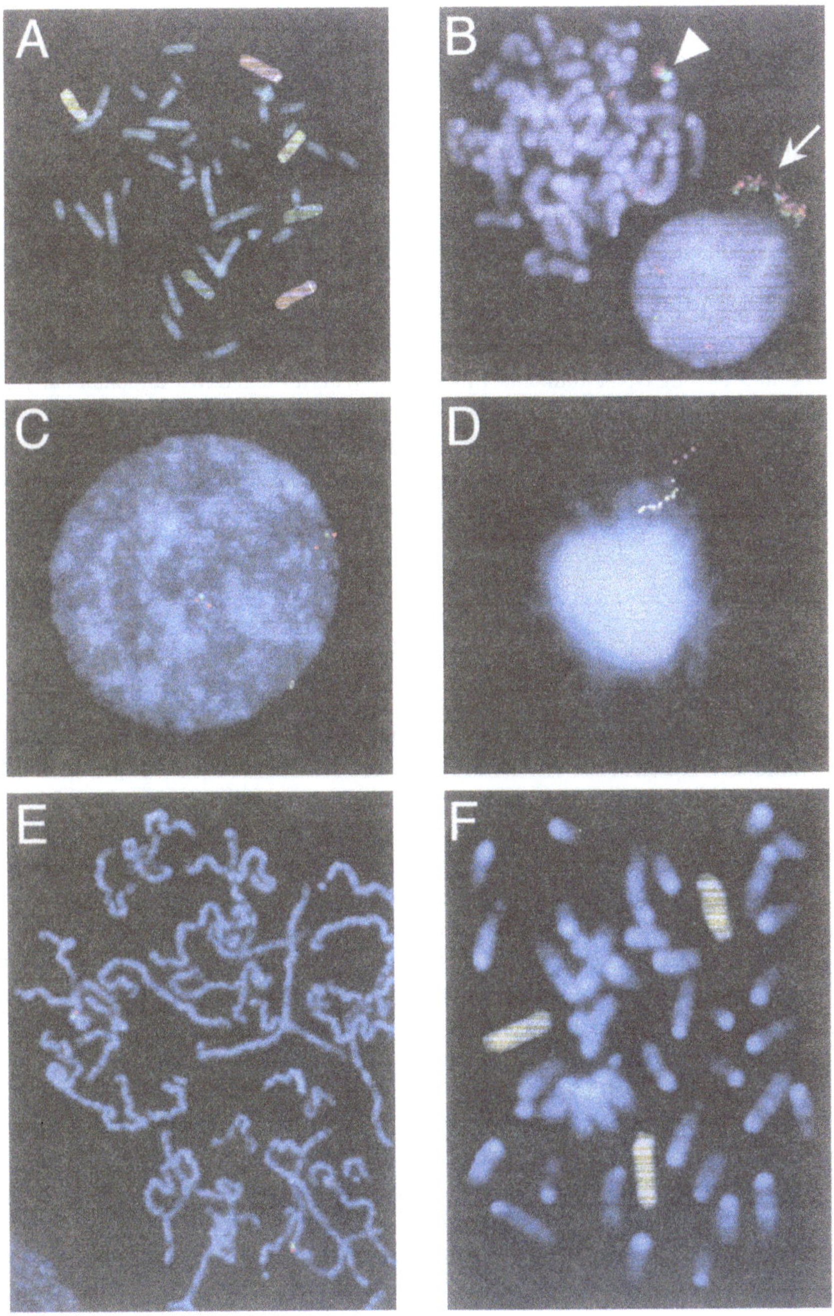
A
B
C
D
E
F

was first described in 1986 when Cremer *et al.*[8] used radiolabeled probes to detect the copy number of chromosome 18 in normal nuclei and nuclei with a trisomy 18. Specimens ranged anywhere from fresh cytological preparations to archived paraffin-embedded, formalin-fixed tissue sections. The replacement of radioactivity with fluorescence made the technique more amenable to use in a clinical setting and also increased the resolution of the assay. Interphase mapping is used to determine the relative order of specific probes with respect to each other (Figure 10.1C). This is typically employed when the distance between a number of probes is below the limit of detection using metaphase chromosomes. Another technique with the same application, but higher resolution, is fiber FISH (Figure 10.1D). In this strategy, the nuclear membrane is perforated using detergents and the uncondensed genomic DNA "spills" out of the nucleus and spreads along the surface of the slide as long DNA filaments. Since the DNA is arranged two-dimensionally, as opposed to its three-dimensional arrangement in the nucleus, the resolving power in terms of physical distance between two loci is increased.[9–13] Thus, due to the diverse applications of FISH, there are many variables involved in the preparation of specimens and probes as well as the exact protocol used for hybridization and detection of the annealed probe.

10.2.1. Preparation of Cytological Specimens

As indicated previously, there are many different scenarios in which one would like to detect specific DNA sequences. Most techniques involving *in situ* hybridization require the preparation of metaphase spreads as either the reference material (that to which the probes being analyzed are hybridized) or the actual sample itself. This is true for gene mapping, somatic hybrid analysis, determination of orthologous regions between species, SKY, and comparative genome hybridization (CGH). In each case, the cells from which the chromosomes are to be prepared are treated with an agent that arrests the cells in mitosis, thereby increasing the number of cells containing condensed chromosomes. The most commonly used drug is colchicine, or Colcemid®, which interferes with the progression through mitosis by inhibiting spindle fiber formation. There is a delicate balance that must be reached, however, between the number of mitoses and the length of the chromosomes, and these two parameters are inversely related. The longer one leaves the cells in Colcemid®, the more cells will reach the mitotic block. The longer cells are blocked, however, the more condensed the chromosomes become. Why then would one want to prolong the mitotic arrest? Some cultures grow particularly slow and therefore have very few cells entering mitosis at any given time. By increasing the Colcemid® time (and usually decreasing the working concentration) one is able to increase the mitotic index of a particular harvest. If the number of mitoses is more important than the resolution of physical distance on the chromosomes, this may be a reasonable compromise. The length of treatment with Colcemid® must therefore be experimentally determined for each cell type.

Other treatments are added depending upon the cell type and length of chromosomes desired. Human lymphocytes are usually stimulated to divide by addition of phytohaemagglutinin (PHA) to the culture media. Mouse spleen cultures are stimulated with a mixture of lipopolysaccharide (LPS) and concavalin A (ConA). Some protocols use methotrexate (MTX) to inhibit DNA synthesis by interfering with the synthesis of the deoxyribonucleotide thymidine triphosphate (dTTP). This has the effect of synchronizing the cells in S-phase. The cells are then released from the block for a predetermined period of time (usually 5 h) and allowed to cycle until they reach mitosis, where they are blocked with Colcemid® and harvested. Another variation of this approach includes the addition of the nucleotide analogs BrdU and FudR after release from the S-phase block. Incorporation of these thymidine analogs into the newly synthesized DNA results in decreased condensation of the chromosomes and therefore tends to be used for high-resolution gene mapping.

Following mitotic arrest, the cells are resuspended in a 0.075M potassium chloride solution (KCl). Since this solution is hypotonic with regard to the concentration of KCl within the cells (i.e., the cells contain a higher concentration of KCl), water enters the cells in an attempt to restore the osmotic equilibrium. The net effect is that the cells swell, causing the cytoplasmic membrane to burst and leaving only the nucleus intact. They are then fixed with a 3:1 mixture of methanol:acetic acid and dropped onto microscope slides. Although this seems like a very simple cookbook recipe, it is difficult to obtain high quality metaphase spreads. Parameters involved in obtaining a high quality slide include the length of time in hypotonic solution, the freshness of the fixative and the number of times it was changed, the cell concentration, and the humidity and slide drying time (a function of temperature). All of these variables affect the degree of spreading (so that as few chromosomes as possible are overlapping), the amount of cellular debris surrounding the chromosomes, and the "hybridizability" of the chromosomes themselves. Removal of cytoplasmic and nuclear debris by treatment with Proteinase K and/or pepsin decreases the amount of background noise caused by nonspecific adherence of probe to the proteins, lipids, and carbohydrates. Digestion of RNA with RNaseA can also help in reducing background noise. Humidity and drying time can be assessed by examination of the chromosomes themselves. Chromosomes that look "shiny" or "hollow" will most likely give a poor or inconsistent hybridization pattern. This is indicative of high humidity and prolonged drying time. The effects of such variables can be minimized by preparation of the slides in an environmentally controlled unit in which both the temperature and humidity can be tightly controlled.[14]

Another highly variable parameter is the manner in which the slides themselves are stored. All protocols require that the slides be dehydrated after preparation and again prior to hybridization. One difference that arises is in the amount of time between preparation and the first dehydration. The length of time the slides "bake" at 37°C can vary anywhere from one day to one month. The slides are then dehydrated and either used immediately, stored in 100% ethanol at −20°C, or stored with Drierite® or similar moisture absorbent material at −70°C for up to one year. Since water adversely affects the hybridization potential of the chromosomes, all of the above treatments are designed to prevent exposure to moisture.

The slides are then pre-treated prior to hybridization with the labeled probe. This involves exposure to RNase A followed by proteinase K and/or pepsin as indicated above. These treatments can be disregarded if there appears to be little evidence of surrounding cellular debris. Denaturation of the chromosomal DNA into single-stranded molecules is essential, however, and is accomplished by the addition of 70% formamide in 2XSSC (pH 7.0–7.5) to the slide, followed by a coverslip and heating to 80°C for a predetermined period of time. The amount of time is dependent upon the species from which the chromosomes were prepared and the age of the slides, older slides generally needing slightly shorter incubation times. The coverslip is removed and the slides immediately immersed in ice cold 70% ethanol for $\geqslant$3 minutes. This is followed by a 90% and 100% ethanol dehydration series. The slides are air dried at room temperature, at which point they are ready for hybridization with the denatured probe.

The above protocol can also be adapted for adherent cells grown on chamber slides. This is sometimes preferable to trypsinization of the cells and preparation as a cell suspension. One application is the use of nucleic acid probes and fluorescent immunocytochemistry for the simultaneous localization of genomic sequences and particular proteins within the cell under growth conditions without perturbation of the cellular organization. This approach is being used more extensively as live cell imaging becomes an integral part of tumorigenesis studies.

Another variation involves the preparation of extended DNA fibers for ordering of probes with resolution in the range of 2–350 kb. There are at least three different methods for obtaining these extended DNA/chromatin fibers. All three involve depletion of histones and extraction of genomic DNA from interphase nuclei on a microscope slide by high salt, alkaline, or detergent treatment. In one protocol, the DNA forms halos around the nucleus in loops of 60–200 kb, the approximate distance between two matrix attachment sites in genomic DNA (Figure 10.1D). Another method lyses the nuclei and the unattached DNA is allowed to stream down the microscope slide. The third approach involves treatment of intact cells with the topoisomerase II inhibitor m-AMSA to prevent chromatin condensation. Subsequent alkaline treatment of the nuclei permits the DNA to leak out of the nucleus. Chromatin extracted using the latter procedure results in more tangled DNA, however new techniques such as DNA combing[9] and optical mapping[11,15] have greatly improved this technique.

Other specimens utilizing *in situ* hybridization protocols are formalin-fixed, paraffin-embedded tissue sections. These must first be treated to remove the wax and formalin. This is accomplished by heating the slides to 65 C for 2 h followed by incubation in xylene. As with the chromosome spreads, the slides are then run through an ethanol dehydration series, and as for the interphase slides, the cells are permeabilized by treatment with a 0.1% Triton® X-100/4XSSC solution. The slides are then treated with a 1M sodium iso-thiocyanate (NaSCN) solution at 56 C to reverse fixative induced crosslinks between the DNA and protein. Pepsin treatment followed by the standard formamide/2XSSC denaturation and ethanol series is performed prior to addition of the hybridization solution containing the denatured probe. Since the specimen preparation is typically performed the same day as the hybridization, storage of the slide after removal of the paraffin and formalin is not an issue.

The preparation of intact nuclei for interphase FISH is the same as for chromosome spreads from suspension or tissue sections, depending on the source of the material. This includes treatment of the slide with RNase A, pepsin, and formaldehyde as well as denaturation with formamide at elevated temperatures.

10.2.2. Preparation of DNA Probes

DNA probes can be prepared from a number of different sources and their nature is dependent upon the type of the *in situ* hybridization being performed. The principles behind probe preparation are the same, however, the exact protocol used will vary from source to source. The most common DNA probes consists of a fragment of genomic DNA or cDNA subcloned into a specific vector. This vector may take the form of a yeast (YAC), bacterial (BAC), plasmid (PAC), or human (HAC) artificial chromosome, a cosmid, plasmid, or phagemid. These types of probes are usually used for gene or translocation breakpoint mapping and chromosome identification.

The technique of SKY (which will be discussed at greater length towards the end of this chapter) requires specially prepared probe sets. Briefly, this technique "paints" each chromosome with a unique combination of fluorochromes. Thus, a probe corresponding to the entire length of each chromosome must be prepared. The first step in this process involves the use of flow cytometry to separate the individual chromosomes of a particular species on the basis of size and relative nucleotide content.[16,17] Each individual chromosome is then amplified by PCR using a degenerate oligonucleotide primer.[18,19] This increases the amount of working material and serves as the labeling reaction by incorporating conjugated nucleotides during the PCR. Obviously great care must be taken not to cross-contaminate the sorted chromosomes prior to labeling.

Another source of probe is total genomic DNA isolated from cell lines or tissues. This type of probe is necessary for performing CGH. Isolation of genomic DNA from most cells or tissues (i.e., blood, cell lines, and organs) involves an overnight incubation with proteinase K and SDS, followed by phenol:chloroform extraction and isopropanol precipitation. DNA can also be obtained from paraffin-embedded tissues, enabling one to retrospectively analyze archived tumor material. The only difference from the above protocols is removal of the formalin and paraffin prior to extraction of the DNA. This procedure is similar to that outlined in the previous section on metaphase preparation from paraffin-embedded, formalin-fixed tissue sections utilizing xylene and NaSCN.

Good quality DNA preparation is essential for obtaining a successful labeling reaction. Cross-contamination of probes during isolation must obviously be avoided, but this is not the only cause for trouble. Since the labeling reactions require the activity of enzymes, care must be taken to eliminate any contaminants which may be inhibitory to their activity. Many enzymes require the presence of divalent cations such as magnesium in order to function. This means that one must be careful of the concentration of chelating agents like EDTA or EGTA in the isolation steps. Some EDTA is acceptable as long as the final concentration in the labeling reaction is not too great. This can sometimes be overcome by addition of extra magnesium to the labeling reactions. Other potential inhibitors are proteinases or other nonspecific proteins that may adversely affect the enzymes. The amount of protein in a DNA preparation can be gauged by calculation of the Abs_{260}:Abs_{280} ratio. This should normally be around 1.8 for DNA. This is generally the easiest step in any hybridization reaction, but if done incorrectly can result in days of frustration and troubleshooting.

10.2.3. Labeling of DNA Probes

Probes can be labeled with a number of different fluorescent and nonfluorescent conjugated nucleotide analogs. The type of label used depends upon the imaging equipment available, type of experiment, cost, and personal preference. Probes containing nucleotides conjugated with a fluorescent molecule are said to be direct-labeled, whereas those containing nucleotides that are nonfluorescently labeled (i.e., with biotin, digoxigenin, dinitrophenol, etc.) are referred to as indirectly labeled. These latter probes must be detected using a second or third molecule containing the fluorescence as discussed in Section 10.2.5.

There are generally two different ways to label a probe for *in situ* hybridization. The first was briefly discussed in the previous section with regard to the chromosome-specific probes used in SKY. The conjugated nucleotides are incorporated directly into the PCR product. For SKY this usually involves the use of three fluorescence-conjugated and two nonfluorescence-conjugated dUTP analogs. This allows for the simultaneous hybridization and detection of probes that emit fluorescence in five distinct ranges of the spectrum. The importance of this will become apparent in later sections of this chapter dealing specifically with SKY. PCR can also be used to label genomic inserts with either insert-specific, vector-specific, or degenerate oligonucleotide primers. DNA fragments smaller than 100 kb can be efficiently labeled using random primed labeling[20] and nick translation as described in the introduction. Fluorochromes can also be chemically attached to nucleic acid probes.

10.2.4. Hybridization of Probe to Specimen

This step of the procedures deviates very little from one technique to another. The desired amount of labeled probe is combined with an excess amount of competitor Cot I DNA.[21–23] The purpose of the Cot I DNA is to suppress elements in the probe (by complementary annealing) which are highly repetitive and dispersed throughout the genome. Such elements include Alu, LINE, SINE, and B1 repeat sequences. Failure to eliminate these sequences from the probe will result in hybridization to sites on every chromosome, not just that region containing the unique sequence of interest. Cot I DNA can be eliminated, however, if the probe is known to be devoid of any repeat elements, which is often the case with cDNA probes. The combined DNA is precipitated, resuspended in a small volume of formamide/dextran sulfate/SSC solution (referred to as hybridization buffer), and denatured by heating to 80°C. This is followed by either quick chilling (if no Cot I) or incubation at 37°C for at least one hour to allow the Cot I sequences to anneal, thereby making the repeat elements in the probe double stranded and preventing them from hybridizing to the sample.

The chromosomes on the slide are denatured by incubation in formamide at 80 C followed by a cold EtOH dehydration series. The probe is then allowed to anneal to the chromosomes at 37°C in a moist dark chamber. The amount of time required for hybridization depends on the complexity of the probe. The standard protocol is 18–24 h for unique sequence probes (i.e., genes and cDNA) and 40–72 h for complex probe mixtures such as those used for SKY and CGH.

10.2.5. Detection and Visualization

The detection stage of the protocol requires a number of increasingly stringent salt washes at 45°C to remove nonspecifically bound probe. Bovine serum albumin (BSA) is used to nonspecifically bind to the charged silica in the glass microscope slide thereby preventing a nonspecific interaction between the slide and any other fluorescence-labeled molecules used in the detection process. Probes labeled with nucleotides directly conjugated to fluorescent molecules (i.e., FITC-dUTP, Rhod-dUTP, etc.) do not require blocking or further detection and can therefore be immediately analyzed. Indirect-labeled probes, however, require at least one more detection step. Bio-dUTP, for example, can be detected with either avidin conjugated to a fluorochrome (i.e., avidin-FITC) or with a fluorochrome-conjugated antibody directed against biotin (i.e., mouse-anti-biotin-FITC). One can add more detection levels in an attempt to increase the signal strength, however this usually results in an increase in the background level as well and therefore little improvement in the signal to noise ratio.

10.2.6. High Sensitivity Detection Procedures

The smaller the actual target sequence in the genome, the shorter the probe becomes and thus fewer fluorescently-labeled nucleotides to provide a signal. A genomic target dispersed over a large physical distance can also result in decreased signal strength. An example of this is hybridization of a 6 kb cDNA clone. This size is generally sufficient for a nice signal, however, if the gene contains multiple exons dispersed over perhaps 100 kb the density of fluorescence is decreased and therefore more difficult to visualize. In order to circumvent this problem, multiple levels of immunological detection are sometimes used. As mentioned above, this often amplifies the background noise as well and is therefore not always an efficient means of detecting weak signals.

A newer method makes use of a well utilized enzymatic reaction involving peroxidase.[24–26] This method was originally developed for ELISA[27] and has been adapted for immunohistochemistry[28] as well as for FISH.[26,29] The probe is labeled in the usual manner by nick translation incorporation of a hapten-conjugated nucleotide (i.e., Bio-dUTP or Dig-dUTP). The first detection layer uses peroxidase conjugated to either avidin or an anti-hapten immunoglobulin molecule. These are then amplified through the use of a peroxidase substrate conjugated to either biotin or a fluorochrome. Upon cleavage of the substrate, the biotin or fluorochrome becomes deposited close to the site of the reaction. Fluorochromes can be immediately visualized while biotin deposits require detection with an avidin- or anti-biotin-fluorochrome conjugate. Because detection involves an enzymatic reaction, multiple substrates are cleaved and deposited, thereby amplifying the signal.

One problem with the original reported protocol was the loss of spatial resolution. This was due to the free migration of the cleaved substrate before its immobilization on the slide. Later modifications included polymeric substances to increase the viscosity of the reaction solution and thereby decrease the diffusion of the intermediate cleaved products.[30] The net result is an extremely sensitive amplification method for the detection of small or difficult to visualize FISH signals with only a slight sacrifice of the spatial resolution of directly labeled probes. This technique is now available as a kit from NEN™ Life Science Products.

10.3. Applications of FISH

Now that we have established an understanding of how FISH is performed, it is possible to explore the practical applications of this technique to biological problems. A review listing a number of applications came out almost ten years after the first publication describing nonisotopic *in situ* hybridization.[31] Although advances have been made and new techniques developed involving the use of FISH the applications have not changed much. References cited in the following sections are not at all meant to be inclusive, but merely represent examples of the various applications.

10.3.1. Gene Mapping

The explosion in manuscripts involving the use of FISH came along as the human genome project began to take form. FISH proved to be a powerful method for mapping the many genes and genomic clones being isolated by a number of laboratories.[32] The initial goal was to use FISH to physically map sequences along each chromosome at 1 Mb intervals.[32,33] This would allow gene hunters to use these markers as landmarks for linkage studies.[34–36] The use of Cot I suppression, as discussed in Section 10.2.4, was an immense improvement.[21–23] Prior to this, the mapping of large clones resulted in hybridization to multiple loci and often gave a banding pattern to the chromosomes. This was due to the presence of highly repetitive sequences such as Alu, LINEs, SINEs, in the probe and in those regions of the chromosome (heterochromatin) that are known to be relatively gene-poor. Suppression hybridization (for review see reference 37) effectively removed these sequences from the hybridization reaction, thereby allowing only the single copy sequences to anneal to the chromosomes. The result was that FISH could be used to physically map the larger clones isolated from cosmid, BAC, and YAC genomic libraries.[38,39] Not only was FISH an effective means of mapping these clones, but it was also beneficial for determining whether they were pure or chimeric in nature. Chimeric clones are fusion products of DNA fragments from different genomic loci created during the generation of a recombinant library. This was often a problem with large YAC clones that contained from ~200 Kb to ≥1 Mb of genomic sequence.

FISH was, and continues to be useful for physically mapping genes cloned by methods other than linkage where their chromosomal location is unknown.[40–42] This is especially true for identifying the location of homologous genes in different species.[43–47] It is also a useful technique for the identification of pseudogenes or other genes in a gene family[48] (Figure 10.1A).

10.3.2. Somatic Hybrid Analysis

Somatic cell hybrid lines have proved very useful for isolating and mapping disease genes in both mice and humans. It is important to determine the genomic contribution of the species of interest in these lines both during their derivation as well as periodically due to their unstable nature. This was originally accomplished through the employment of Hoechst 33258 or Giemsa stains. Mouse chromosomes could readily be distinguished in mouse–hamster hybrids by the intense Hoechst 33285 staining of the mouse centromeres[49] while color differences with Giemsa staining was useful for the analysis of human–mouse hybrids.[50] Giemsa banding of chromosomes was also used to identify interspecific translocations,[51] but the analysis required skilled knowledge of chromosome bands, was extremely time consuming and still was not sufficient for identification of more subtle rearrangements.

With the advent of FISH it became feasible to unambiguously identify species-specific chromosomes in interspecific crosses. This was first accomplished by nick translation labeling the entire genome of a particular species and subsequent hybridization to metaphase spreads prepared from the hybrid cell line.[23,52–54] In order to identify the origin of a particular chromosome, the cell line genomic DNA would be labeled and hybridized to normal metaphase chromosomes prepared from each of the species involved. In this manner it was possible to determine not only which chromosomes were being contributed from a given species, but how much, and which regions of particular chromosomes were involved[55,56] (Figure 10.1F). Another fluorescent technique was the performance of species-specific banding of the hybrid chromosomes using labeled *Alu* and L1 repetitive sequences as probes.[57,58]

10.3.3. Clinical and Cancer Cytogenetics

Pre- and postnatal screening for cytogenetic abnormalities is extremely important as both a diagnostic and prognostic procedure. This is usually accomplished by Giemsa banding of metaphase chromosomes and karyotype analysis by skilled and certified cytogenetic technologists. Many chromosome aberrations, such as aneuploidy and translocations involving sufficiently large chromosome fragments, are readily identifiable through the use of these techniques. There are instances, however, where identification and determination of the exact defect is not possible by these methods. A prime example involves the identification of marker chromosomes. These are small chromosome fragments that usually contain some of the centromeric material necessary for segregation during cell division, but not enough chromatin to identify their origin by standard staining techniques.[59,60] Identifying the chromosomal origin of these markers is clinically significant because not all extrachromosomal material has phenotypic consequences.[61]

Karyotype analysis of tumors by conventional banding techniques alone is also very difficult. This is due to poor morphology, the number of marker chromosomes, double minutes, and complex chromosomal rearrangements often seen, especially in solid tumors.[62] Karyotype aberrations increase in number as tumors progress from premalignant lesions to metastatic disease (Figure 10.2A, B).[63] Identification of these changes is useful in determining the stage of the tumor for diagnostic and prognostic purposes as well as following the course of regression and reappearance of tumors after treatment. Single chromosome paints proved useful in early endeavors using FISH as a complementary technique to Giemsa banding in order to further characterize tumor karyotypes.[64] This method has been enhanced by the development of new techniques known as SKY and m-FISH[65] which allow for the simultaneous chromosome-specific painting of the human[66,67] and mouse[68] genomes. This technique and its many applications will be discussed in Section 10.4.2.

10.3.4. Chromosome Evolution

Comparative cytogenetics is the study of changes in chromosome number and composition within and between different species as a function of their evolutionary divergence from one another. Studies have been performed looking at the chromosomal changes during evolution.[47,69–74] There is only one difference in chromosome morphology that distinguishes humans from chimpanzees.[75] Human chromosome 2 has been derived by the fusion of two acrocentric great ape chromosomes (chimpanzee chromosomes 12 and 13; gorilla and orangutan

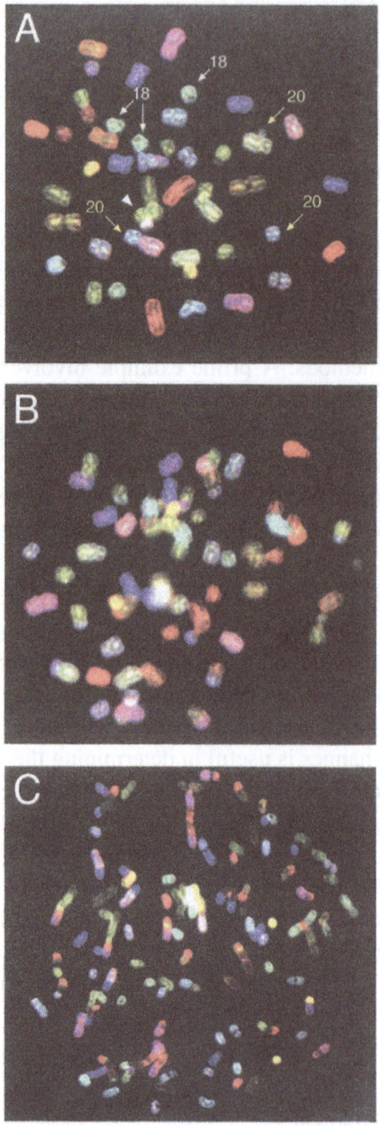

Figure 10.2. A: Immortalized cell line-CPDR-1 from a primary prostate cancer has a hyperdiploid karyotype with gains of chromosome 18 and 20 (arrows) and a t(Y;11) (arrowhead) (courtesy of M. Augustus). B: Highly rearranged chromosomes in a prostate cancer cell line (PC3) derived from a metastasis to the bone. The karyotype is predominantly hypertriploid (courtesy of M. Augustus). C: Differentially labeled human chromosome painting probes hybridized to metaphase chromosomes from the gibbon *Hylobates concolor*. Few chromosomes have not changed throughout evolution as indicated by their uniform hybridization color (arrows). Some chromosomes are simple rearrangements being composed of only two distinct human chromosomes (concave arrows) while the majority, however, have become highly rearranged such that the gibbon chromosomes are composed of pieces from several different human chromosomes (arrowheads), or vice versa depending on your perspective (see reference 66).

chromosomes 11 and 12). The only other difference was a translocation in gorilla between the chromosomes homologous to human chromosomes 5 and 17 (chimpanzee and orangutan chromosomes 4 and 19).[70,71,76] A more striking pattern of chromosome evolution is seen when comparing the chromosomes of great apes to those of the gibbon. Only human chromosomes 11, 14, 20, X and Y hybridized to the entire length of a single gibbon chromosome. All other gibbon chromosomes were involved in interchromosomal rearrangements compared to humans and the great apes[70,71,76] (Figure 10.2C).

Similar studies have been performed to explain the karyotypic differences between the morphologically similar and closely related Indian and Chinese muntjacs.[77–79] The Indian muntjac has the lowest chromosome number known in mammals, with a diploid genome of 2n = 6/7. The Chinese muntjac, however, has a karyotype of 2n = 46. It is curious from an evolutionary standpoint how the genomes of these two related species became so divergent. SKY analysis has also been applied to evolutionary studies and will improve the analysis of genomic relationships.[66]

10.4. Karyotype Analysis

The most commonly used method for whole genome analysis is Giemsa banding, in which the banding pattern of each chromosome is compared to its homologue and the standard banding pattern for chromosomes of that particular resolution. Although this remains an excellent technique for routine cytogenetic applications (i.e., prenatal diagnostics and reasons of infertility), analysis of more complex genomic rearrangements, such as those seen in tumorigenesis, requires great skill and even then subtle aberrations can be missed. The previous two sections have already discussed the use of FISH for analyzing the genomic complement of different individuals and species. Unfortunately, using standard FISH techniques it was only possible to analyze a few predetermined portions of the karyotype in any given hybridization. This was largely due to the limitations in the number of available fluorochromes that could be distinguished using conventional microscope hardware and imaging software. Recent advancements in fluorochromes and photodetection devices, however, have enabled the simultaneous hybridization and detection of probes labeled with multiple fluorochromes. The net result has been the development of combinatorial labeling and hybridization techniques known as SKY and m-FISH. Advancements in quantitative image analysis have themselves lead to the development of a new FISH application for whole genome analysis. This technique, CGH, is a quantitative method for comparing the copy number of genomic regions between control and specimen samples. Each of these two methodologies alone provide significant advances in the way we analyze genomes, but their combination has resulted in a great leap forward in our knowledge and understanding of the chromosome aberrations specific not only to tumors of different origin but also to specific developmental stages of tumorigenesis.

10.4.1. Comparative Genome Hybridization

The complex karyotypes found in solid tumors involve multiple translocations, chromosomal aneuploidy, and the presence of small unidentifiable marker

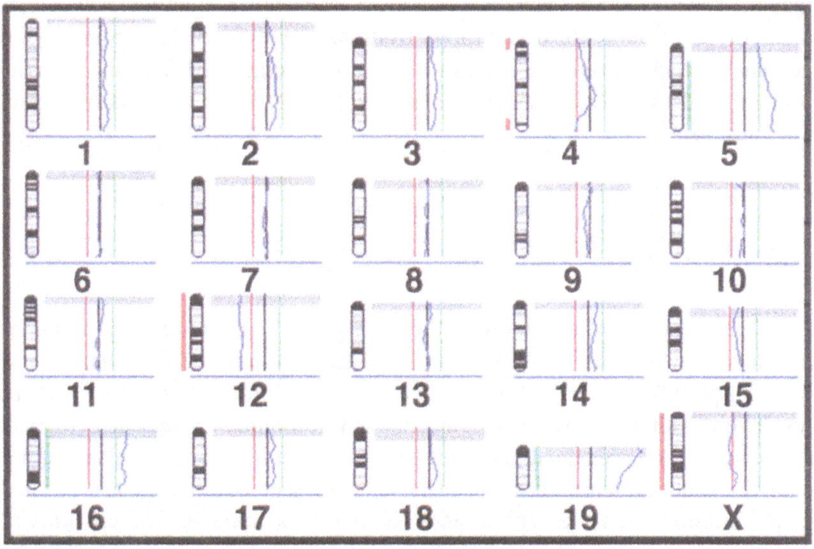

A

Figure 10.3:. A: Comparative genomic hybridization (CGH) analysis of the metaphase in B. Illustrated is the G-banding pattern (ideogram) of each chromosome. To the left and right of the ideogram is indicated losses (thick red bar) and gains (thick green bar) of chromosome regions, respectively. To the far right of each ideogram is a series of lines. The red line indicates the loss threshold, the green line the gain threshold, and the black middle line represents no change in the chromosome content. Superimposed on these threshold is a blue line representing the ratio of sample to control fluorescence (courtesy of C. Montagna). B: Spectral karyotype analysis (SKY) of a liver-derived mouse tumor showing multiple chromosome rearrangements (chromosomes 3, 4, 7, and 10), deletions (chromosomes 12 and X), gains (chromosomes 5, 16, and 19) as well as a strain specific polymorphism (chromosome 8, small centromere on right chromosome). Although a fragment of chromosome 19 has been inserted into the middle of chromosome 3, no material from chromosome 3 appears to be lost at the resolution of CGH (10.3A). Translocation of chromosomes 4 to one another, however, resulted in the loss of telomeric DNA from one chromosome and loss of centromeric DNA from the other. Reciprocal translocation between chromosomes 7 and 10 appears to be balanced since no material from either chromosome was gained or lost. Monosomy of chromosomes 12 and X as well as trisomy for chromosomes 5, 16 and 19 are all reflected in the CGH pattern of these chromosomes (courtesy of C. Montagna) (*continued*).

chromosomes. As a result, it is often impossible to know with certainty whether any genomic material has been gained or lost during the development of the tumor. A copy number change in individual genes of interest could be assessed by interphase or metaphase FISH, but until recently a comprehensive genome screen has not been possible. Comparative genome hybridization, or CGH, is a two-color FISH strategy that does not require growing cells.[80] As such, archived material is amenable to analysis at a later date. The technique involves nick translation labeling of total genomic DNA isolated from tumor and control samples with different fluorochromes. For illustrative purposes let us use rhodamine-dUTP (red fluorescence) for the control and fluorescein-dUTP (green fluorescence) for the tumor. These samples are pooled in a one-to-one ratio, combined with human Cot I DNA to suppress repetitive elements and hybridized to normal human metaphase preparations.

The ratio of fluorescence intensities of the two fluorochromes is measured along the length of each chromosome (Figure 10.3A). Sequences represented at the same copy number in the two genomes will hybridize as an equal ratio of the two colors giving a value of 1.0. Loss of chromosomal regions in the tumor will appear as regions with a ratio of red to green fluorescence greater than 1.0. Gain of material will result in greater hybridization of the tumor DNA (red) and a shift in the red to green ratio to values less than 1.0. The ability to make quantitative measurements requires fluorescence image acquisition using a charge-coupled device (CCD) camera and optical filters specific for the emission spectra of each

B

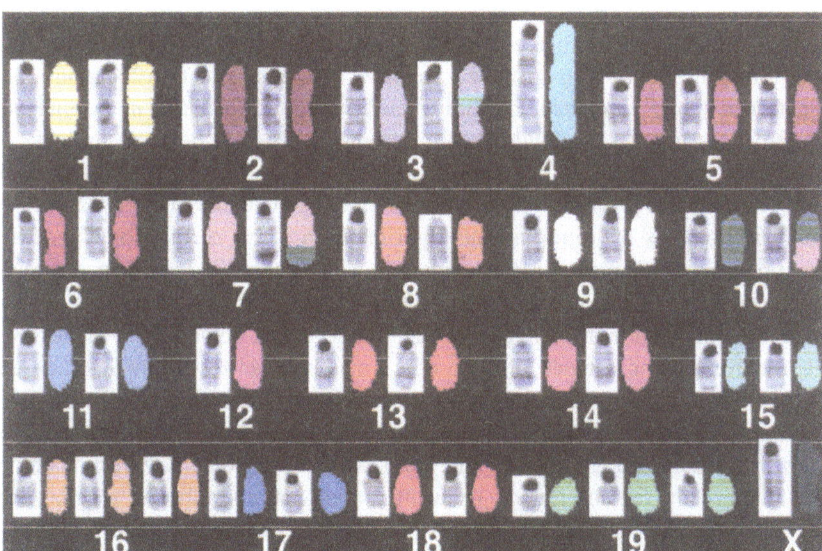

Figure 10.3. (*Continued*)

fluorochrome. Special software packages compare the ratio of fluorescence intensities and output karyotype with pseudocolored chromosomes representing gains and losses. A karyotype can also be produced which displays the red to green ratio along each chromosome as a histogram. This is important for determining how many copies of a particular region have been gained or lost in the tumor sample.

Since the introduction of this technique in 1992, CGH has been applied to nearly all types of human cancers (reviewed in references 81–83). This has resulted in the identification of tumor-specific and stage-specific aberrations that otherwise would have been unnoticed or more difficult to identify by conventional banding techniques. It has now become clear that what was originally thought to be a chromosome mess or a mere reflection of acquired secondary cytogenetic aberrations in solid tumors is actually a very specific gain or loss of genomic material. Determination of this pattern of copy number changes and identification of the genes found in these chromosomal regions will enable us to better understand the sequence of events on the molecular level which leads to the development of tumors in a tissue-specific manner. The ability to determine the particular developmental stage of a tumor based on its pattern of genomic gains and losses will have diagnostic and prognostic value. As we learn more about the different stages of tumorigenesis, we will be in a better position to tailor treatment to individual tumors based on their genetic profile.

10.4.2. Spectral Karyotyping

While CGH is a useful technique for the quantitation of genomic imbalances, it is often useful to also visualize balanced chromosomal translocations. As mentioned above, chromosome banding has limitations, especially in the analysis of tumor karyotypes. Individual chromosome paints can be used to look at chromosomes suspected of being involved in a rearrangement but are not practical for obtaining an *a priori* assessment of genomic organization. Spectral karyotype analysis involves the simultaneous hybridization of all chromosome-specific painting probes to metaphase spreads.[66,68] This technique uses five different fluorochromes. Each chromosome-specific library is PCR labeled using a degenerate oligonucleotide primer (DOP-PCR) and a unique combination of the five fluorochromes. For instance, chromosome 1 may be labeled with rhodamine (red fluorescence), chromosome 2 with fluorescein (green), chromosome 3 with Cy5 (blue), chromosome 4 with rhodamine and Cy5 (red and blue), chromosome 5 with fluorescein and Cy5 (green and blue), and chromosome 6 with rhodamine and fluorescein (red and green). By using a combinatorial labeling scheme, the number of probes that can be distinguished has increased from 3 to 6. Increasing the number of fluorescent molecules to 5 allows the distinction of up to 31 different chromosomes (reviewed in reference 67).

The use of an interferometer attached to the CCD camera allows an analysis of the light with wavelengths between 400 and 800 nm captured at each pixel of the camera chip. A Fourier transformation is performed on the data which results in the generation of an emission spectrum for each pixel. This is translated into an RGB display and the end result is a metaphase spread in which each chromosome has a distinct color. The chromosomes are automatically arranged into a karyotype based on their detected emission spectra and the combinatorial table, which indicates the labeling scheme used. Translocations are seen as the juxtaposition of two or more different colors on a single chromosome (Figure 10.3B; i.e., chromosomes 7 and 10). This technique also enables the identification of marker chromosomes which either had no Giemsa bands or in which the banding pattern did not resemble that of any normal chromosome.[84,85] The combination of SKY and banding patterns (with either Giemsa or the fluorochrome DAPI, which gives a reverse banding pattern) is a potent tool for the band identification of translocation breakpoints. FISH with individual clones in the region is useful for localizing the breakpoint to an individual clone. Designing primers and sequencing the region eventually leads to the identification of the exact breakpoint where the translocation occurred and the genomic reorganization of nearby candidate genes. Direct joining of two different gene segments can result in the formation of fusion proteins. These are seen in a wide variety of lymphoid tumors and are thought to be one of the causative agents responsible for tumorigenesis. The morphological similarity, and often poor quality, of mouse chromosomes makes SKY a particularly useful tool for the analysis of tumors found in murine models of human cancers.[85–89]

References

1. Gall, J. G., and Pardue, M. L. Formation and detection of RNA-DNA hybrid molecules in cytological preparations. *Proc. Natl. Acad. Sci. USA 63*, 378–383, 1969.
2. John, H. A., Birnstiel, M. L., and Jones, K. W. RNA-DNA hybrids at the cytological level. *Nature 223*, 582–587, 1969.
3. Manning, J. E., Hershey, N. D., Broker, T. R., Pellegrini, M., Mitchell, H. K., and Davidson, N. A new method of in situ hybridization. *Chromosoma 53*, 107–117, 1975.
4. Broker, T. R., Angerer, L. M., Yen, P. H., Hershey, N. D., and Davidson, N. Electron microscopic visualization of tRNA genes with ferritin-avidin:biotin labels. *Nucleic Acids Res. 5*, 363–384, 1978.
5. Langer, P. R., Waldrop, A. A., and Ward, D. C. Enzymatic synthesis of biotin-labeled polynucleotides: Novel nucleic acid affinity probes. *Proc. Natl. Acad. Sci. USA 78*, 6633–6637, 1981.
6. Langer-Safer, P. R., Levine, M., and Ward, D. C. Immunological method for mapping genes on Drosophila polytene chromosomes. *Proc. Natl. Acad. Sci. USA 79*, 4381–4385, 1982.
7. Manuelidis, L., Langer-Safer, P. R., and Ward, D. C. High-resolution mapping of satellite DNA using biotin-labeled DNA probes. *J. Cell Biol. 95*, 619–625, 1982.
8. Cremer, T., Landegent, J., Bruckner, A., Scholl, H. P., Schardin, M., Hager, H. D., Devilee, P., Pearson, P., van der Ploeg, M. Detection of chromosome aberrations in the human interphase nucleus by visualization of specific target DNAs with radioactive and non-radioactive in situ

hybridization techniques: Diagnosis of trisomy 18 with probe L1.84. *Hum. Genet. 74*, 346–352, 1986.

9. Monier, K., Michalet, X., Lamartine, J., Schurra, C., Heitzmann, F., Yin, L., Cinti, R., Sylla, B. S., Creaven, M., Porta, G., Vourc'h, C., Robert-Nicoud, M., Bensimon, A., and Romeo, G. High-resolution mapping of the X-linked lymphoproliferative syndrome region by FISH on combed DNA. *Cytogenet. Cell Genet. 81*, 259–264, 1998.
10. Ried, T., Speicher, M. R., Schröck, E., Dietzel, S., Jauch, A., Nagaraja, R., Schlessinger, D., and Cremer, T. (1995). Functional high resolution mapping of DNA-clones on Halo-DNA preparations and in interphase nuclei by multicolor fluorescence in situ hybridization. *Am. J. Hum. Genet. 53* (Suppl.), 1348–1340, 1995.
11. Schwartz, D. C., and Samad, A. Optical mapping approaches to molecular genomics. *Curr. Opin. Biotechnol. 8*, 70–74. 1997.
12. Tocharoentanaphol, C., Cremer, M., Schröck, E., Kilian, K., Blonden, L., Cremer, T., and Ried, T. Multicolor fluorescence *in situ* hybridization on metaphase chromosomes and interphase Halo-preparations using cosmid and YAC clones for the simultaneous high resolution mapping of deletions in the dystrophin gene. *Hum. Genet. 93*, 229–235, 1994.
13. Wiegant, J., Kalle, W., Mullenders, L., Brookes, S., Hoovers, J. M. N., Dauwerse, J. G., van Ommen, G. J. B., and Raap, A. K. High-resolution *in situ* hybridization using D19NA halo preparations. *Hum. Molec. Genet. 1*, 587–591, 1992.
14. Spurbeck, J. L., Zinsmeister, A. R., Meyer, K. J., and Jalal, S. M. Dynamics of chromosome spreading. *Am. J. Med. Genet. 61*, 387–393, 1996.
15. Samad, A., Huff, E. F., Cai, W., and Schwartz, D. C. Optical mapping: a novel, single-molecule approach to genomic analysis. *Genome Res. 5*, 1–4, 1995.
16. Carter, N. P., Ferguson-Smith, M. E., Affara, N. A., Briggs, H., and Ferguson-Smith, M. A. Study of X chromosome abnormality in XX males using bivariate flow karyotype analysis and flow sorted dot blots. *Cytometry 11*, 202–270, 1990.
17. Gray, J. W., Langlois, G., Carrano, A. V., Burkhart-Schultz, K., and Van Dilla, M. A. High resolution chromosome analysis: One and two parameter flow cytometry. *Chromosoma 73*, 9–27, 1979.
18. Telenius, H., Carter, N. P., Bebb, C. E., Norednskjöld, M., Ponder, B. A. J., and Tunnacliffe, A. Degenerate oligonucleotide-primed PCR (DOP-PCR): General amplification of target DNA by a single degenerate primer. *Genomics 13*, 718–725, 1992.
19. Telenius, H., Pelear, A. H., Tunnacliffe, A., Carter, N. P., Behmel, A., Ferguson-Smith, M. A., Nordenskjöld, M., Pfragner, R., and Ponder, B. A. J. Cytogenetic analysis by chromosome painting using DOP-PCR amplified flow sorted chromosomes. *Genes Chromosom. Cancer 4*, 257–263, 1992.
20. Feinberg, A. P., and Vogelstein, B. A technique for radiolabeling DNA restriction fragments to high specific activity. *Anal. Biochem. 132*, 6–13, 1983.
21. Landegent, J. E., Jansen in de Wal, N., Dirks, R. W., Baas, F., and van der Ploeg, M. Use of the whole cosmid cloned genomic sequences for chromosomal localization by non-radioactive *in situ* hybridization. *Hum. Genet. 77*, 366–370, 1987.
22. Lichter, P., Cremer, T., Borden, J., Manuelidis, L., and Ward, D. C. Delineation of individual human chromosomes in metaphase and interphase cells by in situ hybridization using recombinant DNA libraries. *Hum. Genet. 80*, 224–234, 1988.
23. Pinkel, D., Landegent, J., Collins, C., Fuscoe, J., Segraves, R., Lucas, J., and Gray, J. W. Fluorescence in situ hybridization with human chromosome specific libraries: Detection of trisomy 21 and translocation of chromosome 4. *Proc. Natl. Acad. Sci. USA 85*, 9138–9142, 1988.
24. van Gijlswijk, R. P. M., van Gijkswijk-Janssen, D. J., Raap, A. K., Daha, M. R., and Tanke, H. J. Enzyme-labelled antibody-avidin conjugates: New flexible and sensitive immunochemical reagents. *J. Immunol. Methods 189*, 117–127. 1996.
25. van Gijlswijk, R. P. M., Wiegant, J., Vervenne, R., Lasan, R., Tanke, H. J., and Raap, A. K.

Horseradish peroxidase-labeled oligonucleotides and fluorescent tyramides for rapid detection of chromosome-specific repeat sequences. *Cytogenet. Cell Genet. 75*, 258–262, 1996.
26. Raap, A. K., van de Corput, M. P. C., Vervenne, R. A. W., van Gijlswijk, R. P. M., Tanke, H. J., and Weigant, J. Ultra-sensitive FISH using peroxidase-mediated deposition of biotin- or fluorochrome tyramides. *Hum. Molec. Genet. 4*, 529–534, 1995.
27. Bobrow, M. N., and Litt, G. J. Method for the detection or quantitation of an analyte using an analyte dependent enzyme activation system. United States Patent no. 5196306, 1993.
28. Adams, J. C. Biotin amplification of biotin and horseradish peroxidase signals in histochemical stains. *J. Histochem. Cytochem. 40*, 1457–1463, 1992.
29. van Gijlswijk, R. P. M., Zijlmans, H. J. M. A. A., Wiegant, J., Bobrow, M. N., Erickson, T. J., Adler, K. E., Tanke, H. J., and Raap, A. K. Fluorochrome-labeled tyramides: Use in immunocytochemistry and fluorescence in situ hybridization. *J. Histochem. Cytochem. 45*, 375–382, 1997.
30. van Gijlswijk, R. P. M., Wiegant, J., Raap, A. K., and Tanke, H. J. Improved localization of fluorescent tyramides for fluorescence *in situ* hybridization using dextran sulfate and polyvinyl alcohol. *J. Histochem. Cytochem. 44*, 389–392, 1996.
31. Lichter, P., and Ward, D. C. Is non-isotopic *in situ* hybridization finally coming of age? *Nature 345*, 93–95, 1990.
32. Lichter, P., Chang Tang, C.-J., Call, K., Hermanson, G., Evans, G. A., Housman, D., and Ward, D. C. High resolution mapping of human chromosome 11 by in situ hybridization with cosmid clones. *Science 247*, 64–69, 1990.
33. Bellanné-Chantelot, C., Lacroix, B., Ougen, P., Bilaut, A., Beaufils, S., Betrand, S., Georges, I., Glibert, F., Gros, I., Lucotte, G., Susini, L., Codani, J. J., Gesnouin, P., Pook, S., Vaysseix, G., Lu-Kuo, J., Ried, T., Ward, D. C., Chumakov, I., Le Paslier, D., Barillot, E., and Cohen, D. Mapping the whole human genome by fingerprinting yeast artificial chromosomes. *Cell 70*, 1059–1068, 1992.
34. Dixon, M. J., Dixon, M., Houseal, T., Bhatt, M., Ward, D. C., Klinger, K., and Landes, G. M. Narrowing the position of the Treacher Collins syndrome locus to a small interval between three new microsatellite markers at 5q32–33.1. *Am. J. Hum. Genet. 52*, 907–914, 1993.
35. Lichter, J. B., Difilippantonio, M., Wu, J., Miller, D., Ward, D. C., Goodfellow, P. J., and Kidd, K. K. Localization of the gene for MEN 2A. *Henry Ford Hosp. Med. J. 40*, 199–204, 1992.
36. Lichter, J. B., Difilippantonio, M. J., Pakstis, A. J., Goodfellow, P. J., Ward, D. C., and Kidd, K. K. Physical and genetic maps for chromosome 10. *Genomics 16*, 320–324, 1993.
37. Jauch, A., Daumer, C., Lichter, P., Murken, J., Schroeder-Kurth, T., and Cremer, T. Chromosomal in situ suppression hybridization of human gonosomes and autosomes and its use in clinical cytogenetics. *Hum. Genet. 85*, 145–150, 1990.
38. Lengauer, C., Eckelt, A., Weith, A., Endlich, N., Ponelies, N., Lichter, P., Greulich, K. O., and Cremer, T. Selective staining of defined chromosomal subregions by *in situ* suppression hybridization of libraries from laser-microdissected chromosomes. *Cytogenet. Cell Genet. 56*, 27–30, 1991.
39. Ried, T., Mahler, V., Vogt, P., Blonden, L., van Ommen, G. J. B., Cremer, T., and Cremer, M. Direct carrier detection by in situ suppression hybridization with cosmid clones for the Duchenne/Becker muscular dystrophy locus. *Hum. Genet. 85*, 581–586, 1990.
40. Gallagher, P. G., Upender, M., Ward, D. C., and Forget, B. G. The gene for human erythrocyte membrane protein band 7.2 (EPB72) maps to 9q33–q34 centromeric to the Philadelphia chromosome translocation breakpoint region. *Genomics 18*, 167–169, 1993.
41. Otsu, K., Fujii, J., Periasamy, M., Difilippantonio, M., Uppender, M., Ward, D. C., and MacLennan, D. H. Chromosome mapping of five human cardiac and skeletal muscle sarcoplasmic reticulum protein genes. *Genomics 17*, 507–509, 1993.
42. Upender, M., Gallagher, P. G., Moon, R. T., Ward, D. C., and Forget, B. G. Localization of the human alpha-fodrin gene (SPTAN1) to 9q33→q34 by fluorescence in situ hybridization. *Cytogenet. Cell Genet. 66*, 39–41, 1994.

43. Chowdhary, B., Raudsepp, T., Fronicke, L., and Scherthan, H. Emerging patterns of comparative genome organization in some mammalian species as revealed by Zoo-FISH. *Genome Res. 8*, 577–589, 1998.
44. Fronicke, L., and Scherthan, H. (1997). Zoo-fluorescence *in situ* hybridization analysis of human and Indian muntjac karyotypes (*Muntiacus muntjak vaginalis*) reveals satellite DNA clusters at the margins of conserved syntenic segments. *Chromosome Res. 5*, 254–261, 1997.
45. O'Brien, S., Cevario, S., Martenson, J., Thompson, M., Nash, W., Chang, E., Graves, J., Spencer, J., Tsujimoto, H., and Lyons, L. Comparative gene mapping in the domestic cat (*Felis catus*). *J. Hered. 88*, 408–414, 1997.
46. Raudsepp, T., Fronicke, L., Scherthan, H., Gustavsson, I., and Chowdhary, B. Zoo-FISH delineates conserved chromosomal segments in horse and man. *Chromosome Res. 4*, 218–225, 1996.
47. Wienberg, J., Stanyon, R., Jauch, A., and Cremer, T. Homologies in human and *Macaca fuscata* chromosomes revealed by in situ suppression hybridization with human chromosome specific DNA libraries. *Chromosoma 101*, 265–270, 1992.
48. Giordano, S. J., Yoo, M., Ward, D. C., Bhatt, M., Overhauser, J., and Steggles, A. W. The human cytochrome b5 gene and two of its pseudogenes are located on chromosomes 18q23, 14q31–32.1 and 20p11.2, respectively. *Hum. Genet. 92*, 615–618, 1993.
49. Hilwig, I., and Gropp, A. Staining of constitutive heterochromatin in mammalian chromosomes with a new fluorochrome. *Exp. Cell Res. 75*, 122–126, 1972.
50. Bobrow, M., and Cross, J. Differential staining of human and mouse chromosomes in interspecific cell hybrids. *Nature (Lond.) 251*, 77–79, 1974.
51. Kozak, C. A., Lawrence, J. B., and Ruddle, F. H. A sequential stainiing technique for the chromosomal analysis of interspecific mouse/hamster somatic cell lines. *Exp. Cell Res. 105*, 109–117, 1977.
52. Schardin, M., Cremer, T., Hager, H. D., and Lang, M. Specific staining of human chromosomes in chinese hamsters × man hybrid cell lines demonstrates interphase chromosome territories. *Hum. Genet. 71*, 281–287, 1985.
53. Manuelidis, L. Individual interphase chromosome domains revealed by in situ hybridization. *Hum. Genet. 71*, 288–293, 1985.
54. Durnam, D. M., Gelinas, R. E., and Myerson, D. Detection of species specific chromosomes in somatic cell hybrids. *Somat. Cell Mol. Genet. 11*, 571–577, 1985.
55. Boyle, A., Lichter, P., and Ward, D. C. Rapid analysis of mouse-hamster hybrid cell lines by in situ hybridization. *Genomics 7*, 127–130, 1990.
56. Doucette-Stamm, L. A., Riba, L., Handelin, B., Difilippantonio, M., Ward, D. C., Wasmuth, J. J., Gusella, J. F., and Housman, D. E. Generation and characterization of irradiation hybrids of human chromosome 4. *Somat. Cell Molec. Genet. 17*, 471–480, 1991.
57. Lichter, P., Ledbetter, S. A., Ledbetter, D. H., and Ward, D. C. Fluorescence in situ hybridization with Alu and L1 polymerase chain reaction probes for rapid characterization of human chromosomes in hybrid cell lines. *Proc. Natl. Acad. Sci. USA 87*, 6634–6638, 1990.
58. Nelson, D. L., Ledbetter, S. A., Corbo, L., Victoria, M. F., Ramirez-Solis, R., Webster, T., Ledbetter, D. H., and Caskey, C. T. Alu polymerase chain reaction: A method for rapid isolation of human-specific sequences from complex DNA sources. *Proc. Natl. Acad. Sci. USA 86*, 6686–6690, 1989.
59. Sachs, E. S., Van Hemel, J. O., Den Hollander, J. C., and Jahoda, M. G. Marker chromosomes in a series of 10,000 prenatal diagnoses. Cytogenetic and follow-up studies. *Prenatal Diagn. 7*, 81–89, 1987.
60. Warburton, D. *De novo* balanced chromosome rearrangements and extra marker chromosomes identified at prenatal diagnosis: Clinical significance and distribution of breakpoints. *Am. J. Hum. Genet. 49*, 995–1013, 1991.
61. Blennow, E., Nielson, K. B., Telenius, H., Carter, N. P., Kristoffersson, U., Holmberg, E., Gillberg,

C., and Nordenskjøld, M. Fifty probands with extra structurally abnormal chromosomes characterized by fluorescence *in situ* hybridization. *Am. J. Med. Genet. 55*, 85–94, 1995.
62. Heim, S., and Mitelman, F. *Cancer Cytogenetics*. New York: Wiley-Liss, 1995.
63. Ried, T., Heselmeyer-Haddad, K., Blegen, H., Schrock, E., Auer, G. Genomic changes defining the genesis, progression, and malignancy potential in solid human tumors: A phenotype/genotype correlation. *Genes Chromosomes Cancer 25*, 195–204, 1999.
64. Carter, N. P. Cytogenetic analysis by chromosome painting. *Cytometry 18*, 2–10, 1994.
65. Speicher, M. R., Gwyn Ballard, S., Ward, D. C. Karyotyping human chromosomes by combinatorial multi-floor FISH. *Nat. Genet. 12*, 368–375, 1996.
66. Schröck, E., du Manoir, S., Veldman, T., Schoell, B., Wienberg, J., Ferguson-Smith, M. A., Ning, Y., Ledbetter, D. H., Bar-Am, I., Soenksen, D., Garini, Y., and Ried, T. Multicolor spectral karyotyping of human chromosomes. *Science 273*, 494–497, 1996.
67. Macville, M., Veldman, T., Padilla-Nash, H., Wangsa, D., O'Brien, P., Schröck, E., and Ried, T. Spectral karyotyping, a 24-colour FISH technique for the identification of chromosomal rearrangements. *Histochem. Cell Biol. 108*, 299–305, 1997.
68. Liyanage, M., Coleman, A., du Manoir, S., Veldman, T., McCormack, S., Dickson, R. B., Barlow, C., Wynshaw-Boris, A., Janz, S., Wienberg, J., Ferguson-Smith, M. A., Schrock, E., and Ried, T. Multicolour spectral karyotyping of mouse chromosomes. *Nature Genet. 14*, 312–315, 1996.
69. Ried, T., Arnold, N., Ward, D. C., and Wienberg, J. Comparative high-resolution mapping of human and primate chromosomes by fluorescence *in situ* hybridization. *Genomics 18*, 381–386, 1993.
70. Wienberg, J., Jauch, A., Stanyon, R., and Cremer, T. Molecular cytotaxonomy of primates by chromosomal in situ suppression hybridization. *Genomics 8*, 347–370, 1990.
71. Nash, W. G., Menninger, J. C., Wienberg, J., Padilla-Nash, H. M., O'Brien, S. J. The pattern of phylogenomic evolution of the *Canidae*. *Cytogenet. Cell Genet. 95*, 210–224, 2001.
72. Koehler, U., Arnold, N., Wienberg, J., Tofanelli, S., and Stanyon, R. Genomic reorganization and disrupted synteny in the Siamang (*Hylobates syndactylus*) revelaed by flujorescence *in situ* hybridization. *Am. J. Phys. Anthropol. 97*, 37–47, 1995.
73. Koehler, U., Bigoni, F., Wienberg, J., and Stanyon, R. Genomic reorganization in the concolor gibbon (*Hylobates concolor*) revealed by chromosome painting. *Genomics 30*, 287–292, 1995.
74. Müller, S., O'Brien, P. C. M., Ferguson-Smith, M. A., and Wienberg, J. Reciprocal chromosome painting reveals homologies between human and prosimian (*Eulemur macaco macaco* and *Eulemur fulvis mayottensis*) karyotypes. *Cytogenet. Cell Genet. 78*, 260–271, 1997.
75. de Grouchy, J., Truleau, C., Roubin, M., and Klein, M. Evolution caryotypique de l'homme et du chimpanzé; Étude comparative des topographies de bandes après dénaturation ménagée. *Ann. Genet. (Paris) 15*, 79–84, 1972.
76. Jauch, A., Wienberg, J., Stanyon, R., Arnold, N., Tofanelli, S., Ishida, T., and Cremer, T. Reconstruction of genomic rearrangements in great apes and gibbons by chromosome painting. *Proc. Natl. Acad. Sci. USA 89*, 8611–8615, 1992.
77. Yang, F., Carter, N. P., Shi, L., and Ferguson-Smith, M. A. A comparative study of the karyotype of the muntjacs by chromosome painting. *Chromosoma 103*, 642–652, 1995.
78. Yang, F., O'Brien, P., Wienberg, J., Neitzel, H., Lin, C. C., and Ferguson-Smith, M. A. (1997). Chromosomal evolution of the Chinese muntjac (*Muntiacus reevesi*). *Chromosoma 106*, 37–43, 1997.
79. Yang, F., O'Brien, P. C. M., Wienberg, J., and Ferguson-Smith, M. A. A reappraisal of the tandem fusion theory of karyotype evolution in the Indian muntjac using chromosome painting. *Chromosome Res. 5*, 109–117, 1997.
80. Kallioniemi, A., Kallioniemi, O.-P., Sudar, D., Rutovitz, D., Gray, J. W., Waldman, F., and Pinkel, D. Comparative genomic hybridization for molecular cytogenetic analysis of solid tumors. *Science 258*, 818–821, 1992.

81. Ried, T., Liyanage, M., du Manoir, S., Heselmeyer, K., Auer, G., Macville, M., and Schröck, E. Tumor cytogenetics revisited: comparative genomic hybridization and spectral karyotyping. *J. Molec. Med. 75*, 801–814, 1997.
82. Forozan, F., Karhu, R., Kononen, J., Kallioniemi, A., and Kallioniemi, O. P. Genome screening by comparative genomic hybridization. *Trends Genet. 13*, 405–409, 1997.
83. Zitzelsberger, H., Lehmann, L., Werner, M., and Bauchinger, M. Comparative genomic hybridization for the analysis of chromosomal imbalances in solid tumors and haematological malignancies. *Histochem. Cell Biol. 108*, 403–417, 1997.
84. Veldman, T., Vignon, C., Schröck, E., Rowley, J. D., and Ried, T. Hidden chromosomes: Abnormalities in hematological malignancies detected by multicolour spectral karyotyping. *Natire Genet. 15*, 406–410, 1997.
85. Coleman, A. E., Schrock, E., Weaver, Z., du Manoir, S., Yang, F., Ferguson-Smith, M. A., Ried, T., and Janz, S. Previously hidden chromosome aberrations in T(12;15)-positive BALB/c plasmacytomas uncovered by multicolor spectral karyotyping. *Cancer Res. 57*, 4585–4592, 1997.
86. Difilippantonio, M. J., Zhu, J., Chen, H. T., Meffre, E., Nussenzweig, M. C., Max, E. E., Ried, T., and Nussenzweig, A. DNA repair protein Ku80 suppresses chromosomal aberrations and malignant transformation. *Nature 404*, 510–514, 2000.
87. Liyanage, M., Weaver, Z., Barlow, C., Coleman, A., Pankratz, D. G., Anderson, S., Wynshaw-Boris, A., and Ried, T. Abnormal rearrangement within the alpha/delta T-cell receptor locus in lymphomas from Atm-deficient mice. *Blood 96*, 1940–1946, 2000.
88. Weaver, Z. A., McCormack, S. J., Liyanage, M., du Manoir, S., Coleman, A., Schrock, E., Dickson, R. B., and Ried, T. A recurring pattern of chromosomal aberrations in mammary gland tumors of MMTV-cmyc transgenic mice. *Genes Chromosomes Cancer 25*, 251–260, 1999.
89. Xu, X., Weaver, Z., Linke, S. P., Li, C., Gotay, J., Wang, X. W., Harris, C. C., Ried, T., and Deng, C. X. Centrosome amplification and a defective G2-M cell cycle checkpoint induce genetic instability in BRCA1 exon 11 isoform-deficient cells. *Mol. Cell 3*, 389–395, 1999.

Index

MIX
Papier aus verantwortungsvollen Quellen
Paper from responsible sources
FSC® C105338

If you have any concerns about our products,
you can contact us on
ProductSafety@springernature.com

In case Publisher is established outside the EU,
the EU authorized representative is:
Springer Nature Customer Service Center GmbH
Europaplatz 3, 69115 Heidelberg, Germany

Printed by Libri Plureos GmbH
in Hamburg, Germany